"*Flat Earths and Fake Footnotes* offers a comprehensive and compelling demolition of the tired myth of an enduring conflict between science and religion. Peterson not only exposes the historical bankruptcy of this familiar story, but also shows how it became a foundational narrative for Western modernity and why it persists. Beautifully written and impeccably researched, this book deserves a wide readership."

—Peter Harrison
Author of *The Territories of Science and Religion*

"Derrick Peterson combines painstakingly detailed historical research with a delightful writing style to tell the story of a famous war that never actually happened. Through primary source after primary source Peterson uncovers the neglected truth that the supposedly eternal conflict between religion and science is a myth, not only in the technical sense of a symbolic story that people tell to express their worldview, but also in the popular sense. Despite being something that everyone knows, it never happened. Perhaps because the story of something that didn't happen is hard to tell compellingly, this truth that is known to many historians, scientists, and theologians is little known to the wider public, and is even unfamiliar to some who ought to know the truth. Peterson shows himself a gifted storyteller as well as scholar, combining true accounts of famous events (which prove no less interesting than the legends that have grown up around them and in some cases have replaced them) with the story of how those events were overlaid and refashioned into the myth so many treat as common knowledge today: the untrue history of the war between religion and science. In an era full of so much untruth, Peterson's book is a breath of fresh air."

—James F. McGrath
Butler University

"'What if everything you ever believed about the *obvious* conflict between science and Christianity was wrong?' This is how Peterson's book could have begun. By the end of the book one is treated to an exceptional, lively, humorous, informative, and compelling account of how the fictitious idea that there ever was an all-out war between science and Christianity became fact, largely by means of the fun game many of us played in our youth known variously as whisper down the lane, broken telephone, grapevine, or gossip. This work on the historiography of science and Christianity is must-reading for high schoolers and college students, along with their parents and professors, and will, if heeded, change the way future generations will see the world. It is not easy to debunk a history that never happened, but Peterson has done precisely that, and achieved it admirably. The history of science is littered with stellar figures of immense importance, erudite thinking, and deeply Christian convictions. A new generation of Christians needs to be reacquainted with these scientific saints and Peterson's work is a sure guide to this task."

—Myk Habets
Laidlaw College, Australia

"Derrick Peterson is seeking to wake us from our dogmatic slumbers. Historians of science, scholars of religion, and theologians often plough separate furrows, paying little attention to each other's work. But in *Flat Earths and Fake Footnotes*, Peterson has brought them all into conversation, condensing a truly vast amount of scholarship. He has shown, moreover, that scholars ignore each other at their own peril. Despite over a hundred years of scholarship debunking the so-called 'conflict thesis,' the idea that science and religion are at war, perceptions of conflict persist. The only way forward from our scholarly impasse is to combine these fields of scholarship to paint a more comprehensive picture of what is going on. Impeccably researched and thoroughly readable, Peterson offers the reader a tour de force of the best research in the history of science, religion, and theology."

—James C. Ungureanu
Author of *Science, Religion, and the Protestant Tradition: Retracing the Origins of Conflict*

"Those who set the historiographical terms of debate frame the narrative involving the alleged conflict of Christian faith and science. Derrick Peterson's learned interdisciplinary study tracing the formation and deconstruction of the erroneous though ever-popular warfare thesis carefully sets the stage for future constructive work recounting the complex developments of scientific history involving Christianity. *Flat Earths and Fake Footnotes* provides the kind of critical analysis and creative catalyst so greatly needed today if we are to build nuanced understanding and trust between the scientific and faith communities for the sake of human flourishing."

—Paul Louis Metzger
Multnomah University & Seminary
Author of *The Word of Christ and the World of Culture: The Sacred and the Secular Through the Theology of Karl Barth*

"Is the Christian faith locked in inevitable conflict with modern science? Historians of science repeat their answer with a resounding 'No!' There was, however, a grand mythology of conflict woven, from Da Vinci to Darwin, through Galileo, Columbus, and the Scopes Trial. A false history was fabricated, a grand myth of conflict. Historians are frustrated. Why won't this stubborn pseudohistory die? To the rescue comes Peterson, a historian extraordinaire with many stories to tell. Exuding a palpable glee, he quests to debunk the grand pseudohistorical myths of conflict. His book about books leads the reader in an adventure across centuries. Hacking through the webs of false references and out-right fabrications, the payoff is a glimpse of what really happened. The truth is far more hopeful than the fiction. Rather than inevitable conflict, the true arc of science and religion might be dialogue, maybe even friendship. May this book get the wide readership it deserves. Maybe then, finally, historians of science might no longer need to debunk, yet again, the stubborn myths of the warfare thesis."

—S. Joshua Swamidass
Author of *The Genealogical Adam and Eve*

"Historians of modern science have time and again debunked the old saw that Christian faith and scientific reason have stood in deeply entrenched conflict with one another, and yet that myth remains etched into the popular imagination. In hopes of sanding away these fixed paths in our collective memory, Derrick Peterson offers us more than simply another genealogy debunking the warfare thesis. Rather, *Flat Earths and Fake Footnotes* offers us a deeply learned and absorbing *meta-genealogy*: a story of why and how the historiography that manufactured the myth of faith/science conflict came to be superseded by the alternative and debunking historiography that we find today. The result is among the clearest and most bracing articulations I've read on the complex historical interplay between religion and science. Peterson's narrative corrects our retellings of that interplay in the past. But more than that, *Flat Earths* challenges us to make explicit our present interests in such retellings and encourages us to imagine what sort of future for science and religion we are projecting with our historiographies. A remarkable first book by a formidable young scholar."

—**Sameer Yadav**

Author of *The Problem of Perception and the Experience of God: Toward a Theological Empiricism*

"In this remarkable volume, Peterson collates and contributes to a quiet revolution in the world of scholarship that is just now being disseminated to the masses. In your hands lies not just a book but the intersection of more scholarly threads than I thought possible in a single volume. Like a relentless detective, Peterson removes the dramatic curtain that has been put over our collective consciousness for so very long, and what remains is a tale of mere mortals, behaving very much as they do today. None will leave without enduring the slow dissipation of myths that we didn't know we believed. The cosmos that remains is, of course, so much more interesting and grand. This should serve as the definitive nail in the coffin of the warfare thesis for a generation to come. Even more than that, and especially delightful, it is a model of intellectual curiosity—of what scholarship ought to be."

—**Joseph Minich**

The Davenant Institute

"Peterson's detailed and well-researched description and argument ought to dispel any notion that the earliest 'scientists' were hindered by religion in their pursuit of understanding the natural world. A must read."

—**Mike L. Gurney**

Multnomah University

"Studying geology at Oxford in the 60s we heard of Ussher's 4004 BC date for creation, and so we thought science and religion were in conflict. As I started on the history fifty years ago, I realized the story didn't make sense and slowly found the conflict of science and faith to be a fiction. . . . I put in my pennyworth, as did many fine authors over the past few decades, but it still crops up including from leading scientists and theologians who should know better. Peterson's interesting and unique perspective in *Flat Earths and Fake Footnotes* is another well-written nail in the coffin."

—**Michael Roberts**
Author of *Evangelicals and Science*

"Even though I am a very hopeful agnostic, I have a keen interest in the relationship between science and faith, which by necessity includes the history of their relationship. This book is probably the most delightful exploration of that history I have ever read. In an erudite yet entertaining manner, Peterson explores the history in a way that undoubtedly makes the case that the enmity between science and religion, although undoubtedly present, has been exaggerated. Peterson acts as an ideal ambassador between these two disciplines. In all honesty, how could I not enjoy a book that links science and religion while including Indiana Jones allusions? Using the obligated—yet fully felt—remark, I must say: highly recommended."

—**Oné R. Pagán**
Author of *The First Brain* and *Strange Survivors*

Flat Earths and Fake Footnotes

Flat Earths and Fake Footnotes

The Strange Tale of How The Conflict of Science
and Christianity Was Written Into History

By
DERRICK PETERSON

 CASCADE Books · Eugene, Oregon

Cascade Books
An Imprint of Wipf and Stock Publishers
199 W. 8th Ave., Suite 3
Eugene, OR 97401

www.wipfandstock.com

PAPERBACK ISBN: 978-1-5326-5333-9
HARDCOVER ISBN: 978-1-5326-5334-6
EBOOK ISBN: 978-1-5326-5335-3

Cataloguing-in-Publication data:

Names: Peterson, Derrick, author.

Title: Flat earths and fake footnotes : the strange tale of how the conflict of science and Christianity was written into history / Derrick Peterson.

Description: Eugene, OR: Cascade Books, 2021 | Includes bibliographical references.

Identifiers: ISBN 978-1-5326-5333-9 (paperback) | ISBN 978-1-5326-5334-6 (hardcover) | ISBN 978-1-5326-5335-3 (ebook)

Subjects: LCSH: Religion and Science—History. | Naturalism—Religious aspects—Christianity. | Science—History.

Classification: BL245 .P47 2021 (print) | BL245 (ebook)

FEBRUARY 12, 2021

To Mom and Dad. Here is something to put on the fridge.

Leopards break into the temple and drink the sacrificial chalice dry; this occurs repeatedly, again and again; finally, it can be reckoned upon beforehand and becomes part of the ceremony itself.

Franz Kafka, "Reflections in Sin, Pain, Hope, and the True Way"

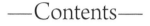

—Contents—

—Preface—

HABENT SUA FATA LIBELLI—BOOKS have their own fate. What shall ours be? In many ways this book is about the fate of books. They say write what you know; they also say write the book you could not find when you needed it. This book is certainly both, and yet also neither. As to what I know, despite the countless hours of pathological research over the last few years, the hundreds of books, perhaps thousands of articles read, all spanning a variety of disciplines, I climbed a mountain only to find the peak was the proverbial iceberg's tip. Like a dream of doors where more appear as you turn one way or perhaps the other, hallways, more hallways always sprouted everywhere. What was beyond them? I always wondered. With a sigh and a grin, off again I would set. Many (too many) are still unopened. My publisher insisting the book be finished this side of eternity, I had to leave many agonizingly shut, or perhaps a few merely kicked ajar as I screamed past. Nor is it totally true that I could not find a book like this when I needed it. As a summary intended to distill the great and nearly unimaginable garden of delights that is the new historiography of science and religion into a workable meal or two for a general audience, there are any number of books at least vaguely similar to this one. With titles such as *Galileo Goes to Jail*, or *When Science and Christianity Meet*, or *Bearing False Witness*, or *Unbelievable,* or *Atheist Delusions*, or *Dominion*, a new cottage industry of brilliant and accessible summaries debunking the smelly dumpster-fire versions of Christian history many of us encountered at universities or while we grew up has begun to form a relatively small albeit happy little fire brigade.

And yet, there was in my estimation also a great deal missing. Very often a few myths are debunked, or the historical methods responsible for the myopia that caused the notion of a perennial struggle between Christianity and science are spoken about—but rarely together, and rarely in a way that makes their inner historical and logical and even (quite gossipy) personal connections apparent. Far from unrelated oddities, there is quite the story to be told about how the warfare of science and Christianity came to be a settled, matter of fact bedtime story in the minds of so many. This is the book I have set upon myself to write: where did the notion of the warfare thesis come from? How did it take hold? What myths were used, and why? How was the falsity of this historiography uncovered, and eventually overturned in academia? Why are Christians not talking about it more? So many in the church who do not have time for these things are often forced into false dilemmas from false facts, and so are burdened with false albeit wily solutions. Whether or not I am perpetuating some of this myself is yet to be seen, but my ultimate intention is to share a bit of the beauty, joy, and, indeed, complex weirdness that has attended the history of Christianity and science. It is my hope that there are many an interesting story or two to tell along the way. Whether I have done the task justice is, of course, another matter entirely. At times it proved overwhelming and intractable. It was a pile that grew exponentially no matter how hard I dug at it. Having finished the manuscript now, if anything the work yet to be done seems even more immense, and so my call for fellow adventurers, all the more urgent.

There is another famous trope, especially popular amongst the old saints, to claim that this book is not the one they wanted to write. It is an elegant strategy that solves any number of problems—it makes you seem humble, popular, a recognized source of authority to be asked, and perhaps just as important relatable in your laziness. Because when it comes down to it, who wants to write a book, really? *Having* written one is the sexier option. Nonetheless, as much as I may have overworked myself to make my way through the mountains of reading and synthesis, the blame is entirely my own. In those bleary hours when I could hardly remember my own name, let alone read one more page, I was my own worst enemy. I have had to make many hard decisions in the course of research about what to include and what to leave out. With little exaggeration, the present book represents something of a third of what I originally wanted to include. And in the strange delirium one gets when culling parts that your deceitful heart says everyone surely must read, I have no doubt made bad and at least questionable decisions here and there that will be pointed out by scholars and generalists alike.

I should have liked to have a chapter devoted to the rise of the notion of the "God of the gaps," and contemporary creationism, for example, and a chapter devoted to how much theodicy has silently impacted what we today mistakenly call the conflict of science and Christianity. A chapter on how the proofs of God have changed over time, and perhaps a look at how speaking of God through nature has changed even in the twentieth century alone by looking at the famous Gifford Lectures through the decades (that last one, to my chagrin, has already been done quite well at least on two separate occasions by Stanley Jaki and Larry Witham). Another chapter sparked in the back of my mind on how the categories of nature and supernature emerged. Surely, I thought, I could squeeze in a few paragraphs (okay, a chapter) on the theological history of astronomy up through present-day hypotheses of the multiverse. I wanted to—somehow—add material on the theological turn in French philosophy in the twentieth century, alongside a (brief, I swear, ever so brief) history of the rise of Christian philosophers in the Anglo-American tradition. Perhaps pepper in a few references to how Trinitarian and christological dogma shaped modern economic discourse. A blocky footnote here and there on dinosaurs. But, I had to exclude these and much more for space consideration (I'll pause here for the sighs of relief). My machinations had an entire section devoted to the rise of atheism, as well, perpetuating the powerful but still contentious thesis of those like Michael Buckley, James Turner, Charles Taylor, Alan Kors, Amos Funkenstein, D. Leech, and others that atheism is a negative image of the theism it is rebelling against. Paying more attention to shifts in the theological landscape (hitherto often ignored) becomes then all the more important as atheism was often a response to ill-considered (perhaps even heterodox) changes in the theological winds rather than Christianity full stop. As those chapters grew, I came to realize I could not afford them the space they needed, either. As they also say about books, though: always leave room for a sequel. I took solace in that.

At the criticisms that will surely come regarding the competence of my judgments, the limitations of my reading, the mania of my prose, it is most likely that I can only nod in agreement and perhaps blame my own broad and interdisciplinary approach. This is neither a work of theology nor of philosophy nor of history properly speaking, though it is bits of them all and more. In the academic world such interdisciplinary work is often frowned upon because it is such a breeding ground for dilettantism. No doubt I have not avoided this entirely, and I look forward to learning more from those who will point such things out to me (hopefully as kindly as possible!). If in researching and writing this book I have condemned myself to many mistakes, I can only plead that I wanted to show the world something beautiful, something mysterious, something interesting, and went about it the best way I know how: by writing a story. This book is not intended as an end, but rather a new beginning, written in the hopes that others more intelligent and well-read than I am can salvage what it was I attempted to do. Theology and Christianity have been an enormous and integral part of that human project of coming to know the world over the last centuries, indeed millennia. This has been forgotten, erased, and distorted—often by narrow apologetic and theological works most committed to defending some notion of Christianity or other. Christ will in the end summarize all things into himself (Eph 1:10), and putting theology and Christianity back into the stories we tell about the grand adventures of ideas and practices is but a small, anticipatory glimpse.

I have accrued many happy debts along the way. There is hardly room enough in this whole book let alone a page or two, but let me try: a deep word of gratitude to Dr. Paul Louis Metzger and Dr. Gregory Oltmann, who helped this project get off the ground with encouragement and a few well-timed threats, as well as to Dr. Jon Robertson and Dr. Michael Gurney who imbued in me a love for historical and philosophical theology. I would like to thank Dr. Peter Harrison of the University of Queensland, Dr. Conor Cunningham of the University of Nottingham, Dr. Sameer Yadav of Westmont University, Dr. Kendall Soulen of Emory University and the Candler School of Theology, Dr. James McGrath of Butler University, Dr. Rod Stiling of Seattle Pacific University, and Dr. Stephen Lloyd of the Oregon Health and Sciences University, who all looked over my book proposal and the manuscript in different stages and lent their expertise to help get it underway. I was truly blown away by their kindness, and their willingness to trust my abilities and my vision for this work despite of its somewhat ludicrous ambition, which it only partially fulfills. Of course, all mistakes in the present volume remain solely my property, but I hope I have done some justice to their trust and good will. I would like to deeply thank my friend and colleague (the soon-to-be Dr.) Sara Mannen, who not only provided friendship and encouragement, but lent her own considerable talents looking over the manuscript for readability, and in helping me with the many edits. Again, all errors remain my own, but without her help this book would have been considerably worse for wear. In addition, I would like to thank several friends and acquaintances in the theological world whose scholarship, work ethic, and encouragement have all helped me greatly at one time or another. Some of them will no doubt be surprised to find themselves named here, but even if we only seldom interact they have been a constant inspiration (which is not to say they agree with anything written in this book, and some may even find things to part with me on quite extensively!). It is thus my pleasure to include Dr. David Congdon, Dr. Benjamin Myers, Dr. Derek Rishmawy, Father Al Kimel, Father Kenneth Tanner, Dr. Fred Sanders, Dr. Myk Habets, Dr. Artur Rosman, Bobby Grow, Ryan Hurd, Sharad Yadav, (soon-to-be Dr.) Kait Dugan, Steven Wedgeworth, Peter Escalante, and several enlightening conversations (and more, of course) about our mutual fascination with the phenomenology of atheism with Dr. Joseph Minich.

I have truly been blessed with best friends in this life as well. And though I can be a bit of a ghost, for whatever reason they keep me around and no doubt would be surprised how often they are in my thoughts as they make my life meaningful amidst what can often be the drudgery of research: Greg and Elisa Oltmann, Eric and Nell Doll, Ira and Sarah Lucia, Brian and MJ Beckman, Zach and Alyse Rathbone, Lauren Foster, Annie Dugas, David and Liz Johnson, Griffin and Laura Williams, Justin and Morgan Warren, Emily Gariepy and Christian Roscoe (soon to be Emily and Christian Roscoe at the time of writing!), John Inskeep, Blake Ehr, Andrew Johnson, Andrew and Hannah Barrett, Kendyl and Josh Shambaugh, and Scottie and Kayla Halgren. I wouldn't trade my memories with them for anything. I fear I am leaving off many others who deserve to be here, but I am deeply grateful for you all. And finally, of course, I would above all like to thank my family. My mom and step-dad Gwendolyn and Kelly Dean, my dad and step-mom Jim and Tamara Peterson, my sisters Tiffany, Janessa, and Tracy, and brothers George and Jacob, they have all been so supportive of this weird profession of mine. I hope I have finally produced something worthy of such kindnesses and love. But even if not, I know they will let me get up and try again.

—Introduction—

Clever Metaphors

History is the product of a technology. It does not simply lie around like stones or apples, ready to be picked up by anyone who pleases. It must first be produced.
—Constantin Fasolt, *The Limits of History*

WE ARE, ALL OF US, haunted by histories. We carry them about with us, bear them in ourselves, often unthinkingly; they sit upon our shoulders as clothing, and fold into the tresses of our hair as fashion; they shape the way we walk—with limps of old wounds, or a strutting gait born from the confidence of a blessed life. We unconsciously choose an eatery because in the glimmer of our memory it reminds us of a lover now long lost; a smell, in itself innocuous, may rekindle happiness in the midafternoon because it swoons with a scene from childhood, and in the evening it may dash us against the rocks of a loneliness that, like darkness, looms up suddenly. The deep ancestries of our words and our concepts drip with history, long roots soaked in the yawning earths of forgotten times, whose forces are ornamented by a phantom glamor that now often escapes our notice. At a less individual and more sociocultural level, what the philosopher and sociologist Charles Taylor has called our "social imaginaries"[1] are born from our collective habits, our theories, our experiences, our histories. The past—both real and imagined—is not ethereal, but cuts through rock and bone and asphalt, shaping our assumptions, our hopes, our fears, our loves, but even our cities whose layouts are lost arguments shaped now in stone.[2] Our biggest mistake is to assume that once the book closes, or the documentary ends, that so too history goes back to where it belongs. No thought is more dangerous. Everything we

---

1. Taylor, *Modern Social Imaginaries*.
2. Ward, *Cities of God*.

encounter, use, indeed everything we *are*, is history given form. "The past is never dead," wrote William Faulkner. "It's not even past."[3]

All of this, of course, can become hopelessly complex. Pasts might linger with us, but to uncover just what this might mean requires the hard labor of memory known as history. Indeed, histories are often packed so densely in some things that to unfold them and see what lay inside would be to drape a single concept or two over the whole world. Taking a look at something like C. S. Lewis' overlooked masterpiece *Studies in Words* can confirm this, when something so intuitively obvious to us (until anyone asks, that is) like "nature" or "sad" or "wit" or (ironically enough) "simple" get the in-depth historical treatments one would expect from the Cambridge Chair in Medieval and Renaissance Literature. But to do this exhaustively or, sometimes even adequately, can be an impossible task. So, most of us decide rather to just carry on with the simple maps we have been passed down, given by those whom we trust—friends, teachers, pastors, reporters, scientists, professors. All the nodes of our social imaginaries. But, just so, again in the words of C. S. Lewis, we often become unknowing prisoners of a recent past.[4] In terms of science and religion, this means most of us walk about haunted by rumors of a long war.

Truth, they say, is stranger than fiction. But it is indeed a certain fiction that has, for science and religion, undergirded what many of us have taken as truth. "Clever metaphors die hard." So opens James Moore's groundbreaking book, *The Post-Darwinian Controversies*. Moore is referring in this instance to the tenacious persistence of "military metaphors" used in literature—both academic and popular—to describe the historical relationship between science and religion: "Through constant repetition in historical and philosophical exposition of every kind, from pulpit, platform, and printed page, the idea of science and religion at 'war' has become an integral part of Western intellectual culture."[5] One of the most "remarkable developments in the nineteenth and early twentieth centuries," concurs theologian and scientist Alister McGrath, "has been the relentless advance of the perception that there exists a permanent, essential conflict between the natural sciences and religion."[6] "The secular public," writes the historian of science Ronald Numbers, "if it thinks about such issues at all, *knows* that organized religion has always opposed scientific progress. . . . The religious public *knows* that science has taken the leading role in corroding faith."[7]

And what is it that "everyone knows"? In their wake, the warlike metaphors drag a stock set of historical examples. Christians believed in a flat earth. Through the zeal of blind faith, they destroyed Greek and Roman science—burned its books, broke its monuments, abandoned its pluralism. They murdered, in a particularly gruesome

3. Faulkner, *Requiem for a Nun*, 73.

4. Lewis, "*De Descriptione Temporum.*"

5. Moore, *The Post-Darwinian Controversies*, 19.

6. McGrath, *The Twilight of Atheism*, 79.

7. Numbers, "Introduction," in Numbers, ed., *Galileo Goes to Jail*, 6.

fashion, the renowned female philosopher Hypatia for being both a woman in authority, and a scientist. They burned down the Library of Alexandria, the great wonder of ancient knowledge (supposedly run, as it happens, by Hypatia's father, Theon Alexandricus). Christians brought about a one-thousand-year age of darkness in the West. Experiments were seen as infringing upon God's jealously guarded secrets. Christians valued pain, and forbade pleasure. They laughed at Columbus for thinking the world round. Censored Copernicus. Demanded—against all evidence—that the earth was special and so at the center of all things. Tortured, imprisoned, and executed Galileo, having refused even to look through his telescope. Abhorred medicine, and while millions died from plague advocated only for prayer. Acted something like an upset troupe of apes when Darwin published *The Origin of Species*. They . . . well, you get the idea. The rest, it might be said, are but a series of footnotes to this bedeviled threshing of the Western mind.

These examples are bracing, and together form a narrative crescendo as harrowing as any drama. The rise of science is momentous on its own, but now with such grand stories it can be likened to the humble origins of a peasant who, becoming a knight, slayed a dragon and lived happily ever after with the rescued maiden. Science is never just science, but now a full history conjured up, one of heroes and villains, one of demons and genius. And examples of such epics are everywhere—from popular science gurus like Carl Sagan or Richard Dawkins, Stephen Hawking or Neil DeGrasse Tyson. They form the premise for best-selling novels like *The Da Vinci Code*, entire best-selling video game franchises like *Assassin's Creed*; show up in popular shows like *Family Guy* or *The Simpsons*. And this before our tour bus even arrives at the bristling sea of memes and the grimy comment forums of keyboard warriors who with untainted confidence topple Christianity with half-remembered history lessons. As I will be wont to quote several times, far from mere annoyance, internet forums in fact bring us close to the heart of the "problem over secularization. When we come down to the axioms which intelligent boys of fourteen years learn from less intelligent schoolboys of fifteen, we come near to the point where the cloudy apprehensions of what is known as intellectual history . . . can be shown to affect the attitudes of a whole society."[8] Not merely historical set pieces, stories of the war between religion and science have become cultural artifacts. Not merely historical trivia, they are the very marrow of our worldview's bones. They shape our presuppositions and ballast our judgments. In many ways any attempt to set science and religion in dialogue is lost at the outset. For every one of us, because of the great expanse of ready-made stories, has an inkling of just how terribly things have gone in the past and so carry in our hearts as well an estimate of their somewhat dismal future outcome. "As a first step toward correcting these misperceptions," writes the historian of science Numbers, "we must first dispel the hoary myths that continue to pass as historical truths."[9] And so,

8. Chadwick, *The Secularization of the European Mind*, 164.

9. Numbers, "Introduction," 6.

we come to our story. In the latter half of the twentieth century, scholars set out to do just that.

Historiography, for those unfamiliar with the somewhat ungainly term, bears no easy definition. For our purposes we might summarize it as philosophical and practical reflections regarding the fact that how we retell history very much shapes and reshapes the content of the history being told. Historiography is as such an attempt to look at the tools brought to bear upon history and ask: how have they changed the story (for better, or for worse)? Specifically, historians discovered as the latter half of the twentieth century marched on that the methods that had crafted our notions of the perennial warfare of science and religion were using a particularly blunt stock of tools. As the literary critic George Steiner once wrote, "method is metaphor made instrumental."[10] And if war was the guiding metaphor, its methods shaped and reshaped history into its image. As the historian of science Peter Harrison has recently observed (and to which we will frequently turn in this book), in other words, "once the constructed nature of the categories [of science and religion] is taken into consideration, putative relationships between science and religion [like their historical conflict] may turn out to be artifacts of the categories themselves . . . determined by exactly how one draws the boundaries within the broad limits given by the constructs."[11] A quiet revolution in the historiography of science and religion has occurred on the back of this observation and others like it in the twentieth century, not just overturning the warfare thesis but revealing the myriad of complex ways Christianity and science benefitted the other, and how this was overshadowed by the peculiar ways the history of science was retold. In fact, in the last thirty years or so not only do historians of science not speak in terms of "Warfare," they have "mount[ed] a sustained attack on the thesis."[12] The concept of warfare really only exists now insofar as its dregs linger "in the cliché-bound mind" of some popular works of polemical history.[13] The Eastern Orthodox theologian David Bentley Hart with his typical panache notes these findings in summary by stating that for quite some time we have been

10. Steiner, *Real Presences*, 100.

11. Harrison, "'Science' and 'Religion,'" 39.

12. C. Russell, "Conflict of Science and Religion," 4. Some of the best works summarizing the primarily historical scholarship revising and deconstructing the warfare thesis include Brooke, *Science and Religion*; Dixon, Cantor, and Pumfrey, eds., *Science and Religion*; Ferngren, ed., *Science & Religion*; Grant, *God & Reason in the Middle Ages*; Hannam, G*The Genesis of Science*; Hardin, Numbers, and Binzley, eds., *The Warfare Between Science and Religion*; Numbers, ed., *Galileo Goes to Jail*; Harrison, *Territories of Science and Religion*; Harrison and Roberts, eds., *Science Without God?*; Hooykaas, *Religion and the Rise of Modern Science*; Klaaren, *Religious Origins of Modern Science*; Lightman, ed., *Rethinking History, Science, and Religion*; Lindberg and Numbers, eds., *When Science & Christianity Meet*; Lindberg and Numbers, eds., *God & Nature*; Livingstone, Hart, and Noll, eds., *Evangelicals and Science in Historical Perspective*; Moritz, *The Role of Theology in the History and Philosophy of Science*; Stark, *For the Glory of God*, 121–200; Ungureanu, *Science, Religion, and the Protestant Tradition*; Yerxa, ed., *Religion and Innovation*.

13. Numbers, "Science and Religion," 65.

left with little more than "attitudes masquerading as ideas, emotional commitments disguised as intellectual honesty . . . ballasted by a formidable collection of conceptual and historical errors."[14]

Yet such is not a story you will often hear in pop culture, among the New Atheists, or on internet forums. Why is this? Aside from the typical biases that stall the flow of information, the manner in which the infelicities and fictions of the warfare thesis came into being not only make for an interesting story, but also allowed the notion of conflict to sink to a level akin to a prereflexive attitude of antagonism between religion and science in histories of the West.[15] Our tale to be told in this book is as such a rather slippery thing: a story within a story, stories within stories folding one over the other in an origami of memory and forgetfulness, of forgetfulness that appears as memory. *Flat Earths and Fake Footnotes* is the story of how the story of the warfare of science and Christianity was built through smaller stories supposedly exemplifying this war; how these smaller stories were built through the use of very particular historical methods; and ultimately how this perfect storm of method and meaning and myth started to be deconstructed through the labor of many scholars through the latter half of the twentieth century.

Without wanting to test the patience of the reader too much, however, there is a sort of positive thesis here: there is no such thing as the "history of the conflict of science and Christianity," and this is a book about it. Or, put in less provocative but slightly more accurate terms: once, there was no warfare between science and Christianity. It was not latent, merely waiting to erupt when the power of Christendom began to fade, or when men and women began to turn to science; rather it had to be instituted and imagined, in both theory and practice. Ultimately, even scientific atheism is not a perennial option to humanity simply waiting for science to emerge and the opportunity for godlessness to finally be enacted, but has a history; its time emerged, was narrated, and can be grasped only as a historically contingent phenomenon of self-fashioning declarations by the storytellers.[16] Or, to try a third way to describe it: this history of war between science and religion—one so common that "everyone knows" it—is over. Or, rather, if the scholarship recounted in these pages is to be believed, it never really even began.

---

14. Hart, *Atheist Delusions,* 19.

15. Unfortunately this book will indeed, like so many others, confine itself to the history of the West and because of space considerations leave the East to its undue neglect. For a few works shedding light on the fascinating history of Eastern Christianity and science, see: Nicolaides, *Science and Eastern Orthodoxy*; Obolevitch, *Faith and Science in Russian Religious Thought*; Nesteruk, *Light From The East: Theology, Science, and the Eastern Orthodox Tradition*; Miller, *The Birth of the Hospital in the Byzantine Empire.*

16. Some readers may pick up on the fact that this paragraph was built as a variation of three relatively famous lines from Shapin's *The Scientific Revolution*, Milbank's *Theology and Social Theory*, and Jüngel's *God as the Mystery of the World.*

And so, a second thesis of sorts: the quiet revolution in historiography in the twentieth century has revealed that the warfare thesis is primarily an artifact of earlier historians reconstructing events, of textbooks and tales rather than titration or test tubes; a product of recollections and reception rather than the historical actors themselves (though of course historians, as creatures of time, are historical actors themselves). "The path of a fact or factoid," writes Anthony Grafton, "from archive to notebook to footnote to book review is, in short, often anything but straight"[17] and it was this meandering path frittered with footnotes along which war was often summoned. Alongside Darwinism as promoted by the recently professionalized class of "scientists" (a word of much contention and debate when it was still of recent coinage in the mid-nineteenth century), "history" as an independent discipline arose for the first time in the early nineteenth century in France, then in England, Germany, and America as with the humanities in general. It turns out that it was here where warfare began to insert itself into our collective memories, precisely through the rise of these new professions. The births of history, in other words, were where the seeds of the grand conflict myth were also incubated; it was because they were born alongside and within the new discipline of history that they could be given the illusion of an ancient and noble pedigree. "In what we usually regard as the Age of Science," write historians Jon H. Roberts and James Turner for example, "it was not the natural or the social sciences that provided the greatest novelty in the academe, but the new humanities."[18] In fact, "the determined attempt of historians in the late nineteenth century to ally themselves with the natural sciences could be seen at every turn . . . [because they] viewed Darwin's work as the primary impetus for a 'science of history.'"[19] Likewise, historian Peter Novick astutely observes that "no group was more prone to scientific imagery, and the assumption of the mantle of science, than the historians."[20] Indeed, during the debates over the professionalization of science "history was the master narrative of the period." Not only was it the basis "from which judgment—moral, political, and philosophical—could be made" it was also the fulcrum by which men and some women could self-consciously become "adjudicators of science" but also of society, of law, of religion. In an age of wondrous and controversial discovery, history was actually the meta-discourse running things behind the scenes, defining sides and even deciding the broader meaning of these discoveries for posterity.[21] And, of course, of the many outcomes to this invention for us one stands out: "the notion of a conflict between religion and science was not so much the work of scientists, as of general historians."[22] Held quite often today by those advocating for the advance of human

17. Grafton, *The Footnote*, 14.

18. Roberts and Turner, *The Sacred & the Secular University*, 75.

19. Roberts and Turner, *The Sacred & the Secular University*, 45.

20. Novick, *That Noble Dream*, 33.

21. Yeo, *Defining Science*, 147–51; Young, *Darwin's Metaphor*, 164–248.

22. Gilley, "The Huxley-Wilberforce Debate," 326.

knowledge, in truth proponents of the warfare thesis remain in any number of ways stubbornly Victorian.

Flat Earths and Fake Footnotes is not just a snappy title, but is an attempt to provide an entryway into the massive array of scholarship that has over the last half decade unwound the knots that tangled to form an image of perennial war (let's be clear though, it is a snappy title!). Unfortunately, despite many introductory works making their way to market, the price of entry for the juiciest tidbits and the most interesting cutting-edge analysis remains gated behind paywalls, intimidating monographs, and complex jargon. Moreover, even when one has access to all of these things how to organize it all into useable and story-like information can lead one quickly into madness. Or, at least, an endless series of interesting but ultimately aimless rabbit trails. Indeed, even among the rising number of excellent works on offer debunking myths, these remain an almost unrelated collection of overturned eccentricities. Galileo may no longer go to jail in our memory, but why that was related to the flat earth beginning in the nineteenth century, or again why both of these are conjured alongside Darwin, is left untold. "Flat earths" is meant to indicate that, just like the myth that Christians believed in a flat earth, any number of memorable images, set pieces, and historical anecdotes are levied to exemplify warfare and Christian irrationalism. Just as importantly, these set pieces are often misunderstood either through mistakes, fabrications, or perhaps simply false angles that illuminated the wrong things, while keeping other information in the dark. Indeed, take any one of the examples from above (Galileo, the Library of Alexandria, ad infinitum)—and here we have but a tiny subset of the usual stories—and in each, recent scholarship has uncovered that these mythologies stay afloat upon questionable historical categories and methodologies doing too much heavy lifting, combining everything into a sort of super-myth self-perpetuated by its own critical mass. As historian Jeffrey Burton Russell summarizes the matter:

> Fallacies or "myths" of this nature take on a life of their own, creating a dialectic with each other and eventually making a "cycle of myths" reinforcing one another. For example, it has been shown the "*The* Inquisition" never existed, but that fallacy, like the flat earth fallacy, is part of the "cycle" that includes The Dark Ages, the Black Legend [of the Inquisition trials], the opposition of Christianity to science, and so on. The cycle becomes so embedded in our thought that it helps to form our worldview in ways that make it impervious to evidence. We are so convinced that medieval people *must* have been ignorant enough to think the world flat that when the evidence is thrown in front of us we avoid it, as we might, when driving, swerve around an obstacle in the road. Thus our worldview is based more upon what we think happened than what really happened. A shared body of myth can overwhelm all evidence . . .[23]

23. J. B. Russell, *Inventing the Flat Earth*, 76.

"Fake footnotes" on the other hand is meant to reference errors—both of fact, and of method, that likewise accompanies the set pieces both to make them exemplify a war between science and religion, and to knit them into a broader story. To be sure, we will have the opportunity to see that straightforward misinformation has deeply affected the perpetuation and invention of the warfare or conflict thesis. So often, in fact, that as I made my way through recent scholarship on the matter it became rather astounding how frequently what has been taken for granted as the building blocks for the warfare thesis were born quite literally from fake footnotes, bad sources, garbled reception histories. To be sure, it would be reduction of the highest order to say that warfare was created by nothing but mistakes. "Flat Earths and Fake Footnotes" remains a convenient heuristic for telling of the new histories emerging in the twentieth century, we must constantly remind ourselves, but a heuristic that contains a great deal of truth nonetheless. There are, of course, an innumerable variety of invisible structures and elements that must be narrated to be "seen"[24] as one retells a historical tale. Think even upon all the intricate nuances of a conversation you had the other day, or some background context that hovers over an event like a storm cloud but that, to later audiences, is lost. Or, on the flip side, the sheer cacophony of events that hit us at any moment need to be bracketed out, put aside, in order for stories to avoid being a vast and shapeless sea of happenings where it is "just one damn thing after another" (to pluck a partial quote from Arnold Toynbee).[25] In many ways what is *not* said in a story is often just as important as what is. Positivism as a historical method often controlled both the said and the unsaid in such a way that religion and theology were either cut out of the picture or, if left in, only appeared as antagonists, hapless mythologizers, cranks.[26] From a host of historical methodologies—again, in particular by the movement known as positivism—history was colored in ways that many frequently anti-Christian historians wanted, but which have thankfully been thoroughly exposed as deficient (and more often than not by "secular" historians to boot).[27]

Our journey moves through three major sections comprised of several chapters each. The first section, titled "Deleting Theology," will tell the story (or at least, *a* story) about how much of the historical mischief that went in to creating the warfare thesis was uncovered. To do this "one need only point to religion," which is an area that has

---

24. For an excellent overview of this problem in historiography, see Clark, *History, Theory, Text*.

25. Toynbee, *A Study of History*, 195.

26. See, e.g., de Certeau, *The Writing of History*, 31: "theological . . . writings [and ideas] have been virtually erased from the works of the great 'savants' of the sixteenth and seventeenth centuries [and so on] as vestiges of epochs long since over, esteemed as unworthy of interest to a progressive society. Within the tissue of history, analysis therefore chooses 'subjects' conforming to its place of observation."

27. For different angles on this story see Toulmin, *Cosmopolis*, 5–44; H. F. Cohen, *Scientific Revolution*; I. B. Cohen, *Revolutions in Science*; Osler, ed., *Rethinking the Scientific Revolution*; Harrison, *Territories*; Josephson-Storm, *Myth of Disenchantment*; Zammito, *A Nice Derangement*; Shapin, *Scientific Revolution*; Shapin, *Never Pure*; Pannenberg, *Theology and the Philosophy of Science*; McGrath, *Theory*.

"largely [been] neglected by modern historians, to see what benefits the abandonment of a priori notions of modernity might mold," writes John Gillis, a European historian. "This recognition [of religion] offers us the opportunity to explore aspects of European [and American] history that have been treated only condescendingly as mere survivals," or antiquities for some reason not yet abandoned. "It may be that we will end up questioning the distinction between the pre-modern and the modern, [the theological and non-theological, the religious and the secular] that defines the way Europe [and America] have been approached."[28]

Chapter 1 provides an entryway into the revisions in our understanding of history by telling the story of the French physicist and historian Pierre Duhem who found an actual "Da Vinci Code" running through history (and actually involving da Vinci no less). Chapter 2 moves on to show how Duhem's discoveries began branching out in the twentieth century as an entire industry—that of the history of science—was completely (though quietly) overturned.[29]

We then turn to section two: The Lords of Time. Here, after we have uncovered some of the themes we will be dealing with in section one, we loop back to the primary moments and mythmakers who created the notion of warfare. What were their own contexts? How did they go about writing history? The point here is not simply to debunk the way they told their stories, but also to enter sympathetically into their situations to understand how it was that they came to write them this way in the first place. This is not to justify what they did, but to better understand their motivations and so in a clearer and more efficient way undo how they pressed themselves upon the historical record. Though I am not without my own clear biases, what I do not want this book to be is yet another tale of heroes and villains. Here we will walk among men and women of understandable agendas and human situations who often made bad but understandable decisions regarding how history was to be retold. Chapter 3 looks at Thomas Huxley and the debates over Darwin, while chapters 4 and 5 turn respectively to Andrew Dickson White and John William Draper. Together, Huxley, White, and Draper form the major trifecta of the mythmakers who not only crafted the understanding of the long historical war between science and religion, but often created quite bespoke historical categories of "religion" and "science" that were in this creation primed to be turned into easy enemies.

In the final section we turn to the Legendarium, and look at a handful of the actual myths used in the warfare thesis as they were combined, created, or catalyzed in the nineteenth and early twentieth centuries from a potpourri of mythology that began its life not too much beyond the seventeenth century. Here, all of the themes from the first two sections come to a head as we look at the flat earth, two chapters on the dark ages, the Galileo affair, and the Scopes trial. There we will see how these

28. Gillis, "The Future of European History," 6.

29. For one brief albeit relatively thorough overview on some of the literature involved, see: Clark, "Secularization and Modernization: The Failure of a 'Grand Narrative,'" 161–194.

familiar scenes have been revised by the new waves of scholarship overturning what we thought we knew. Far from exemplifying battles between religion and science (or, more precisely, Christianity and science), things turn out to be much more complex, and dare I say much more interesting.

This of course does not solve everything. Christianity is not magically proven true by new realizations of how intricately tied to the rise of science it was, nor are some very real difficulties set aside about what we are to do today regarding their interaction. But some false problems definitely are eliminated. Many complications that we often feel the need to deal with today (even as seemingly simple as couching things in terms of "religion and science") in turn are shown to be false problems bred from short or distorted historical memory. As was mentioned, this book is primarily deconstructive in its goals: if the reader finishes the book having learned a portion of the story of the construction and deconstruction of the warfare thesis, our goal will be complete. It is our hope that some clever and intrepid readers will be able to turn some of this into their own constructive work one day. All in all, what this book claims by being an ambassador for the immense amount of scholarship out there on the history of Christianity and science, is to convey the exciting wonders of history in all their ugliness and all their glory. While not all of this revisionist history paints Christianity in a good light (just the opposite, in some cases) and is too complex to be wielded as an easy apologetic tool, nonetheless by removing some of the common obstacles assumed by nearly everyone about how religion and science have related in the past, a new horizon for their future does hopefully appear. Also, it's great at parties.

But beyond this deconstruction, the ultimate goal here is to pave the way for something that I am quite shocked does not yet exist: a retelling of philosophical and scientific history where Christianity is once again put into the picture in the myriad of its complexities. Not as with the more apologetic projects, which are often too sclerotic and narrow for their own good. What we truly need is a bright-eyed retelling of the adventures of the human spirit free of Christianity as cartoonish villain or angelic savior.[30] Whether this task will fall to me or (more likely) to someone else, our present book is but one stone or two on the plotted path toward such a revivification.

---

30. Tom Holland's recently released book *Dominion* is close to what is envisioned here. It is a beautiful example of even-handed history with Christianity playing a role in all of its beauty and ugliness.

Part One

Deleting Theology

In the face of increasing specialization and the growing autonomy of scientific discourse, explicit discussion of theological issues often—but not always—vanished. Nevertheless, differences in scientific practice, approaches to scientific controversy, and the occasional methodological comment reveal their [theological and religious] *ancestry* . . .

—Margaret Osler, *Divine Will and the Mechanical Philosophy*

It's easy to suppose that almost all that is vital in seventeenth-century philosophy can be attributed to the Scientific Revolution, but [it has recently been rediscovered] . . . *the most interesting developments can be traced to the demands of theology.*

—Nicholas Jolley, "The Relationship Between Theology and Philosophy"

The constructing of tales such as [those representing scientific advance as an emancipation and triumph over theology] *should be understood as a symptom of a widely shared—almost subconscious—view that one of the primary tasks of twentieth-century early modern studies was to illustrate the history of secular modernity: that is to say, construct it.*

—S. J. Barnett, *Enlightenment and Religion*

It is still a metaphysical belief on which our belief in science rests—we knowers today, we godless ones and anti-metaphysicians, we too still take our fire from that great fire that was ignited by a thousand-year old belief, that belief of Christians, that was also Plato's, that God is truth, that truth is divine . . . [and] *the highest authority.*

—Friedrich Nietzsche, *Genealogy of Morals*

—Chapter 1—

A Da Vinci Code Is Found

Pierre Duhem and the Rediscovery
of the Christian Contribution to Science

Science had thought that it was seeing religion out the back door. But while it was distracted by questions that lay beyond its ability to resolve, religion . . . quietly walked back in through the front door.

—William R. Shea, "Assessing the Relations Between Science and Religion"

IN THE INTRODUCTION TO his masterpiece, *The Everlasting Man*, G. K. Chesterton opens with a short story he imagines he would one day like to write. In it, a young boy has become terribly bored, as young boys tend to become. Wanting to find adventure, to discover "the effigy or grave of some giant," some wondrous thing, he sets out from the comfort of his hearth for that elusive horizon line to peer over the span of the world. Having packed his things, he starts his long journey downward from the hillock he lived upon, ramshackled with an oversized pack of pots and pans and (one might imagine) a hand-drawn map whose edges are lined with dragons. Once a good distance away, as the story goes, he turns back, pans clanking, for one long, last look at his homestead. This backward glance, to his amazement, reveals home anew. As the panorama opens now to his gaze he discovers what he thought was his homely and boring hill was actually an exotic thing, "shining flat on the hill-side like the colors and quarterings of a shield, [which] were but parts of some gigantic figure, on which he had always lived, but which was too large and too close to be seen." He had, in fact, been living on a giant the whole time. The exotic, the weird, the uncanny, were not merely "out there"; they were the very bones that built his seemingly ordinary hillock. The extraordinary, Chesterton concludes, is always with us but is typically too close to see. To realize this, "[is], I think . . . a true picture of progress of any really independent intelligence today."[1]

Scripture, of course, exhorts us to have faith like a child, but even in the history of science and religion the childlike wonder of Chesterton's boy adventurer has come to the fore. Over the last half-century, the discipline of the history of science and religion was turned on its head. The New Atheists might bellow the defeat of religion at the hands of science, but amongst professional historians things have begun to depart from such tales in rather dramatic fashion. Like detectives in a movie coming to that key moment where a collage of sources meld together in a spark of realization, historians of science have for the last thirty years collectively sat back in their chairs in wonder:

> The issue of the proper interpretation of our scientific heroes has been most pressing of all. . . . When we look at them a little more closely . . . [to our astonishment] they are not like us at all . . . but have [among other things] metaphysical and religious commitments . . . and horror of horrors, they take seriously such misbegotten ideas as astrology, alchemy, magic, the music of the spheres, divine providence, and salvation history.[2]

The "they are not like us" here is of course meant to indicate the stereotypical ideal of a modern, secular scientist who does not believe in God, and perhaps has at best a benign indifference to all things religious. They say never meet your heroes. When many younger historians did meet them while bringing new methods and new ways of understanding to the table, they began to realize how deficient older narratives of the rise of science were. In particular, religion—for our purposes more precisely, Christianity—was once again seen to have played a major role in the advance of what we today refer to as science. Slowly, the typical pictures we have inherited of science finally winning its long war against religion were again and again revealed to have been shaped by the manner of telling histories crafted by their predecessors. This startling realization began to soften the hard lines and contours many wanted as an absolute separation for the Scientific Revolution from that strange, obtuse religious past that came before it. Most historians will not go so far as Steven Shapin, who announced in the infamous opening to his book *The Scientific Revolution*: "There was no such thing as The Scientific Revolution, and this is a book about it."[3] Nonetheless, reconsidering the Scientific Revolution as a distinct period of history by reconsidering the socio-religious and theological roots contributing to scientific change has emerged as a quiet revolution, one taking the discipline of the history of science in the latter half of the twentieth century by storm. Unfortunately, despite the massive scale of the change—and despite Christians usually flocking to secular scholarship that seems more or less friendly to them—this sea change has largely gone unnoticed. We remain prisoners in

1. Chesterton, *The Everlasting Man*, 9.
2. Dobbs, "Newton as Final Cause and First Mover," 34–35.
3. Shapin, *The Scientific Revolution*, 1.

our homely little disputes as such, unaware that great things lay lurking, sleeping just beneath us waiting for discovery.

On the other hand, these same strands of scholarship have impacted what we thought we knew of historical episodes that typically populate what is usually referred to as the "Warfare" or "Conflict" thesis enumerating the less-than-friendly history-spanning relationship between Christianity and science. Much again to the surprise of many scholars, these newly discovered details not only did not confirm or reinforce the "Warfare thesis," but began to actively dismantle it. Even amongst so-called "secular" scholarship—in fact, primarily there—there has been a quiet revolution producing a "new historiography"[4] over the past half century or so, where the "Warfare thesis" of science and Christianity has not only become passé amongst professional historians, but they have mounted "a sustained attack on the thesis."[5] So thorough has this dismantling of the "warfare" or "conflict" thesis been, that it exists now only in "the cliché-bound mind" of popular works of polemical history.[6] The persistence of any such sustained model of "warfare" in our textbooks, amidst our classrooms, or on the television and the internet, remains historical "only in the sense that it is the creation of historians."[7]

The story that tells the tale of these historiographical changes reads something like a detective novel. A hint here, a clue there, were the first bread crumbs on a trail that led into the great depths of the story of human knowledge itself. One of the first intrepid explorers of these labyrinthine depths was a French Catholic physicist turned historian, Pierre Duhem. By following Duhem who, while certainly not solely responsible for the historiographical revolution,[8] was a brilliant catalyst,[9] we can watch as a strange new world of lost texts and forgotten—even fake—footnotes fell into place like puzzle pieces. In fact, it is something of a (quite real) "Da Vinci Code." Through his studies on Leonardo da Vinci, spurred on after da Vinci's lost notebooks were rediscovered at the end of the nineteenth century, Duhem stumbled upon a lost world of medieval theological and physical speculation that, much to Duhem's own surprise, indicated that not only Leonardo, but Galileo, and even Newton, were reliant upon a long predecessor culture of science intimately connected with Christian practice, theology, and metaphysics.

From there, the trails uncovered kept spilling onward and onward. Glancing over Duhem's shoulder provides an entryway into the great reconsiderations of history of

---

4. This phrase "new historiography" was coined by Kuhn, "The History of Science."

5. C. Russell, "The Conflict of Science and Religion," 4.

6. Numbers, "Science and Religion," 65.

7. Livingstone, *Darwin's Forgotten Defenders*, 2.

8. Recent historical investigations have revealed that Duhem was the inheritor of many sophisticated trends in French historiography that were beginning to emerge by the middle of the nineteenth century. See Bordoni, *When Historiography Met Epistemology.*

9. Oliveira, "Duhem's Legacy for the Change in the Historiography of Science."

science and religion that swept through scholars in the twentieth century, as we will see in the next chapter. The new portraits of da Vinci and others revealed how far the nature of how historical figures were represented had been controlling our grander notions of historical change and their meanings. And since historians have typically assigned a relatively low priority to the history of theology and religious ideas, half-formed histories have wandered our textbooks like wraiths, whispering of a centuries-long secret slumbering beneath the usual tools and assumptions of the academy. Duhem's rediscovery of the Christian contribution to science through the figure of Leonardo and his rediscovered notebooks began to turn many typical historical conventions and procedures on their heads.

A (Real) Da Vinci Code Is Found

In May 1913, the Catholic physicist and mathematician Pierre Duhem (1861–1916)[10] signed a contract tethering him to a truly Herculean obligation: he was to write his magnum opus on the medieval origins of modern physics, *Le System du Monde* [*The Order of the World*] by sending to his publisher every year for the next ten years "a manuscript equivalent to 500 printed pages."[11] It is a testament both to the furious pace at which Duhem had been researching since at least 1905, and to his talent for producing written work at a gallop, that though he only lived three years after the initial contract signing, enough notes to finish ten volumes had been produced—120 notebooks of 200 pages each, filled to the brim with notes handwritten by Duhem himself as he had no research assistants. This is all the more remarkable when one realizes the many documents Duhem was combing through had often not been properly indexed by the libraries in which they were housed. Frequent asides in his notes mention the constant struggles he had with file clerks and librarians as his various requests for documents were either denied or declared unable to be completed at the time, as no one knew exactly where the documents Duhem requested resided in the bowels of the archives. It is a jungle, he said, and one could be thankful only that there were no snakes haunting the jigsaw forests of unmarked documents. We can only imagine Duhem's spasms of joy should he have lived to see photocopying or digitization. Though the resultant volumes of *Le System du Monde* have yet to be translated into English, we are in the debt of Roger Ariew, who has skillfully given us a selection in *Medieval Cosmology: Theories of Place, Time, Void, and the Plurality of Worlds*.[12] Ariew's selection concentrates on Duhem's findings specifically related to medieval cosmology, and even as a selection, it weighs in at just over 600 pages. Despite the ludicrous amount of planning that preceded the creation of so many 500-page manuscripts all of this work of Duhem's was spurred originally almost as an accident. Or, one might say, the work of providence.

10. The best introduction and biography of Duhem remains Jaki, *Uneasy Genius*.
11. Jaki, *Scientist and Catholic*, 64.
12. Duhem, *Medieval Cosmology*.

Duhem had even by 1903 already gained an enormous reputation as a physicist and mathematician. At the same time, he had done little to no historical work on his discipline. Never one lacking for ambition, however, he set out to write *Les origins de statique* or "the origins of statics" (that is, broadly put, the analysis of force on physical systems) and intended to include a sustained foray into the deep past. In chapter 5 of his new book, after outlining some of the more theoretical aspects of the project, Duhem turned to this history. In anticipation of the switch into a genre of work he had previously left unexplored, Duhem had been devouring a number of books treating the development of statics, and that of science in general. This was easier said than done. As curious as it may seem to us today, the history of science was a very newborn genre—indeed history as an academic discipline was itself in its infancy, and exceptional works on the topic were few and far between.

It was here in the wilds of this new discipline that Duhem began to run into a curious problem: the gloriously detailed discussions of statics in the modern period were clearly represented in these summaries as going through periods of evolution—theory and technique continuously honed and polished as they were by a network of scholars going back and forth about method and math and mechanism. This was not unusual. What was unusual was that when traced backwards, the historical reviews of science Duhem was reading as preparation simply stopped around the sixteenth century. Despite the obvious development over the years up to Duhem's own time, right before Galileo without fail a large portion seemed simply to leap out like Athena, fully formed from the head of Zeus from an age where there was, apparently, no prior hint of what was to come. "Was there science in the Middle Ages?" asks Duhem's biographer Jaki. "A mere look at the best histories of science available in 1903 would have cured him of any illusion [that there might be]."[13] In these tales of human knowledge science ever so suddenly walked out of the dark with its many brother and sister sciences, plucked from the air or from whatever storehouse of the mind from which geniuses receive their flashes of insight.

Little lamps then suddenly sprouted to slowly illumine the halls of history—names like Copernicus, Galileo Galilei, Leonardo da Vinci. But always in the background of these newborn stars, whether it was merely passed over like a memory so worn by time one no longer even knew it had gone missing; or whether it was spoken of in that quite awe reserved for the monster lurking in some fairy tale, the great abyss of time known as "The Dark Ages" continued to yawn. To be sure, many acknowledged hints and swells of science in the far deeps of the Greek past, an Archimedes here, a Ptolemy there—glimmers now regathered and magnified by thinkers suddenly enlightened by some uncanny muse called "human genius"—a newly minted but now archetypal figure of historical progress invented by eighteenth-century thinkers and historians.[14] This concept of "the genius" allowed an even further separation of sci-

13. Jaki, *Uneasy Genius*, 381.

14. On the history of the evolving concept of "genius," see McMahon, *Divine Fury*.

ence from its medieval past—science like a great magnitude passing from giant to giant like a prodigious relay; or perhaps, depending upon who was writing, like an underground railroad ferrying knowledge through the shadows and away from the inquisitorial torches of priests and bishops.

Of course, all of this was in some sense a well-worn trope. The image of a dark age, by Duhem's time, had slummed around history for long enough in one form or another to become merely an assumed part of the landscape.[15] Even back when the French *philosophe* Jean d'Alambert spoke of the Dark Ages in his preliminary discourse to the infamous *Encyclopédie,* he was merely repeating a prized cliché of both the Reformers and the Humanists.[16] In France, (where Duhem was situated) the Enlightenment, when perceived in broader historical context, was a victory neither of method nor epistemology per se, but was a triumph in "narratological"[17] key. Against the presumed backdrop of a barbaric darkness, humanity's coming of age for the French *philosophes* was driven by a story that allowed men and women to perceive their place in the grand scope of things: "The narrative of 'new science' progressively dismantling all remnants of superstition and Scholasticism in its way was . . . central to the self-perception of the *philosophes*," as Dan Edelstein puts it.[18] Seeing themselves as emerging from an age of scholastic darkness, both to recover lost ancient wisdom and move beyond it, "offered a seductive account of the discoveries in [their] past century, in conjunction with a more overarching history of human civilizations."[19]

This is in part both a cause and a product of how the genius of someone like Leonardo da Vinci has been presented in history. As such these narrative "encodings" of how science is presented are also important for how science has been related as a historical phenomenon to religion, specifically Christianity. As Jaki notes, "beyond Leonardo there was [in these works] by all appearances nothing to look for."[20] Though now in the key of scientific progress, this too was not a totally new conviction about the uniqueness of the man. Even within a generation of his death, the singularity of Leonardo became an archetype for epochal change. When Raphael produced his famous fresco *The School of Athens* in the Vatican, for example, the figure of Plato was in actuality painted in Leonardo's likeness. Raphael adorned the figure with a rose-colored toga, a shade favored by da Vinci, who dressed flamboyantly in his youth. And with index finger raised upward to the heavens, one of the two central figures of the fresco is not just Raphael's apt summary of Plato's philosophy (with the contrasting

15. See chapters 8 and 9.

16. For a survey of early pejorative Protestant uses of "the Middle Ages," see Fergusson, *The Renaissance in Historical Thought,* 46–58.

17. Edelstein, *The Enlightenment,* 2.

18. Edelstein, *The Enlightenment,* 117.

19. Edelstein, *The Enlightenment,* 2–3.

20. Jaki, *Uneasy Genius,* 382.

paired figure of Aristotle pointing downward to the earth), but also produced a characteristic gesture contained in Leonardo's own artwork.

With the 1881 publication of Leonardo da Vinci's notebooks, the perception of his intense and mysterious depths of insight swelled even further. Over 7,000 extant pages of relentless curiosity were now again to see the light of day, winding labyrinthine and revealing layer upon added layer of thought and image as Leonardo focused himself on any given problem over time. Much as Matteo Bandello described the process that went into painting *The Last Supper*—where Leonardo would from sunup to sundown not once lay down his brush, forgetting to eat or drink—so too the notebooks reveal a genius that seems to sit patiently in eternity imperious and unperturbed, gazing from subject to subject as their secrets played before him. The notebooks were among many key documents that Napoleon eventually acquired as he made his way through Italy to France—others being previously unreleased records of the Templar, Inquisition documents, and indeed the unseen records of Galileo's trial. Napoleon had seized them as part of his war payments from the *Biblioteca Ambrosiana* in Milan in 1796.[21]

The journey by which the notebooks had initially arrived in Milan is itself an extraordinary tale. It can be traced in part because of memoirs written by Giovanni Ambrogio Mazenta, an Italian architect who designed the cathedral of San Pietro in Bologna. After Leonardo died in 1519, his trusted friend and pupil Francesco Melzi was bequeathed possession of the notebooks. And when in 1570 Melzi died, so began a great odyssey. Divided into sections here and there by Melzi's less scrupulous heirs, they were scattered across Europe through both legitimate trade and black market interest. Several of the codices ended up the hands of one Pampeo Leoni, who decided Leonardo's musings were too haphazardly strewn across the notebooks. So, he disassembled them and pieced them back together by subject in what has come to be known as the *Codex Atlanticus*. This is the version that would eventually be taken by Napoleon as a spoil of war, and then returned to Milan in 1815 at the intervention of the Vatican. Far from covering up da Vinci, as in Dan Brown's novel *The Da Vinci Code*, we actually have Catholics in part to thank for returning da Vinci's works to the public. A large part remained in Paris, however. Known as "Manuscript B," these documents had been separated by Leoni when he was creating the *Atlanticus*. Some of Manuscript B ended up in the hands of art collector Charles Fairfax Murray between 1859 and 1864; while several pages were removed from Manuscript B by the mathematician Gugliemo Libri. Yet other portions of Manuscript B were purchased by Theodore Sabachnikoff, a Russian student of the Italian Renaissance, who desired to piece together a subsection known now as *The Codex on The Flight of Birds*, and give it to Queen Margherita of Italy who subsequently deposited the gift to the *Biblioteca Reale* in Turin. That these documents should have ever been reunited is nothing short

21. Our account of the journey of the notebooks in this and the next paragraph is quite indebted to the wonderful retelling given by the Chief Curator of the Smithsonian, Peter Jakab, "An Extraordinary Journey."

of a miracle. The last several pages took another monarch to discover them, when King Victor Emmanuel III of Italy managed to secure their purchase and add them to the collection Queen Margherita had previously deposited at Turin.

Such a piecemeal distribution of da Vinci's works is also a good metaphor for how his reputation has changed through time, depending on what aspects are appreciated most about his work in a given era. As Barrie Bullen points out, Leonardo's reputation went through something of a major transition in the middle of the nineteenth century.[22] While many of us now take for granted that Leonardo's artistic and scientific endeavors are taken together as constituting his greatness, the magnitude of his initial reputation was earned despite of, not because of, what we now call his "scientific" musings. The image of a scientific genius that Duhem was working with was itself a product very specific to Duhem's day. Many took Leonardo's adventures in mechanics and other matters to be representative of a haphazard and easily distracted approach to life, his passions burning themselves lukewarm as they diffused across too vast a landscape. He would have burned brighter even than Michelangelo and Raphael, wrote Matthew Pilkington as late as 1852,[23] would he have avoided his "negligence" by way of versatility. Earlier in 1801 in lectures delivered at the Royal Academy, Henry Fuseli accused Leonardo of "having wasted life, insatiate, in experiment."[24] In other words, our picture of the "Renaissance man" combining art and science was fairly late born. As was our notion of "the genius." Both of these categories are very convenient locations to describe the fact that *how* one represents a historical figure, and what categories are used, can deeply affect the content of what is being represented. What was the source of this new picture of Leonardo's scientific acumen? Curiously, though "the political upheavals in France and the struggles between Church and state," involving Napoleon and others seem far removed from the serenity of the *Mona Lisa*, this is exactly where and when the interpretation of Leonardo as a light against the religious dark bloomed.[25] It is in this broader setting that da Vinci became baptized as a scientist heroically subverting religious credulity was formed. He was, to use the now common image, a "genius."

The Genius and the Light

We call this model of [scientific history] *heroic because it made scientific geniuses into cultural heroes.*

—Appleby, Hunt, & Jacob, *Telling the Truth About History*[26]

---

22. Bullen, "Walter Pater's 'Renaissance.'"

23. Quoted in Bullen, "Walter Pater's 'Renaissance,'" 269.

24. Quoted in Bullen, "Walter Pater's 'Renaissance,'" 269.

25. Bullen, "Walter Pater's 'Renaissance,'" 271.

26. Appleby, Hunt, and Jacob, *Telling the Truth About History*, 15.

But was Leonardo da Vinci a genius? Even to ask such a question today will no doubt sound the alarms of sacrilege. For few figures apart from Albert Einstein, Amadeus Mozart, or Isaac Newton so clearly embody the ideals generally associated with the term. The use of genius in this case actually evidences a largely forgotten change in how we view human agency and the contours of history. It is one example among many illustrating how the categories we have come to use in historical investigation can often hide as much as they reveal. In its first usages "genius" was a guardian spirit, an act of God or the gods. *Genii* were seen in Rome and Greece as guardians, spirits, *daemons*, oracles that mediated between the divine and the human.[27] Socrates (d. 399 BC) for example spoke of his ideas as "signs" (*semeion*) given to him by *daimonion* which, while the ancestor term for demon, had not yet taken on the evil spiritual connotations that Christians would place upon the term. It was indeed upon the charge that Socrates had brought "new demonic beings" (*daimonian kaina*) into the city that he was put on trial and sentenced to suicide by hemlock. Homer, similarly, opens both *The Iliad* and *The Odyssey* with invocations to the muse: "Sing, O'Goddess, of the anger of Achilles" he writes in the *Iliad*, and "sing, muse, of the man of twists and turns," in the *Odyssey*. This is followed by an invocation to "sing in me, Muse, and through me tell a story."[28] Cicero (106–43 BC) in his *On the Nature of the Gods* codified this and turned it into a definition: "No man has ever been great, but by the aid of divine breath. . . . Genius is that god within [*sacer nobis inest Deus*]." The Christian traditions of the guardian angel, and the patron saint, themselves were born and evolved from this line of thought.[29]

In this manner to speak of "the genius of Homer," or even "the genius of Shakespeare" would not be affected even if it turned out to be true that the corpus of both men was in fact written by several individuals. For their "genius" would not, in this way of thinking, be located in the man Homer, or the man Shakespeare, but in the muse, *locus genii*, or *daimon* that inspired (literally, "breathed into") them. Indeed, the result of this inspiration was often ascribed to a sort of madness, an ecstasy (*ekstasis*) or "standing outside oneself." To be sure, there were traditions stemming from the poet Pindar and the philosopher Heraclitus that in some sense identified the *daimon* with the individual through which it manifested. In this sense these "men of great soul" (*megalaphues*) are one with their *daimon* in character (*ethos anthropoi daimon*).[30] But it has been only two centuries or so since the notion of genius transformed into the incalculable creative capacity of an individual. "With a bias against religious authority, the [French] philosophes looked back at the discoveries of the previous hundred years and marveled at the trials through which science had been forced to pass. As they angrily surveyed the constraints set by the religious and political authorities, they

27. McMahon, *Divine Fury*, 1–33.
28. McMahon, *Divine Fury*, 10.
29. McMahon, *Divine Fury*, 33–66.
30. McMahon, *Divine Fury*, 16–17.

concluded that only genius, uninhibited by superstition and prejudice, could account for the wondrous discoveries that began with Copernicus and ended with Newton."[31] And only since World War II has the notion of genius further specified to be associated with intelligence or IQ. It is, for example, no mere coincidence that the rise of copyright laws coincided with the individualizing of genius.[32] When the complex patronage networks that supported men like Leonardo and Galileo collapsed, the need to fend for oneself by asserting the very notion of "oneself" as creative surfaced all the more. Retroactively, the "heroic" model of science—where science leaps from genius to genius—was born.[33]

While we will have more to say about this in coming chapters, our point for now is to look at how this transformed notion of "the genius" affects how historical figures have been represented, especially in relation to their surrounding contexts, and not least of all religion. Once, the heroic intellectual and practical feats of humanity would have been viewed as the evidence of God's love or generosity, and so far from a world where God seems continuously absent, he was present everywhere, pervading the splendors of the creativity of men and women. But now our lenses to see the world have changed, and genius became almost the opposite, signifying men and women waving their Promethean torches aflame with humankind's will and ability to conquer not just this world, but any of the infinite cascade churned up by the engines of imagination and the will to conquer. Genius is the heroic loneliness of humanity standing before a world where God is silent, indifferent, or dead. But here, as so often, no straightforward account of "secularization" where the wisdom of man inevitably conquers the irrationality of religion will do. Indeed, some of the rejection of traditional notions of genius actually came at the hands of religion itself, attempting to quell its more zealous components. Already in 1650 Thomas Hobbes rejected the "enthusiasm" of many—enthusiasm meaning literally "to be indwelled by (a) God" (*en + theos*). This was followed by the philosopher John Locke, who described the mind as "tabula rasa" or a blank slate. Human knowledge was not an indwelling, but the accumulation of empirical observation. This was not just a theory about how we come to know things (that is, epistemology), but also a remark against enthusiasm. Yet, despite how they are occasionally represented neither Hobbes nor Locke were anti-theological in their reformatting of the notion of genius, but rather followed the implications of certain interpretations of the biblical fall of Adam, and how it was argued in different quarters to affect the human capacities for knowledge.[34] Nonetheless, it was a small step to take from here to turn genius against religion full stop.

---

31. Appleby, Hunt, and Jacob, *Telling the Truth About History*, 17.

32. McMahon, *Divine Fury*, 72.

33. On the "heroic model" of science, see Appleby, Hunt, and Jacob, *Telling the Truth About History*, 15–91.

34. Harrison, *The Fall of Man and the Foundations of Science*.

The big picture we need to see for now is how "genius" in its reformulations helped produce the image of a figure who was paradoxically totally separate from, and yet represented, the course of whole epochs. Leonardo, for example, now represented the unleashing of the human mind from the dogmatic slumbers of theology, the chains of the teaching magisterium of the church. "Novelty—originality—was central to the genius's self-presentation and reception."[35] As we approach the nineteenth century, "the new embrace of genius places a premium on individuality and uniqueness, heightening the growing eighteenth-century appreciation of the self." But also, the growing depreciation of Christianity—in particular Catholicism and establishment Anglicanism amidst the emerging class of scientists in Victorian Britain.[36] In this sense the great French polemicist Voltaire, who was not totally unqualified in his positive judgments on the creative newness of genius, noted that "however perfect [an artist] may be[,] . . . if [they] are not original, [they] are not considered a genius." So too, the great philosopher Immanuel Kant in his *The Critique of Judgment* wrote that "everyone agrees that genius must be considered the very opposition of the spirit of imitation." Which was to say, as Kant went on, that the spirit of genius must contain "originality [as] its foremost property."[37] His motto of Enlightenment—*sapere aude* (think for oneself!)—in many ways constituted the new approach to history that occurred in the Enlightenment. Anything that was considered valuable was evidence of a future-oriented mode of thinking, one that had little to nothing to do with the past.[38]

As Duhem continued his research, he realized more and more how much Leonardo was considered such a figure because of presupposition of the category of genius, and this played a heavy hand in how the rediscovery of his magnificent notebooks were initially interpreted. As a corollary, this interpretation of the notebooks and da Vinci shaped the emerging idea of a Scientific Revolution as a "genius-driven" novel break from the past and, most particularly, the Christian religion. "During [the notebooks'] successive rediscoveries in modern times," write historians Frances and Joseph Gies, "these astonishing collections were mistakenly perceived as sketches of original inventions, products of an individual 'Renaissance' genius."[39] In part, they note, this misinterpretation comes from the startling clarity of the "aesthetic quality of the drawings," and in another part to the "exaggeration of the contribution of the individual 'inventor'" in the history of science with an underappreciation of the "social nature of technical invention." What this model did, however, was initially blind historians to the deep debts of da Vinci, and ignore the long draughts he drank from the wellsprings of those who came before him.

---

35. McMahon, *Divine Fury*, 89.

36. McMahon, *Divine Fury*, 89.

37. McMahon, *Divine Fury*, 90.

38. Brewer, *The Enlightenment Past*, 7.

39. Gies and Gies, *Cathedral, Forge, and Waterwheel*, 238.

Ultimately Gies and Gies agree with Duhem and scholarship since his time, and note that "the historical value of Leonardo's 'notebooks' . . . lies less in their author's own contributions to engineering than in their incomparable illustrations of the atmosphere in which he lived . . . opened by the discoveries of [his] medieval predecessors."[40] Indeed, Leonardo's thoughts consistently "echo those not only of his peers but of his predecessors, sometimes of a much earlier era." Startlingly, this applies even to things like his famous sketch of the ornithopter, a flying machine similar to it having actually been built and tested in prototype by the eleventh-century English Benedictine monk Eilmer of Malmesbury (c. 1005)—who may in fact be the first person in history to take flight. His fellow monk William of Malmesbury, records the discovery (or debacle, depending upon one's perspective):

> [Eilmer] was a man learned for those times, of ripe old age, and in his early youth had hazarded a deed of remarkable boldness. He had by some means, I scarcely know what, fastened [an apparatus resembling a gliding bird's] wings to his hands and feet so that, mistaking fable for truth, thought he might fly like Daedalus, and, collecting the breeze on the summit of a tower, he flew for more than the distance of a furlong [220 yards]. But, agitated by the violence of the wind and the swirling of the air, as well as by awareness of his rashness, he fell and broke his legs, and was lame ever after. He himself used to say that the cause of his failure was his forgetting to put a tail on the back part.[41]

Leonardo's magnificent creativity and insight notwithstanding, to set him up as a figure that not only stands apart but even contradicts the theologians and scholastics that preceded him because he is a "scientist," or even a tamer label like "innovator" or "freethinker," is not accurate. The "Da Vinci Code," so to speak, reveals that hidden within Leonardo's notebooks are evidence of an entire array of influences that lead to not just the mathematics and inventions, but also the theological view of the world embedded in the natural philosophy of the Christian scholastics and beyond. All of this runs like a separate stream from the tangled question of Leonardo's own personal religious beliefs. If, as some represent, Leonardo was himself not a Christian, this would hardly affect the material perpetuation of ideas born in Christian scholasticism. But even here things are more complicated than they first appear. Our earliest reports of Leonardo's beliefs come from the "Lives of the Artists" of Giorgio Vesali. Leonardo's "cast of mind was so heretical he did not adhere to any religion," Vesali remarked. In the definitive second edition of the text, though, Vesali expunged this judgement, having reevaluated the artist's positions in regard to the Almighty. Leonardo may not have been exceptionally pious in terms of ritual, but there is, it seems, no question regarding his belief in God. He was disgusted by the commercial exploitation of religious symbols like relics, and found special proof for God's omnipotence in

40. Gies and Gies, *Cathedral, Forge, and Waterwheel*, 238.
41. Quoted in Gies and Gies, *Cathedral, Forge, and Waterwheel*, 238–39.

nature through examination of the play of light and color, the great festival of plants and trees and flowers marching through the world. Above all, of course, God was revealed in the extraordinary composition of the human body, represented so famously in his sketch of the *Vitruvian Man*. "Oh you who look upon this, our machine [of the body], do not be sad that with others you are fated to die, but rejoice that our Creator has endowed us with such an excellent instrument as the intellect," wrote da Vinci. Far from scientific anatomy as we understand it today, Leonardo and the Renaissance at large understood anatomical investigation as a particularly direct route to understand God's glory.[42] Again like Prometheus bringing fire down from the gods to light the gloom, however, Leonardo was in spite of his debts set up by historians as one key figure dragging an entire era out of the religious dark.

This leads to a second aspect—the diorama of the "Dark Ages" in which many place Leonardo to shine all the brighter.[43] We will see in more detail later how the "Dark Ages" were constructed, and several of the myths that play their part in this. It is important to note, however, that Leonardo plays a role here, too, with the figure of Jules Michelet—who was one of the primary intellectuals responsible for initiating this picture of Leonardo as the first to lead captivity captive, striding out from the dark. Michelet was also the first to coin the term and periodization of "The Renaissance," in a series of lectures given in 1840 at the *College de France*, and was a full-fledged participant in the French struggles of the State against the Church. In turn, the Church, in a mode of high alert because of their memories of Napoleon (who was, it should be noted, now himself considered a great genius who stood astride history to separate it into parts)[44] and the other French grievances briefly outlined in chapter 5, saw Michelet's criticism of ultramontanism and medieval Christianity as sounding alarms threatening the Church's security.[45]

Ironically, as a young historian, Michelet saw the Middle Ages as a "period of light and creativity," as the historian Jacques le Goff puts it. But, after the death of his first wife in 1839, the tone changed completely. His grief was like an overturned ink bottle, spilling upon the Middle Ages and casting them in the sudden shock and soak of the ink tide. They are now full of "gloom, of obscurantism, petrification, and sterility." With his lectures at the *College de France*, the Middle Ages are abandoned as a wasteland, and "we have come to the Renaissance through the phrase 'return to life' . . . thus we come into the light."[46] Though we do not have the 1840 lectures in their original form, according to one eyewitness they were a "field of battle," against the Church and the Middle Ages they inflicted upon humanity, and Leonardo was Michelet's chosen champion. His "genius" was bidden to ride forth. The lecture spawned

---

42. A. Cunningham, *The Anatomical Renaissance*.

43. For more on the "Dark Ages" see chapters 8 and 9 in this book.

44. McMahon, *Divine Fury*, 90.

45. Le Goff, "The Several Middle Ages of Jules Michelet," 3–28.

46. Le Goff, *Must We Divide History*, 31, 32, 34.

violent debates that spilled across the campus. As a warning to Michelet about the unrest he and another lecturer Quinet were continually causing with their work, in 1848 the government closed the lecture halls to them and both men were not allowed to teach for three months. "When they returned," says Bullen, "they were greeted as conquering heroes."[47]

To really send his point home, Michelet represented Leonardo's vision as "Protean, Faustian, even demonic in inspiration."[48] This, of course, is meant as a compliment, and is symbolic of Michelet's opinion that Leonardo is little else than the incarnation of genius providing the antithesis to medieval theological and dogmatic attitudes. By invoking Leonardo's "genius" in this way, of course, Michelet can also infuriate the Church by appearing to name the devil. Michelet seems to affect both the new use of the concept of genius as a heroic lone individual standing against the credulity of faith, and the old sense, where the muse of Leonardo is imbued into him by a devilish realm likewise opposed to Rome. Michelet, who considered his work as a historian quite explicitly in terms of a necromancer wresting the dead from their tombs,[49] sent Leonardo forth as a pale rider on a skeletal horse, galloping onward to scourge not the world but the Church. In this way, Michelet also identified Leonardo with the persecuted witches of the Middle Ages, who, as agents of Lucifer, a "Satanic Middle Ages" conjured forth "from Michelet's depths of despair." At this stage, then, Michelet's elegant work becomes a somewhat frantic attempt to topple the canopy of Catholicism still hovering over Europe. The historian Le Goff puts this into glorious prose: "It is as if the winged Michelet could not free himself from the suffocating darkness of a long tunnel. His beating wings struck against the wall of a cathedral shrouded in shadow."[50] Though fancying himself a necromancer, Michelet came to perceive in the Middle Ages a sort of "anti-nature" that contradicted what he perceived to be the scientific spirit of his contemporary France and its struggle against the Church.

Though perhaps not so melodramatic as Michelet, similar portraits of Leonardo emerged from other influential thinkers of the nineteenth century. For William Whewell, an Englishman who coined the term *scientist* in 1833 (and a Christian, mind you), "the dark ages" in good Protestant form represented an uncanny valley of 1,000 years without progress because of the Catholic Church: "We have now to consider more especially a long and barren period, which intervened between the scientific activity of ancient Greece, and that of modern Europe; and which we may, therefore, call the Stationary Period of Science."[51] In the second edition of his *Philosophy of the Inductive Sciences,* Whewell incorporated his encounter with the newly rediscovered manuscripts of Leonardo da Vinci in Paris, and so the new sciences began with da

47. Bullen, "Walter Pater's 'Renaissance,'" 274.
48. Bullen, "Walter Pater's 'Renaissance,'" 274.
49. Le Goff, "The Several Middle Ages," 4.
50. Le Goff, "The Several Middle Ages," 6.
51. Whewell, *Philosophy of the Inductive Sciences,* II:181.

Vinci himself, inching the start of progress back slightly from Galileo. Regardless, for Whewell the medieval period still represented a "noonday slumber" between the Greeks and then Leonardo and Copernicus, with whom the great dawnsprings of reason finally broke over this sleepy world of shadows.[52] Perhaps even more influentially, Europe's leading naturalist Alexander von Humboldt in his five-volume work *Cosmos* (which according to historian William Langer was the most widely read book in Europe outside of the Bible)[53] called Leonardo a "scientific Columbus" exploring the unmapped terrain of anatomy and anticipating mathematics and mechanics in the nineteenth century. "Before Galileo and before Bacon" Whewell, Humboldt, Michelet, and others claimed that Leonardo was the first "who placed experience before received opinions, and human reason above superstition."[54]

This portrait of Leonardo was soon to be beatified and placed in the inner sanctum of the newly emerged discipline of the history of science. George Sarton, one of the first to argue for the history of science as a legitimate discipline, and rightly called its father and archon, set forth the historiographical molds for what was to follow. In many ways he followed the general pattern in Michelet, Whewell, and Humboldt, though the contrast between da Vinci and the dark ages of the Catholic church altered to become a contrast between scientific progress and theology, even religion, as a whole. For the 400th anniversary of Leonardo's death Sarton penned an article in the May 1919 edition of *Scribner's Magazine* and noted that Leonardo da Vinci was the first true scientist.[55] To be sure, Sarton (whose specialty was medieval science) acknowledged science taking place in the Middle Ages. The view that "everything was wrong and dark in the Middle Ages," says Sarton, "was a childish view long exploded." Indeed, many of the medieval schoolmen were geniuses. However, as part of that same thought he notes regardless "their point of view was never free from prejudice," that is "theological" prejudice.[56] This made them "cocksure . . . they knew everything except their own ignorance."

Elsewhere, this same strategy is used by Sarton to reap the rewards of the medieval discoveries while distancing himself from the theology that went hand in hand with it.[57] For Sarton's "New Humanism" Leonardo was a typological figure, who could as such not be too closely associated with the theology of the Christian medievals from whom he learned so much. When Leonardo read thirteenth- and fourteenth-century figures "his mind was already proof against the scholastic fallacies." He was able "to filter through his own experience whatever medieval philosophy reached him either in

---

52. Whewell, *Philosophy of the Inductive Sciences*, I:14.

53. Quoted in Gregory, "Continental Europe," 89.

54. Quoted in Bullen, "Walter Pater's 'Renaissance,'" 270.

55. Sarton, "The Message of Leonardo."

56. Sarton, "The Message of Leonardo," 537.

57. Sarton, *The History of Science*, 100.

print or by word of mouth."[58] Indeed, more properly than inheritors of scholasticism, Sarton characterized the debts of Galileo, Kepler, and Leonardo to the scholastics only to the extent that the scholastics preserved the spirit of the Greek Archimedes. What is most curious of all in Sarton's review is that though Sarton had personally reviewed the first volume of Duhem's *Le System du Monde* in 1914 for the journal *Isis*, and was well aware of Duhem's Leonardo studies, not once is Duhem's work on Leonardo mentioned in Sarton's article even to be refuted. This peculiar silence would unfortunately be a fairly characteristic feature in Duhem reception, for one reason or another.

In the end, what Duhem emphasized was needed is not the heroic or genius-driven model of history, but one focusing upon broader contexts, and the long duration of innumerable patient laborers. "Brilliant and solid as Leonardo was," writes Duhem, "he takes his place in the chain of scientific tradition"[59] that goes back deep into the Christian Middle Ages. In fact, Leonardo "sums up and condenses so to speak in his person all the intellectual conflict through which the Italian Renaissance becomes the inheritor of . . . Parisian scholasticism."[60]

The Labor of Centuries

The history of science is distorted by two prejudices, so similar to one another that they could be fused into one: the current thinking is that scientific progress is made by a sequence of sudden and unforeseen discoveries. It is, according to general belief, the work of geniuses, who have no precursors at all.

—Pierre Duhem, *Studies on Leonardo Da Vinci*

Despite the length and pedigree of the "Dark Ages," the rising notion of the genius, and the sudden eruption of science from the darkness imposed by Christianity in the literature of his time, Duhem was not convinced—though early on he could not yet say exactly why. In fact, he himself had been a general supporter of the idea that the Middle Ages was not but a barren wasteland. Even by 1903 there would have been little to summon from his pen about medieval science, precisely because he still believed there had not really been any such thing.[61] He noted, however, that despite all of this "it was easy to recognize [in the summary works Duhem had been reading] that most of them were very condensed and lacking in detail."[62] Yet, in one of those peculiar connections of history, by an incidental reference in a postcard from his friend and fellow scholar Paul Tannery—outlining a text fragment of Euclid on

58. Sarton, "The Message of Leonardo," 537.
59. Quoted in Jaki, *Uneasy Genius*, 390.
60. Quoted in Jaki, *Uneasy Genius*, 393.
61. Martin, *Pierre Duhem*, 147–63.
62. Quoted in Martin, *Pierre Duhem*, 150.

the measurement of weights—Duhem became dimly aware of a man he eventually termed "the enigmatic" Jordanus de Nemore (c. 1250).[63] In 1904, when researching *Les origines de la statique* another unusual reference to de Nemore in a sixteenth-century book solidified Duhem's curiosity. When Duhem began tugging at this little thread of Ariadne, this hint of science where it ought not to be, he found that he had unwittingly stumbled into a labyrinth. One that he would spend the rest of his life exploring.

Duhem's publisher, Jules Thirion, SJ, noted the first sign of Duhem's unexpected discoveries in a letter replying to Fr. Henri Bosmans. Sections of Duhem's book were set to be published one after the other in the journal *Revue de questiones scientifique*, and having been enamored by the first published section, Bosmans wanted to peek at the manuscript for the upcoming January release. Given Duhem's reputation for an astonishing pace of work, the reply Thirion sent to Bosmans no doubt shocked both men: "I haven't got it yet. Duhem hasn't finished it. He says he still has some reading to do."[64] Duhem himself seems to have been taken aback at this. When he finally wrote the preface to his finished text, he remarked "when we turned to the study of the texts [books on the history of statics] referred to, we anticipated having to add or alter many details, but nothing led us to suspect that the history of statics in its entirety would be upset by our researches."[65] Duhem had "discovered that his sources were so inadequate he had to start again, and he was now discovering the facts, barrowloads of them."[66]

In fact, Duhem's work led him to discover a great series of historical plagiarisms one on top of the other. Or, put more charitably: a series of historical "discoveries" that had been forgotten, or perhaps had merely remained stubborn with their secret that they had been discovered previously. Before Steven Simon (1548–1620) and Galileo Galilei (1564–1642), Niccolo Tartaglia (1500–1557) determined the apparent weight of a body on an inclined plane, for example, along with many other precursors to modern method. But here, Tartaglia was accused by a contemporary, Ludovico Ferrari, of merely erasing the name of the prior discoveries of our previously mentioned mystery man, Jordanus de Nemore, a mathematician of the thirteenth century. Another similar discovery has been termed the "de Soto Enigma" after another transitional figure discovered by Duhem—and later confirmed by historian William A. Wallace—Domingo de Soto, whom Duhem argued was the missing link connecting Galileo to the fourteenth-century scholastics, and even earlier to thinkers no less prestigious than Thomas Aquinas.[67] This will no doubt seem inconsequential to most of us. To Duhem, it again reinforced the paradox that demanded to be solved: even more lost precursors

---

63. Quoted in Martin, *Pierre Duhem*, 149.
64. Quoted in Martin, *Pierre Duhem*, 149.
65. Quoted in Martin, *Pierre Duhem*, 150.
66. Martin, *Pierre Duhem*, 157.
67. Wallace, "The Enigma of Domingo de Soto," 384–401.

to modern science cried from the dust, signaling discovery where, according to the textbooks, none should be.

Just nine months after he began his initial descent into study, Duhem had not only fully abandoned his original opinion regarding the sterility of the Middle Ages. He had, in fact, come to what was at first dimly perceived, but now fully realized idea that a complete historiographical rewrite of the emergence of modern science was on order.[68] In November of 1903 he had accepted Tannery's invitation to contribute a paper on some aspect of the history of mechanics. When he turned in the much-discussed results, which channeled into the tight space of an essay the oceanic sweep of his initial efforts, Duhem's tone was that of a man caked in dust, but whose eyes were still bright and shining as he emerged from a treasure cove with his prize. Not unscathed, but triumphant with gems in hand, Duhem announced:

> To establish this detail, we have had to impose laborious drudgery on our-selves: we have had to examine and analyze the many manuscripts relating to statics held in the *Bibliothèque Nationale* and the *Bibliothèque Mazarine*. This analysis has allowed us, we believe, to discover more than one spring, un-known or misunderstood until now, whose waters have copiously contributed to the formation of modern science.[69]

What the documents revealed in this case—among their many treasures—were precursors for the theory of inertial motion put forward by Descartes, Galileo, and then in its ultimate form by Isaac Newton. Newton's famous "first law" of motion notes that bodies at rest tend to stay at rest, and bodies in motion tend to stay in motion, unless either are acted upon by an external force. This no doubt seems commonsense to us today, being as we are at the far end of so much intellectual labor. But for the reigning paradigms of motion at the time—mostly variations of Aristotle—theories of motion and projectiles had always been something of a weak point.

More to our point, suddenly the intense historical singularity of Leonardo da Vinci—and in turn others like Descartes, Galileo, even Newton himself—found precedent in the forgotten depths of time. Here was a renaissance before the renaissance; a light—not just *in* the dark—but one that shone all around it making all it touched, light. And this noon flaring amidst a place of dusk was from the medieval scholastics no less, who were despised as useless pedants arguing about how many angels could dance on the head of a pin, if they were remembered at all. "Science does not know spontaneous generation," wrote Duhem, "Not even the most unforeseen discoveries have been made in all details in the mind that discovered them." This was no doubt a dark saying for Duhem's contemporaries, who relished in the novelty of science as a method to chastise the Church in its relationship to new knowledge, especially as Duhem's words crept closer not just to da Vinci but to Galileo. Soon no doubt, many

68. Patapievici, "The 'Pierre Duhem Thesis.'"
69. Martin, *Pierre Duhem,* 157.

realized that even that lonely god among men Isaac Newton was to be held in the gaze of Duhem's newfound historiography. Duhem chastised what he called the philosopher René Descartes's "prodigious arrogance," to see only errors in the past, when in fact "in [Descartes's] brilliant essay on statics [he] said nothing that would not have been known long before him in the school launched by Jordanus."[70] Descartes no doubt is also something of a stalking-horse for Duhem's chronologically snobbish contemporaries:

> The science of mechanics and physics, of which modern times are so rightfully proud, derives in an uninterrupted sequence of hardly visible improvements from doctrines held in medieval schools. The pretended intellectual revolutions were all too often but slow and well-prepared *evolutions*. The so-called renaissances were often but unjust and sterile reactions. Respect for tradition is an essential condition of scientific progress.[71]

Duhem's use of "doctrines" is here well advised. These were not figures who merely "happened" to be Christians *also* interested in what we would now label scientific pursuits. Such is a riposte meant to concede the historical point of the connection of Christianity and science, but weaken it by suggesting the relationship was merely one of happenstance, unrelated by anything else except for the fact that both pursuits were undertaken by the same individual. Undoubtedly, there are certain instances where this is true—though even at such a basic level it rings as a concession that already complexifies the notion that Christianity is essentially adversarial to science. Duhem's claim here is, nevertheless, stronger: Christian theology and practice provided resources that, at multiple different levels, both removed certain impediments to science and provided theoretical underpinnings that gave direction and justification to certain methods, principles, and theory selection.

In this manner Duhem's historical work plays directly into his constructive approaches in the philosophy of science. An advocate for what later became known as the "underdetermination of data" in which a variety of theories seem to explain the data at hand, the question becomes how theories emerge and are selected from among a menu of explanations each of whose power is roughly equivalent. "The thesis of historical continuity [of the Christian theological framework of physics] is one part of [Duhem's] epistemology, with which Duhem attempts to resolve the problem of the choice of hypotheses,"[72] as Robert Maiocchi points out. Theology and religious beliefs can impact science and other disciplines in a variety of ways, and not just in specific findings or ideas. The preference for simplicity in explanations of the natural world for example—sometimes referred to as Ockham's Razor after William of Ockham (1287–1347), a scholastic and one of the players in Duhem's new story—can hardly be

---

70. Quoted in Jaki, *Uneasy Genius*, 386–87.

71. Quoted in Jaki, *Uneasy Genius*, 386–87.

72. Maiocchi, "Pierre Duhem's *Aim and Structure of Physical Theory*," 395.

justified except that in terms of explanatory heft and aesthetic beauty, such simplicity is what we have historically come to expect because of belief in the One Creator God. "There is no reason—in the absence of independent belief in the simplicity of nature," writes philosopher of science James McAllister, "why [the preference for simplicity] should result in hypotheses that are true more often than would any other."[73] Despite mostly failed attempts to make it an independent criterion, it still bears the marks of its theological origin. For that which is simple need not be beautiful, nor the beautiful true, nor all expected together.

In a classic 1960 paper, for example, the mathematician and Nobel Laureate Eugene Wigner remarked upon "the unreasonable effectiveness of mathematics," and how often mathematics are linked both to simplicity, but also, remarkably enough, to beauty and aesthetic judgments. Not only the notion that human minds reflect reality—and so that reality itself is open to human minds—but that simplicity, unity, and beauty are all frequent and indispensable components of this "is something bordering on the mysterious, and there is no rational explanation for it."[74] This idea has been such a powerful assumption that ironically some have come to critique its lingering presence it in a most unexpected place: that of natural selection in evolutionary theory. The Yale evolutionary ornithologist Richard Prum, in an insightful recent work on the evolution of beauty, takes no less than the vocal atheist Richard Dawkins to task for his stress on natural selection as a single, all-encompassing explanation. "Contemporary adaptationists should question why they *feel* it is necessary to explain all of nature with a single powerful theory or process." Though not a historian, his tentative answer is that "the historic monotheism" surrounding Darwin's theory and its reception "predisposed them . . . to replace a single omnipotent God with a single omnipotent idea." He closes this thought with a pointed question: "Is the desire for scientific unification simply the ghost of monotheism lurking within contemporary scientific explanation?"[75] The answer stemming from Duhem's work would seem to be a resounding *yes*. Just so, it is not merely that figures like Jordanus, Jean Buridan (c. 1300–1361), or Nicole Oresme (c. 1325–1382) provided mathematical, practical, and theoretical precedent for later figures, which they did in abundance. In explicit defiance of Mach and other positivists whose positions we will see in the next chapter, theology and metaphysics are shown by Duhem to play a key role in providing the very conditions for understanding the world scientifically.

The Duhem Scandal

Unfortunately, a dismissive response to the implications of Duhem's newfound data for the history and historiography of science was hardly uncommon even among

73. McAllister, "Truth and Beauty in Scientific Reason," 32.

74. Wigner, "The Unreasonable Effectiveness of Mathematics," 2.

75. Prum, *The Evolution of Beauty,* 52–53.

those who appeared to have learned a great deal from him.[76] Duhem—as one example—noted that Christian theology in many ways dismantled the scientific straitjacket of Aristotelianism when the so-called "Condemnations of 1277" proclaimed that the Christian God was not subject to the physical necessities demanded by Aristotelianism:

> If we had to assign a date to the birth of modern Science, without doubt, we would choose the year 1277 when the Bishop of Paris solemnly proclaimed that several Worlds could exist, and that the entirety of the heavenly spheres could, without contradiction, be animated by rectilinear motion.[77]

Putting aside our shock that the Church declared that many worlds could exist in the thirteenth century, it is nonetheless quite easy to scoff at Duhem's suggestion regarding a new date for the birth of modern science. A great admirer of Duhem, the historian and philosopher of science Alexander Koyré, is reputed to have remarked that "if Duhem had never made any mistakes, we would have had no great jobs to do. We have lived on his mistakes."[78] And indeed, the notion of science starting with the 1277 condemnation is perhaps too boldly stated, and open to misunderstanding. Nonetheless, many criticisms of Duhem including Koyré's "rest upon misapprehensions of Duhem's work" that stem from their own, fundamentally different metaphysical commitments.[79] As Ariew and Barker put it, "[Koyré's] metaphysical commitment to revolutionary accounts of science obliges [him] to read Duhem as a bad revolutionary." Yet, "in Duhem's accounts there can be no revolutions, so he can hardly be blamed for getting the dates wrong."[80]

Duhem wrote in the third volume of *le System du Monde* that "never was any science invented at any particular time, but from the beginning of the world knowledge has grown slowly, and is still not complete at this very age."[81] In focusing upon the condemnations of 1277, Duhem meant rather that a series of thresholds were needed to remove chaff that were inhibiting scientific growth. New possibilities undreamt of—or literally defined as impossible in Aristotle (such as the existence of voids)—with the condemnations suddenly became thinkable, even if it would take a good deal of time to realize this. Duhem writes, "New thoughts passed through [the 1277 condemnation of Aristotelianism] many of which can be rediscovered, barely modified, in the writings of our contemporaries who philosophize about the principles of science."[82] As we will see in the chapter on Copernicus and Galileo, "the shift in

---

76. Patapievici, "The Discovery of the Physics of the Middle Ages by Pierre Duhem."
77. Quoted in H. F. Cohen, *The Scientific Revolution*, 52.
78. Quoted in Oliveira, "Duhem's Legacy," 135.
79. Ariew and Barker, "Duhem and Continuity," 324.
80. Ariew and Barker, "Duhem and Continuity," 337.
81. Quoted in Jaki, *Uneasy Genius*, 402.
82. Duhem, *Medieval Cosmology*, 369.

human self-understanding [to an infinite universe was] inseparably bound up with a changing understanding of God and of God's relationship to man and to nature."[83] The condemnations placed a newly recharged emphasis on the infinity of God and his omnipotence to create and change conditions as he pleased, leaving it up to investigators to empirically determine just what it was he had made.[84]

Far from a fun bit of speculation to do at the pub, this led to the notion of an infinity of perspectives, variables, and frames of reference for finite observers to take into account. The mind was sent reeling before this abyssal God in medieval thought experiments, to which Galileo, and even Newton were directly indebted for formulating things like the laws of inertia, frictionless planes, the theoretical isolation of variables, and other ideal conditions necessary for the formulation of laws of nature.[85] Thus, for example, space could not be defined as that which contained a body, as Aristotle held, for in theory God could destroy the whole world around the apple that Eve took from the tree, and yet the apple would remain as it is, including its dimensionality. Rather than a container, space, it was thought, was a notion relative to bodies that were all created by God simultaneously and, as created, could ultimately be defined only relatively to one another. It would be too much to say this anticipated Einstein's notion of relativity, and yet some of the conceptual similarities, as pointed out by theologians like T. F. Torrance, are there.[86] As such, when this and a host of other changes are kept in mind, "the revolution to which [historians of science like Alexander] Koyré calls our attention seems much less revolutionary. We are more nearly right when we speak of modern science, and thus also of our own culture, as a product of the self-evolution of the Christian culture of the [Patristic and] Middle Ages."[87]

Others have taken Duhem's searching even further. William A. Wallace's work for example has criticized Duhem, but only to the extent that he underplayed even earlier sources outside the Parisian masters post-1277, such as "Robert Grossetest at Oxford and Albertus Magnus, Thomas Aquinas, and Giles of Rome at Paris, all who did their work before the condemnations of 1277, or in essential independence of them."[88] Perhaps one of the earliest examples of such precursors for Duhem's notion of theological change was discovered, for example, in the work of John Philoponus, sometimes called "John the Grammarian" (c. 490–570), who is notable among other things for his work on astrolabes. In contrast to Aristotle—who described vision as a ray coming out from the eye itself—the notion of the light of grace as gift, along with his own observations, led Philoponus to reorient entirely the framework for optics

---

83. Harries, *Infinity and Perspective*, 15.

84. Osler, *Divine Will and the Mechanical Philosophy*; Funkenstein, *Theology and the Scientific Imagination*; Riskin, *The Restless Clock*; Harries, *Infinity and Perspective*.

85. Funkenstein, *Theology and the Scientific Imagination*, 152–84.

86. Torrance, *Divine Meaning*, 343–73.

87. Harries, *Infinity and Perspective*, 15.

88. Wallace, "Pierre Duhem," 304.

as receptive.[89] Nor could matter be eternal, said Philoponus, and the heavens by the same chain of reasoning were created and so not divine, immutable, or a separate system to terrestrial mechanics.[90] Rather, both the glittering wheels of the heavens and the brooding, damp stone of the earth were brought into being by the creation and continued preservation of God, subject to singular principles straddling both. This was a remarkable moment in history. Even Galileo remarked concerning his own dependence upon many of the anti-Aristotelian arguments found in Philoponus.[91] And so Wallace can finally remark: "Thus we too are advancing the continuity thesis first proposed by Pierre Duhem, only enhancing it now to show a fuller dependence on medieval [and patristic] thought than has hitherto been proposed by historians of science."[92]

Unfortunately, the general response to Duhem's work was only a very loud silence. Herbert Butterfield—himself not a historian of science but an especially gifted and thorough generalist in history—could write that "the work of Duhem . . . has been an important factor in the great change which has taken place in the attitude of historians of science to the Middle Ages."[93] Nonetheless, what should have been the centenary celebration of Duhem in 1961 passed without comment. In 1977 no less a figure than Thomas Kuhn—author of one of the most influential books of the twentieth century, *The Structure of Scientific Revolutions*, and the coiner of the now *ad nauseum* phrase "paradigm shift"—could write "Pierre Duhem's search for the sources of modern science disclosed a tradition of medieval physical thought which, in contrast to Aristotle's physics, could not be denied an essential role in the transformation of physical theory that occurred in the seventeenth century. . . . The essential novelties of seventeenth-century science would be understood only if medieval science had been explored first . . . More than any other, that challenged has shaped the modern historiography of science."[94] Strangely, if the case is as Kuhn said, it remains thoroughly mystifying why he did not cite Duhem in his main and most famous work, *The Structure of Scientific Revolutions*. So conspicuous is this absence that Kuhn's own commentators picked up on it and noted that "Kuhn ignored his debt to Duhem while respecting his [that is, Duhem's] leading followers."[95] Indeed, another goes so far as to say that there is "nothing of real relevance to this particular issue in *The Structure of Scientific Revolutions* that was not raised already in Duhem . . . many of the Kuhnian

---

89. Sorabji, ed., *Philoponus and the Rejection of Aristotelian Science*; McKenna, "John Philoponus."

90. Schukin, "Matter as Universal."

91. Torrance, "John Philoponus," 98.

92. Wallace, "Pierre Duhem" 304.

93. Butterfield, *The Origins of Modern Science*, 15.

94. Kuhn, "The History of Science," 108.

95. Agassi, "Kuhn's Way," 409.

theses that have created such a stir in philosophy of science seem at most to be (often rather less clear) restatements of Duhemian positions."[96]

Silence nonetheless reigned. Even early on in his career, before he had even turned to the history of science, Duhem had made powerful enemies that hindered his reception. Chief among them was the enormously influential French scientist Marcellin Berthelot (1827–1907), whose ideas had received devastating criticisms at the hands of Duhem's doctoral work on thermodynamic potential in physics and chemistry. We need not detain ourselves with the specifics. This criticism led Berthelot in a panic of anger to declare that "this young man will never teach in Paris." It was no idle threat, for the well-placed academics who owed the advancement of their careers to Berthelot's paternal favor were legion. Others with no particular debt to Berthelot nonetheless were well aware enough of the order of things to do nothing about the blackballing of Duhem. In the same savage gesture Berthelot informed major publishers in Paris that they would become the targets of his considerable ire should they ever release anything written by Duhem. Duhem's daughter Hélène laments the situation in somewhat melodramatic terms that nonetheless seem to fit the afflicted circumstances of Duhem's academic career:

> "This young man will never teach in Paris," declared Berthelot. . . . Then began that thirty-year struggle between the Sorbonne on the one side and Pierre Duhem on the other. He will be the enemy, the man never to be spoken of, all of whose productions will be ignored, whose discoveries will all go unmentioned, whom by this silence and oblivion they will hope to discourage, whom even today they affect not to cite, even when a sentence in a work seems to be taken verbatim from one of his books.[97]

Berthelot was not the only obstacle to the eventual reception of Duhem's work. For one reason or another the second half of *Le System du Monde* (comprising five of the ten volumes), was delayed in publication for nearly forty years after Duhem's death. It is notable that the torturous delay in publishing only ended when a lawsuit was leveled against the publishing company.[98] It did not help that because of his Catholicism, Duhem was also taken to be an ultramontane Catholic zealot—an ultramontane, that is to say, is one who believes the pope's authority transcends all borders, geographic or national—who was seen as a deep threat to the stability of the French state. Later, in the 1960s after the full breadth of *Le System du Monde* had been published, many who read Duhem also dismissed him as nothing more than a Catholic apologist disguised as a historian. Quite ironically, on the other side of things, since Duhem considered himself as something of a subversive in terms of his disagreements with the reigning Catholic neo-scholastic theology of the day—which was officially sanctioned by

96. Worrall, "'Revolution in Permanence,'" 77.
97. Quoted in Martin, *Pierre Duhem*, 199.
98. Jaki, *Scientist and Catholic*, 123.

the Church and consisted largely in variations and commentary on Thomas Aquinas—Duhem also found few supporters among the theologians of his time. As Eugene Klaaren noted in the 1970s, Duhem's findings regarding the contribution of the medievals to the sciences has often been neglected for—quite ironically—very *theological* reasons: "many Christians have tended to slight the fourteenth- and fifteenth-centuries" so that "Thomists, Calvinists, and Lutherans alike have not found their heroes in these periods."[99]

The journey of Duhem's historical work to publication was, against these forces, a long and arduous quest. One, in fact, undertaken tirelessly over nearly four decades by his daughter Hélène, and whose twists and turns were incredible enough to warrant an entire book written on this strange limp toward publication.[100] Part of the difficulty was, initially, simply pragmatic. After the First World War the *franc* was severely devalued and the cost of printing paper skyrocketed, and so the expenditures associated with the original contract signed by Duhem were seen to be unworkable by the current publishers. Moreover, as Hélène herself noted, publication was costly "because of the frequent footnotes," which included lengthy citations of Greek and Latin and technical asides into mathematics; and so, by necessity, involved the lengthy and expensive procedure of switching typefaces and hiring editors with the proper and quite divergent sets of expertise.[101]

Many of Duhem's friends who desired his work to be published nonetheless remained optimistic, and Hélène was, for example, informed in a letter from the new publisher Freymann after several frustrating years with no progress in publishing that the current director of the publishing house "Paints. He adores Leonardo. He said he is resolved to publish Duhem, because of Leonardo. Long live Leonardo! [Yet] the old father-in-law and his son [the previous directors of the Hermann publishing house] have left behind the finest disorder." Freymann ends on a somewhat ominous note, however, asking Hélène if she had a copy of the original contract that the current publishers were unable to locate.[102] This would not be the last time she was asked to send the original contract. Hélène promptly sent it via registered mail and waited for a reply, or for the new contract to be issued. Neither came, and Hélène was left adrift in a somewhat eerie and discomforting silence, one that simply continued on with no obvious signs of reprieve. "I fear," wrote close family friend and a colleague of Duhem's, Albert Dufourcq, "that something contrary to your interest is at work."

99. Klaaren, *Religious Origins of Modern Science,* 32–33. Klaaren is not specifically referring to Duhem (in fact, Duhem is curiously absent from Klaaren) but the point Klaaren is making is directly applicable.

100. Jaki, *Reluctant Heroine.*

101. Jaki, *Reluctant Heroine,* 166.

102. Jaki, *Reluctant Heroine,* 106.

Indeed, "living here in Paris, I hear that certain people are happy . . . thinking that the monument of Duhem [*le System du Monde*] will never see light."[103]

Whether they were related or not, Hélène could not help but see in the Parisian publishers dragging their feet the echoes of Berthelot's own campaign in Paris against her father. "You will recall as I do with deep sadness," she wrote to Dufourcq, "that the official potentates who barred his road to Paris and organized a conspiracy of silence about everything he had written, made that dream impossible to realize."[104] She would, many years later, still reflect that "Certain publishing houses would retreat in front of the official opposition of the Sorbonne." This is the reason hidden behind the many excuses that exaggerate the delay, she wrote to Dufourcq, "[here] is the very reason for which Hermann [the publisher] has, since 1917, never resumed the publication of the work, even when the economic situation was prosperous, not even when the copies in stock were diminishing . . ."[105]

Anecdotally, perhaps the clearest illustration of this silence was from a then just-graduated PhD student, Hélène Metzger-Bruhl, who would later become a very notable figure in the sciences before her destruction by the Nazis at Auschwitz in 1944, for the crime of being a Jew. After no doubt pouring thousands of hours into her thesis on the intellectual development of crystallography, Metzger-Bruhl learned only after submitting to her committee at the Sorbonne, and much to her amazement and consternation, that her conclusions were anticipated to a startling degree by Duhem.[106] She was mortified at this oversight, but in the main both curious and perplexed at how she had overlooked Duhem. The solution, which she was too kind to state later to her advisers, was that the Sorbonne had been, at the leading charge of Berthelot, the main opponent of Duhem's work. Just so, they did not keep copies of his publications for student reference. "The day after the exam," she wrote, "I went out and purchased a copy of his *Théorie Physique* and (if I may say so) rapidly devoured his [other] philosophical and historical works."[107]

Metzger-Bruhl was so impressed by Duhem's work, and so perplexed that she did not previously know of its existence, that she began writing letters with Hélène Duhem, and in fact coaxed her to attend a meeting of the *Academié des Sciences* in 1937. The meeting was to be dedicated to the memory of Duhem, and Metzger said all in attendance would be delighted to hear from Duhem's daughter on the memories of her father, the scientist and historian. Reluctantly, Hélène Duhem agreed, nervous to speak in front of such a learned audience. The first to present that evening was Maurice d'Ocagne, the President of the French Mathematical Society and Inspector General over the engineering of French roads and bridges, who remarked that Duhem

103. Jaki, *Reluctant Heroine*, 109.
104. Jaki, *Reluctant Heroine*, 122.
105. Jaki, *Reluctant Heroine*, 168.
106. Jaki, *Scientist and Catholic*, 79.
107. Jaki, *Reluctant Heroine*, 145.

"had definitely destroyed the legend held for a long time by many authors as an article of faith, about the scientific night of the Middle Ages."[108] D'Ocagne then continued noting that it was not just an insult to Duhem that the second half of his great work had yet to be published, but it was an insult to France itself: "By not making the necessary effort on behalf of the publication of the remainder finished by Duhem, the powers of officialdom will fail in what they owe to the intellectual elite of the country, [and] to its great renown in the world."[109] Metzger-Bruhl was the next to speak, and similarly called for publication:

> In order to draw attention [to Duhem's reputation], I ask you not to leave this meeting before voting with unanimity the position which I now submit to you, with possible modification, of course, as follows: The French group of historians of science associated with the *Académie international de histoire des sciences*, goes on record with its wish that the unpublished work of Pierre Duhem, and in particular the concluding part of his admirable work, *Le système du monde de Platon a Copernic*, be speedily published because it will give the greatest service to all interested in the development of human thought.[110]

The historian Abel Rey was up next, and praised Duhem's "intuition," that had allowed him to see the unity in a history that "on a superficial look appears jostled and diverse, made of branches with no connection but which is nevertheless [as Duhem demonstrated] one in its essence just as the human spirit that created it." Indeed, Rey concludes that Duhem's work has done nothing less than effect "a Copernican revolution in the historiography of science."[111] Several more speakers went, including Hélène, who no doubt was glowing from the warmth of the good will and praise for her father and her own labors. The group closed the day by unanimously supporting Metzger-Bruhl's call for the publication of the second half of Duhem's masterwork.

A key figure, however, was absent from the meeting, despite a standing invitation: the publisher, Freymann. No doubt had he attended, there would have been little recourse to the delaying tactics he had employed through letters up to that point. In some sense, nonetheless, his absence was a portent for how the next two decades would continue to go. Along with this, the historian of science George Sarton—whom we met earlier—signed a document urging publication, along with Madame Tannery (the wife of Duhem's close friend and scientific associate Paul Tannery) in 1939. "To leave volumes VI to IX of Duhem's work unpublished is almost a scandal,"[112] he wrote. In hindsight, one feels the ominous force of the "almost." Despite this declaration it would still be another twenty years before the final five volumes saw the light of day.

---

108. Quoted in Jaki, *Reluctant Heroine*, 144.

109. Jaki, *Reluctant Heroine*, 144.

110. Jaki, *Reluctant Heroine*, 145.

111. Jaki, *Reluctant Heroine*, 146.

112. Jaki, *Reluctant Heroine*, 141.

Many of Sarton's students and interlocutors felt that he held a heavy hand of censorship and redaction over their works, especially in regards to any religious content.[113] Here, with Duhem, it seems little was different. Sarton was eager for his New Humanism to distinguish the science of the medievals from their theology, and this it seems played a factor in his Janus-faced treatment of Duhem's work.[114] The circle around Hélène comprising of several old friends of Duhem and his fellow scientists confided to her in a letter that they felt Sarton had duped them all. "When we were alone again [after speaking with Sarton about the publication of Duhem's *System*]—Cartan, Peres, and myself—all agreed we had been taken for a ride"[115] wrote Dufourq.

And so: it was through this lengthy concatenation of delays, dilemmas, and silences that the full heft of *Le System du Monde* only arrived on printed paper in 1959, a full forty-three years after Duhem had passed away. It comes with perhaps no surprise that the debate regarding the continuity or discontinuity between medieval and modern physics reached something of a head around the same time.[116] Duhem's "Da Vinci Code," through war, dogmatism, and both personal and professional strife, was almost lost. Of course, the ultimate point in these inquiries and historical paradigms is not really claiming the figure of Leonardo, or indeed any given thinker leading up through the so-called Scientific Revolution. Rather, they each serve as particularly poignant references symbolizing an entire sequence and shape—and so meaning—of history. Despite the fact that Duhem's work went through such trials—and certainly we should not be so naïve as to say it was the sole inspiration for later work—Duhem's da Vinci, so to speak, began to leaven the history of science and religion in the twentieth century. "All future work will consist, in large measure, in working in the veins [Duhem] has opened," as one commentator wrote, even well before the final five volumes had been published.[117]

As the noted expert on the historiography of the Scientific Revolution H. Floris Cohen remarks, at minimum "Duhem's work marks the point where [the notion] . . . of *absolute discontinuity* in the birth of early modern science is left behind more or less for good."[118] And yet discontinuity was still the view that was, until the last quarter of the twentieth century, the story that found itself passed down to posterity in footnotes and forged historical memories of the rise of the sciences. This involved not just chronological discontinuity between the Middle Ages and the births of science that only came after, but also the discontinuity between the theological and the scientific. "Which historian will restore us to . . . Duhem?" asked A. Leboeuf in his prescient

113. Ungureanu, "Relocating the Conflict," 1120.

114. Sarton, *The History of Science*, 38, where Sarton acknowledges some good from the theological influence upon the sciences, but argues that in the majority the relationship has been "aggressive."

115. Jaki, *Reluctant Heroine*, 216.

116. McMullin, "Medieval and Modern Science."

117. Durand, "Nicole Oresme," 168.

118. H. F. Cohen, *The Scientific Revolution*, 54.

1919 review of volume 5 of *le System du Monde*. "Is not there [in this work and its early reception] . . . a melancholy image of the fragility of our efforts, of the inexorable slowness imposed on the march of truth" that come in the form of the "brutal rupture," the "long eclipses" between master and disciples?[119] A long eclipse, that is, until the latter half of the twentieth century, when the story began, suddenly, to change. The quiet revolution was starting to rise into a murmur. Scholars began to see, in the almost prophetic early admission of the thoroughly pagan Benedetto Croce in 1949, that not just in culture and practice, but so too now in science, "it is impossible for us to call ourselves completely non-Christians."[120] To that we now turn.

---

119. Quoted in Jaki, *Uneasy Genius*, 432.
120. Croce, "We Cannot Help But Call Ourselves Christians."

—Chapter 2—

Paper and Bone

The Deletion of Theology from the History of Ideas

The emergence of the history of science as an academic discipline was . . . in part conditioned by the fact that it served as a battleground for the wider debate over the significance of science in Western culture. . . . [In other words, the history of science as academic discipline was] *in part an offshoot of a major ideological disagreement over the relationship between science and religion.*

—Peter Bowler, *Reconciling Science and Religion*

So profound has been the impact of Christianity on the development of Western civilization that it has come to be hidden from view.

—Tom Holland, *Dominion*

THE LONG DELAY IN the publication and reception of Duhem serves as an apt metaphor for a broader problem that occurred in the history of science. As one scholar has asked, "to enter the primary sources is to find a very different phenomenon, so foreign to the . . . mythology of secular emancipation [from theology and religion] that we have to consider what evidential fallacy could explain the discrepancy?"[1] The question is much larger than the personal imbroglios like those overshadowing Duhem's work, and there is perhaps no one answer. Nonetheless, through Duhem's work and those that followed a clue emerged: something had crept into the method of science and the history of science that had caused a vast and systematic blindness. As this book argues, such blindness was "a fantastic product of the secularization of scholarship in the [nineteenth] and twentieth centuries, rather than a reflection of any real historical

1. Erdozain, "A Heavenly Poise," 77.

trend."[2] If "genius" as a category could subtly rework history to separate the religious past from an enlightened present, ironically enough it turns out the word *scientist* and by extension "The Scientific Revolution," has done much the same. Let us rewind time for a moment by traveling to the year 1881 for our detective story, when the London Natural History Museum in South Kensington opened. It was not only one of the pinnacles of the Gothic revival in the Victorian period, it also encapsulated the newly minted Victorian spirit of scientific investigation. All the great and recent discoveries of humankind were there, sitting displayed as in a great cathedral of light. But was that light of God, or of man? Such a question would not have scandalized the ancient world as we may think it might, because this juxtaposition would have been seen as ridiculous. Whatever wisdom mankind may have, this was a gift of God. Though the concept of a "scientist" as we now know it is so familiar it may appear ageless in its self-evidence, William Whewell coined the term in the 1830s.[3] The English word *science* came even later. To be sure, the Latin term *scientia* is quite old. It signified, however, any area of knowledge that could admit of some type of systematic investigation appropriate to the nature of its object of inquiry. Indeed, as was the topic of countless works in the Middle Ages even up through the nineteenth century, in this manner theology was often considered a science, a *scientia*. The term *scientia*, as such, indicated not so much a topic of inquiry but rather a type of method and habitual virtue taking root in the individual. It was a form of mind, a mode of knowledge, indeed one could even call it a way of life.[4]

The terms *scientist* and *science* as we have come to use them supposedly inaugurated something quite new, especially in English-speaking circles.[5] This term was meant to distinguish its object of inquiry from theology, metaphysics, aesthetics, natural philosophy, and other matters (including what was to many at the time the annoying social prestige of the Victorian cleric). The *Oxford English Dictionary* records the first use in this limited sense only in 1867.[6] Regardless, at the inception of the Natural History Museum, outside the walls of this house of wisdom, sitting at the apex of its highest gable resided a terra-cotta statue of the biblical first man, Adam.[7] This was no aberration, some vestigial holdover from a quaint religious and theological past that had yet to be discretely removed, now juxtaposed awkwardly with the newly won vistas of legitimate human knowledge. Just as the grand cathedrals of old (in whose style the museum itself was crafted), "The Natural Museum in Kensington . . . encouraged its visitors to view it as a temple of science." Indeed, one commentator even remarked that as visitors came to this "animal's Westminster abbey" with its

2. Sheehan, "The Enigma of Secularization," 1063.

3. See Ross, "Scientist," 66–67; H. F. Cohen, *The Scientific Revolution*, 27–39.

4. Harrison, *Territories*, 1–21.

5. Kelley, *History and the Disciplines*.

6. Cited in McGrath, *Nature*, 25.

7. Yanni, *Nature's Museums*, 142f.

"stained glass windows, and church-like atmosphere," they were known to "respect-fully remove their hats as they entered the building."[8]

Gazing down from his parapet, Adam represented the interlocked worlds of sci-ence and theology. As the historian of science Peter Harrison records, many at the time saw the figure of Adam as the prototypical investigator of nature. "Much as [Adam] surveyed the creation, named and classified the creatures, and bent them to his ends, those who now labored within the confines of the museum also sought to bring order to the unruly diversity of nature, and to organize the whole of the living world into a kind of material encyclopaedia."[9] This is how Francis Bacon (1561–1626)—often titled "the father of modern science"—justified the turn to empiricism. Because of the corruption consequent on Adam and Eve's expulsion from the garden, humanity, he reasoned, was incredibly likely to err in their opinions. The meticulous empirical col-lection of facts was therefore a strict necessity for the development of knowledge; like bread crumbs they would form an inexorable trail wending back to the lost homestead of knowledge forfeited in the fall. This vision for scientific advance was also driven by a particular eschatological visualization of the final end and goal of all things. The epitaph to Bacon's *Great Instauration* reads: "Many shall go to and fro, and knowledge shall increase." You may recognize this as coming from the futuristic vision of Daniel 12:4. As Harrison notes, in Bacon's time this was not to lead to a passive piety lying in wait, rather "Godly individuals were to be active participants in history, directing their efforts towards the establishment of those conditions that would usher in the final age of the world"[10] and hence to do "science" was to participate in Christ's resto-ration of what Adam lost.

Now, whatever one makes of this type of reasoning, it should at least be clear when it is put this way, that the question: "was he doing theology, or science?" or "was this an act of religion, or scientific method?" does not seem to register so neatly into our contemporary categories. As Margaret Osler puts it, "concepts formed in each domain [of early modern theology and natural philosophy] can be found to be deeply embedded in the other."[11] We will look more at how "science" was rhetorically con-structed in the next chapter, and shall turn in chapter 5 to look at how "religion" was likewise rhetorically constructed around the same time by anthropologists like Edward Burnett Tylor and James Frazer. For now, what must be recognized is that far from some enterprise justified in terms of "knowledge for knowledge's sake," or indeed, as something antagonistic to the faith, the rapidity of scientific advance—and the unity of a bewildering variety of new disciplines—were all often justified and circulated in explicitly theological and religious terms. "How ironic it is to read in popular histories of the 'antagonisms of religion and the rising science.' That was precisely what the

---

8. Livingstone, *Putting Science in its Place*, 39, picture caption 9.

9. Harrison, *The Fall of Man*, 245.

10. Harrison, *The Fall of Man*, 187.

11. Osler, "Mixing Metaphors," 93.

problem was not!" writes historian Michael Buckley. "These sciences did not oppose religious convictions, they supported them. Indeed, they subsumed theology, and theologians accepted with relief and gratitude this assumption of religious foundations by Cartesian first philosophy and Newtonian mechanics."[12] As history unfolded, the problem was not warfare but more often the various competing harmonies that had been struck too deeply between science and religion. Thus, when aspects of the science changed, the religion that was so deeply wedded to them in many instances appeared to become obsolete as well.

Nonetheless, to continue our present theme, knowledge was seen in all its variety as restoring the primal wisdom of Adam, to pick up but one theological thread of the time. Theology and religion sanctioned, justified, and unified scientific pursuits, while also providing frameworks for selection of both method and theory. Even a list of concepts which science uses—nature, cause, space, time, matter, experiment, motion, contingency, law, and so forth—invokes a series of complex histories that deeply involved theological matters that survive—even if only in an implicit way—in scientific discourse today:[13]

> [Historical investigations reveal how] each of the sciences get philosophical [and theological] as it nears its theoretical source—where it did once regard itself as *natural philosophy*—because each at its source and in its most comprehensive theoretical articulation embodies an aspiration to ultimacy or universality that is simultaneously obscured in the mundane work of the specialists and operatives within it. The closer one gets to these original sources, the closer one gets to indispensable assumptions about the meaning of nature, place, body, causation, motion, life, explanation, and truth. In short, one gets closer to the indispensable assumptions about being qua being and therefore being in relation to God [the theological and philosophical explanations of which] remain axiomatic within science in its more mundane practice at the experimental level.[14]

For us today, puzzlement no doubt arises to the extent this interplay disrupts the purity of typical notions of science and religion. Our attention is here called to the fact that our categories are not transhistorical norms, but ones whose shapes bear all the marks of history. While the fluid traffic between theology and natural philosophy was thought to be stopped by the nineteenth century's walling off of the categories

---

12. Buckley, *At the Origins of Modern Atheism*, 347.

13. Brooke, *Science and Religion*, 19–33; for a lengthier look, see Brooke, "Religious Belief and the Content of the Sciences." See also Moritz, *The Role of Theology*, 51–90; Hanby, *No God, No Science*, 9–48.

14. Hanby, *No God, No Science*, 11–12; Pannenberg, *The Historicity of Nature*, 30: "As it often happens, the philosophical problem-horizon of the respective themes [of the natural sciences], along with the history of the problem in philosophy, is not adequately considered. It is then a task of theology, in dialogue with the natural sciences, to recall the philosophical problem-horizon of the themes in question and, within that framework, to bring to bear the specifically theological accent on these themes."

of science and religion, historians have recently become more and more amazed to discover just how many concepts and practices previously flowing to and fro in translation between fuzzy or at least quite different borders marking different types of knowledge, now often find themselves locked on the wrong side of the gates within these two new kingdoms. There are, in other words, still survivals of religion in science, and the new science in religion, calling into question just how discrete even our contemporary categories really are. But we are again getting ahead of ourselves. Despite this pervasive intertwining in the modern period—what some scholars have gone so far as to call an unprecedented "fusion," or "triangulation" between disciplines with shockingly "porous borders"—opening any given textbook on the history of science, such interconnections would not be the conclusion one would draw. "Put bluntly," write two contemporary scholars, "the significance of God to modern thought has gone missing in action."[15] To take but one example at this point, the "Father of Modern Economics" Adam Smith, about whom Jacob Viner famously wrote in his *The Role of Providence in the Social Order*:

> Modern professors of economics and ethics operate in disciplines which have been secularized to the point where the religious elements and implications which were once an integral part have been painstakingly eliminated. . . . [Scholars] either put on mental blinders which hide from their sight these aberrations in Smith's thought, or they treat them as merely traditional and in Smith's day fashionable ornaments to what is essentially naturalistic and rational analysis. I am obliged to insist that Adam Smith's system of thought, including economics, is not intelligible if one disregards the role he assigned to the [theological] elements.[16]

Concepts of human nature, freedom, liberty, selfishness, private property, the invisible hand, and so on in early economic thought—not just in Smith but the whole range of modern economists including Hugo Grotius, John Locke, Thomas Malthus, and beyond—all deeply incorporated Adam and Eve into their theories in a nice coincidence that allows the picture of the Adam statue to continue to pull weight in this chapter. And this was but one interwoven theological thread. Yet, though such theologies "indelibly stamp" economic theories with their signatures through time, just as Viner mentioned its erasure from memory other scholars have begun likewise to note a great ensemble of disciplines that have gone through a "process of de-theologization."[17] Indeed, visitors to the Natural History Museum in Kensington will not today come upon the terra-cotta Adam, gazing down as a unique symbol unifying the variety of disciplines, discoveries, and data held within the towering spires and dizzying hallways

15. Firestone and Jacobs, *The Persistence of the Sacred in Modern Thought*, 10.

16. Viner, *The Role of Providence*, 81–82. Cf. as well Hilton, *The Age of Atonement*; McCarraher, *The Enchantments of Mammon*.

17. Boer and Petterson, *Idols of Nations*, 1.

yawning beneath Gothic archway stones. "Intentional or not," writes Harrison, "some time after the end of World War II [the Adam statue] was toppled from its commanding position." This, in turn, serves as a powerful metaphor for our theme:

> This particular fall of Adam might also be vested with symbolic significance, for the twentieth century witnessed the final stages of the secularization of scientific knowledge along with the development of a degree of historical amnesia about the role of religion in its early modern origins.[18]

The depth that we have been affected by this symbolic fall of Adam goes a long way to explaining why the reception and influence of Duhem's work was so glacially slow, or why it was needed in the first place. We can refer to this in general as secularization. This term is notoriously slippery, admitting of nearly innumerable definitions. For this story, it refers to "the tendency of interpreters after the modern period to downplay, extract, hinder, or otherwise work contrary to the religious and theological dimension of . . . [historical] thinkers." While it is often thought of as a sociological commentary upon society, its decline in religious adherents, or the movement of religious forces to the periphery of societal influence, it is overlooked that secularization has just as often been a revisionist historical program where thinkers and movements are "recast and repackaged as agnostics, skeptics, or atheists, antithetical to religion and theology, regardless of whether this portrait captures the actual contours of the given figure's" thought and practice.[19] In de Certeau's analysis, "theological ... writings [and ideas] have been virtually erased from the works of the great 'savants' [read: geniuses] of the sixteenth and seventeenth centuries [and beyond] as vestiges of epochs long since over, esteemed as unworthy of interest to a progressive society. Within the tissue of history analysis therefore chooses 'subjects' conforming to its place of observation."[20] We saw this already in part with Duhem's findings. And this is not limited to the sort of cheap pamphleteering on gets on the internet, but can occur in very sophisticated ways such as the overall stellar work of Peter Gay and Jonathan Israel.[21] But again and again as other currents of forgotten theology and religion were slowly surfacing in the twentieth century, this pattern that was being uncovered of selective emphases and amnesias that had built the story of the rise of science started to raise serious questions about past historical methods and conclusions describing the rise of science as also, necessarily, the fall of religion. Those promoting such judgments typically had methods that made invisible anything to their minds that did not resemble

---

18. Harrison, *The Fall of Man*, 245.

19. Firestone and Jacobs, *The Persistence of the Sacred in Modern Thought*, 3; Chapp, *God of Covenant*, 11–13.

20. De Certeau, *The Writing of History*, 31.

21. For a brief analysis of Israel and Gay from this perspective, see Minich, *Bulwarks*, 16–36.

contemporary science. Such things were in turn bracketed out, dismissed, or simply overlooked as they crafted these quite bespoke tales of history.[22]

Through continuous acts of hindsight, a golden narrative pathway of scientific development was carved out as a cumulative, relatively straight line. All of the contingent, messy clouds of mediations leading to today were lost, downplayed, cleaned up. In this way many who were thoroughly religious figures who saw their "science" as being of a piece with and aiding them in their spiritual vision of the world, and whose religion likewise impacted their science in a variety of ways, were rewritten. To anticipate slightly: "The most significant issue that mid-twentieth century scholars did not really probe deeply," writes Victorian scholar Frank Turner, "were the actual, specific, concrete character of nineteenth-century religion and nineteenth-century secular developments. Rather they allowed a positivist concept of science derived from [Auguste] Comte and [John Stuart] Mill to provide the undergirding for their definition of the secular."[23] More on that momentarily. While positivism died as a movement in the mid-twentieth century, the relics of the histories it had rewritten and left behind were often left unexamined. As ghosts they wandered, living a shadowy afterlife haunting textbooks, popular histories, and pop culture at large. These wraiths of men and women, the phantasmagoria of movements and institutions, were held up in memorial as exemplary of the scientific course of history, and so in turn conditioned our expectations about the pathways of science and religion. Even more than that, they set expectations for what—surely—must have happened given the essences of these things. Nonetheless, there is a deeply "uncomfortable irony" in these works, which seek to "expunge theological motivations from the history of philosophy [and science] while leaning upon theological sources to make the case."[24]

In terms of our own more historically minded quest, this opens our eyes and fixates them upon the historical fact that the irrelevance or even antagonism between scientific and theological disciplines has been produced by an act of writing that selectively unmade the very thorough intermeshing of theological and scientific ideas and practices in their mutual historical development. While the call among theologians is often regarding a "dialogue" between disciplines, latent theologies linger hidden in many nooks and crannies, waiting to again be conjured and brought forth. Duhem's findings, fascinating on their own, were one of many catalysts that led scholars into a vast world—what has been called "a quiet revolution" in the history and historiography of science and religion—a pathway that led far beyond the overlooked scholastics into

22. It should be noted that there were often many theological reasons for these deletions as well. See: Gregory, *Nature Lost?*, who describes this phenomenon as "de-natured theology" (15). Nonetheless, "the breakdown of the older positivistic understanding of science and the history of science characteristic of scholarship in the second half of the twentieth century has made a facile and complete separation of science and religion [through history] far more difficult [for historians] to maintain" (263).

23. F. Turner, *Contesting Cultural Authority*, 9.

24. Erdozain, "A Heavenly Poise," 74.

the theological underside of histories now thought settled and secular. Uncovering the hidden fact that theology and religion were key components in the rise of science that were intentionally and unintentionally deleted from the breadth of our memory, such deletions brought historians to realize that the warfare narrative had been written into history using these (intentional and unintentional) deletions to drive the stark contrasts of war like science vs. religion, or reason vs. faith.[25] The sea change that occurred in the twentieth century was one in no small measure seeded from the fruit of the tree that Duhem initially planted—though other independent movements arose as well, such as the historical work of Robert Merton, and the philosophy of science of Michael Polanyi.[26] To tell the tale in its fullness would require a book in itself. To make the job more manageable, we can enter by way of looking at how Isaac Newton and René Descartes have come to be reconceptualized. Much as Leonardo was in the last chapter, these men represent not just themselves, but stand as shorthand symbols for the scientific and philosophical pillars of the nature of the modern age, respectively. Both of whom, through the selective writing of their memories, "came down to us as coopted . . . Enlightenment [figures] without parallel who could not have possibly been concerned [with theology] or establishing the existence and activity of a providential God."[27]

A Forgetfulness, Which Appears As Memory

The time has . . . come for us to ask why the twin myths of "rational" Modernity and "modern" Rationality, which continue to carry conviction for many people even in our own day, won such an eager response among philosophers and historians of science after 1920. Like any historical tradition, the standard account of Modernity is the narrative of a past episode reflected in a more recent mirror: as such, it can be a source of insight both about the episode itself, and about the writers who held up this particular retrospective mirror.

—Stephen Toulmin, *Cosmopolis*

Like all human endeavors, science relies on "big picture" stories that center it, give it broader meaning, name its essence.[28] Unfortunately, most of us have inherited certain "big picture" stories of the nature and rise of science as purely "internalist"—that is, relying upon no other factors than those we today deem as scientific; and so also either inevitably or simply as a matter of fact driving the decline of religion, and Christianity

---

25. Chapp, *God of Covenant*, 12.

26. See, for example, Nye, *Michael Polanyi and His Generation*; Cohen, ed., *Puritanism and the Rise of Modern Science*.

27. Dobbs, "Newton as Final Cause," 38.

28. Cunningham and Williams, "De-Centring the 'Big Picture.'"

in particular—now seen as completely "external" to this science that is conquering it.[29] These motifs exist in so many works both popular and academic that the shape such assumptions give to history lingers in the back of our minds as a sort of common-sense presupposition, shaping all other information we are given into their image. Imagine, for example, walking down a marble aisle bathed in immortal light, peering upon the stained-glass windows of the icons of science. Who might we see, and how would they appear to us? Here, Archimedes surrounded by mathematical notation; there, Copernicus with his sun-centered universe. A few windows show others like Muhammad ibn Musa al-Khwarizmi, Galileo, or Ibn Sina. In none of them do we see that all of these men and their compatriots were deeply religious. Or, if we are made aware of their religion, little to nothing will tip us off that this had robust connections to their scientific endeavors and discoveries. Here, a particular visage of Newton sits above all, molten in Taboric glory and robed in the semi-circular apse of this church scientific. The "scientist" par excellence. Silent, eternal, he waits in the stained-glass endless under the apple tree in patient observation for its fruit to fall. Science may claim objectivity, but it still relishes its myths. "Most people," writes Patricia Fara, "know little about Newton's physics, but they do know that he watched an apple fall from the tree."[30] As Alfred Noyes mythologizes Newton in his poem after the First World War: "Or did he see as those old tales declare/(Those fairy-tales that gather form and fire)/Till, in one jewel, they pack the whole bright world/A ripe fruit fall from some immortal tree/Of knowledge, while he wondered at what height/Would this earth-magnet lose its darkling power?"[31]

In the story of Genesis, the forbidden fruit of knowledge that beguiled Adam and Eve has often been pictured as an apple, though these are not native in the Middle East and so were undoubtedly not the fruit that broke the world. Rather, the idea of an apple came about because the Latin for evil—*malum*—makes a convenient wordplay for apple tree—*malus*. Pictures of Christ through the Middle Ages and the Renaissance display the infant Son of God holding this piece of Eden, to represent him as the second Adam, the one who will restore what has been lost. But, now with his own apple, Newton became represented as the new secular Adam of the scientific world. Like the figure of Prometheus stealing fire from the gods for the benefit of man, Newton's apple represented man's ascent, not his debasement. "Nature and nature's law lay hid in night/God said 'Let Newton be'/And all was light"[32] as the poet Alexander Pope wrote. Only, in many of the portraits written of the man it seemed Newton had taken from God as he pleased. Newton has received divine revelation, so to speak, but without the divine. Chateaubriand makes the idea all but explicit: Newton was a

29. R. Young, *Darwin's Metaphor*, 23.

30. Fara, *Newton*, 192.

31. Quoted in Fara, *Newton*, 193.

32. Alexander Pope, *Epitaph for Sir Isaac Newton*.

thief "whostol so to speak the secret of nature from God."[33] He was, in other words, a genius.

Newton himself related his supposed story of inspiration by the apple no less than four times.[34] One must wonder at this. As a pious man, it seems unlikely Newton was trying to conjure some of the more blasphemous connotations, yet he no doubt knew the story would take root in the popular imagination in ways that lent themselves quite easily to mythologization. Though Newton has often been represented as the typical aloof academic, he was in fact a fastidious curator of his public persona. Even if the apple was in reality his inspiration, it was not until twenty years after the apple crisped Newton's head that he would publish his theory of gravity. Regardless, the apple story was almost completely unknown until the nineteenth century, where it intersected with the newly magnified figure of "the genius." And so, like Leonardo, Newton began to embody the lonely scientific virtuoso bursting forth from the dark froth of time to cast his glamor on a religiously benighted world. Or, like Galileo, he was a man who used his eyes to see the world as it really is, rather than close them to pray.

All of this is nonsense, of course. Newton—just as Galileo—was zealously Christian, and his work was deeply impacted by this. Moreover, instead of a scientist (which as we might recall was a term yet to be coined) Newton understood himself to be doing natural philosophy when writing his *Principia Mathematica* (indeed it is in the very title of the work), which included reflections upon the place and nature of God.[35] It could have hardly been otherwise. In the nineteenth century nonetheless "reinterpretations of Newton's life were not just ornamental flourishes, but were laden with ideological import." Where the notion of a "scientist" was still in disarray, and was by believers and secularists alike being pried from bureaucratic oversight of Church hierarchy, "anecdotes like the falling apple [indicating direct insight of nature apart from tradition or theological creed] helped to determine what it means to be a scientist. Inconsistencies and absences cloud factual accounts of Newton's own existence, but his life's legendary reinterpretations [often as "pure" secular insight] had a very real impact on the subsequent course of science."[36] This is Newton in the visage of genius, much as we saw Leonardo: Newton the genius, working in the rarified air of a "pure" science. Here, through insight latent in the mighty depths of humanity Newton leads humanity out of religious ignorance as the key figure representing both the scope and nature of what has come to be called "The Scientific Revolution."

As a stage upon which the new category of genius might play, there are in fact few concepts so emblematic and so unifying in the history of science as that of "The Scientific Revolution." It allows a reference point marking a distinct period of time for

33. Shank, *Newton Wars*, 8.

34. Fara, *Newton*, 48–56.

35. A. Cunningham, "How the *Principia* Got Its Name."

36. Fara, *Newton*, 196.

the questions that historians of science apply to their trade. As such it also marks the unique essence that historians think characterize the activities of the period, justifying it as a unique block of time in the first place. As one of the greatest historians of science in the twentieth century, Richard S. Westfall, put it, "[The Scientific Revolution is] our central organizing idea . . . [because] without it our discipline [history of science] will lose its coherence."[37] This is not merely a tool for the historian of science, but marks something central to the very notion of life in the contemporary West. Westfall elsewhere mentions that "for good or for ill, science stands at the center of modern life. It has shaped most of the categories in terms of which we think."[38] Speaking of "The Scientific Revolution" is, in this sense, speaking of the creation story of modern life as we now know it.

Despite how commonplace it is to drop casual references about The Scientific Revolution today, as a historical category it has lived a rather short and contentious life. Surprise will no doubt accompany the reader of I. B. Cohen's *Revolutions in Science*,[39] where he examined at length how the concept only came into circulation with Martha Ornstein in 1913, passed through Alfred North Whitehead in 1923, E. A. Burtt in 1925, John Hermann Randall Jr. in 1926, J. D. Bernal in 1939, and in the later 1940s and 1950s began achieving saturation in the work of Alexander Koyré and A. Rupert Hall. Eventually "The Scientific Revolution" was sedimented in our collective imagination by its use in Thomas Kuhn's explosive 1962 work, *The Structure of Scientific Revolutions*, considered to be in the top 100 academic works of the twentieth century and which—as was mentioned—is in fact deeply, albeit silently, reliant on much of Duhem's work. A curious by-product of this critical mass that the concept had achieved was a new sense of normalcy, that is, that the "Scientific Revolution" was a concept that was more or less "simply there" marking the hinterlands of history. Though it would increasingly be refined and contested, many of its major aspects carried over from the nineteenth century's rewriting of figures like Newton. Westfall above insisted that the Scientific Revolution provides the locus of coherence for the discipline of the history of science. Elsewhere, we get an inkling of what he and others often mean:

> Before the Scientific Revolution, theology was queen of all sciences. As a result of the Scientific Revolution, we have redefined the word "science," and today other disciplines, which once took their lead from Christian doctrine, strive to expand their self-esteem by appropriating the word in its new meaning to themselves. Theology is not even allowed on the premises anymore. . . . A once Christian culture has become a scientific one. The focus of change, the

---

37. Quoted in Osler, "The Canonical Imperative," 4.
38. Quoted in Shapin, *Never Pure*, 379.
39. I. B. Cohen, *Revolutions in Science*.

hinge on which it turned, was the Scientific Revolution of the sixteenth and seventeenth centuries.[40]

Science, and the Scientific Revolution, came not just to signal an increase in knowledge, but a separation from—indeed a victory over—religion and theology. As we have already begun to hint, non/religious and non-theological characterization of science "as defined by twentieth-century philosophers of science," not only "supplied almost all the defining events" like the Copernican revolution in astronomy, the Galilean revolution in mechanics, and the Newtonian synthesis as events characterizing the scientific revolution *par excellence*. It also supplied the character of these events, overlooking their religious contexts by using "an ideal derived from [purely] physical sciences" where, as such, these events were not merely described as scientific discovery, but "historians [also] regarded the main theme of the 'Scientific Revolution' as being the elevation of experience above tradition and authority . . . the rise of research and experiment as against the study of ancient texts." Indeed, because "historians conceived the 'scientific revolution' in terms of the advancement of free thought, they generally saw [all 'extrinsic' factors like] politics, religion, and economic circumstances only as factors impeding progress."[41] Just so, the historians responsible for creating and centralizing a historiography centering on the Scientific Revolution tended also to be the ones "to divorce the important developments in science from religion and theology."[42] A great map had been made; what was left was for the details to be placed as the map's legend had already decided.

Due in major part to trends started by Duhem, for those of us used to images of the sciences slowly but surely triumphing over religion, the route of the history of ideas and practices has become weirder, more theological, and given one's disposition either wondrous or infuriating. Through the act of what has been called the "problematization of the canon" of figures that make up the essence of what is meant by "Scientific Revolution" by recent scholarship, there has emerged an increasing awareness that the exploratory and foundational aspects of Newton's alchemy and theology subsequently were "subtly and deeply" ignored—even deleted—by historians, in exchange for a focus upon the light of Newton's success in the mathematical and physical sciences.

> Thus [theology and alchemy] became a curious anomaly [to historiography—and one to be explained away [by later historians—that Newton's studies in astronomy, optics, and mathematics only occupied a small portion of his time. In fact most of his great powers were poured out upon church history, theology, the chronology of ancient kingdoms, prophecy, and alchemy.[43]

40. Westfall, *Science and Religion in Seventeenth Century England*, 2. Cf. Israel, *Radical Enlightenment*, 4.

41. Cunningham and Williams, "De-Centring the 'Big Picture,'" 410, 412–13.

42. Osler, "Religion and the Changing Historiography of the Scientific Revolution," 77.

43. Dobbs, *The Foundation of Newton's Alchemy*, 170.

Written off as an anomaly (where it was written of at all), theology in fact had a deep impact upon Newton's "scientific" work.[44] Indeed, "the remarkable success of the *Principia* . . . in restoring true natural philosophy" was motivated on Newton's part by a fervent quest where "he sought the border where natural and divine principles met and fused."[45] The tale of how this was rediscovered is a fascinating journey that in many senses parallels Duhem's own plunge into the depths of forgotten archives. Just one year before Hélène Metzger-Bruhl invited Hélène Duhem to the conference at the *Academié des Sciences* dedicated to the titanic exploits of her father, the economist John Maynard Keynes found himself at an auction held by the multinational company Sotheby at New Bond Street in London, on July 13, 1936. Specializing in brokering fine art, on this day Sotheby had a particularly unusual offering for its bidders—a trove of previously unknown private writings by one of the most luminary individuals of all time, Sir Isaac Newton. Curiously, despite Newton's reputation, enthusiasm was only lukewarm for the entirety of the sale. As Sarah Dry puts it, "the reason for the lackluster bidding was that few understood what the dusty papers actually contained." These were no mere journal entries by the man, or perhaps a few grocery lists. Rather the documents contained well over five million previously unread words. Apart from the sheer quantity, however, was their explosive content. These papers had not always been unknown, records Dry. Some of Newton's close associates and admirers who knew them "had worked hard to conceal their contents, and they had been mostly successful." They contained no less than the entire bulk of Isaac Newton's religious, theological, and alchemical musings, revealing how deeply his own thought—even his science—was indebted to these unusual pastimes. In fact, Newton had written far, far more on these topics than he had on what we would today consider science proper. The papers were seen by many of Newton's associates as potentially threatening "to undermine not just Newton's reputation, but, some felt, that of science itself."[46]

And so, after Newton died the papers were obscured through the course of time, quite literally separated out from his scientific writings and hidden with varying degrees of success. Of course, there are many reasons such a clandestine scattering of the Englishman's works went underway after his death. Some of this is merely an English tendency to de-systematize their forbearers, taking only what they think they need like bricks from an old building instead of the entire floorplan. The sequestering of the papers was also due to Newton holding many ideas that orthodox Christianity viewed as heretical—in particular his anti-Trinitarianism. Part of this particular story of deletion is no doubt the fastidiousness of his associates wanting to protect him from overweening churchmen. Nonetheless, in the light cast by these papers science, too, suddenly seemed much stranger, much less clean than many wanted. As such, the singular "scientific" Newton is viewed as such only when "seen through the prism

44. Cf. Oliver, *Philosophy, God, and Motion*, 156–90.

45. Dobbs, *The Janus Faces of Genius*, 170.

46. Dry, *The Newton Papers*, 3–5.

of anachronism." In this way, many could retain their image of "Newton the heroic scientist who has, with other scientific hall-of-famers (Darwin, Einstein) 'charted the course' of modern thought."[47]

So deep did this felt need to ensure that God and science had not cohabited in one and the same mind delve, for example, that one of Newton's first French biographers Jean-Baptiste Biot (1774–1862) postulated that the Englishman had gone quite mad after his fiftieth birthday.[48] A neat dividing line was created, suggesting Newton's decline in mental health and acuity must surely account for the startling proliferation of his religious writing. This became a fairly typical device for dealing with such odd intermixtures as theology and science. Newton had indeed suffered a nervous breakdown—writing several strange and accusatory letters to his friend the philosopher John Locke for example that reveal his fatigued and fragile state of mind dripping with paranoia—and so Biot's contention had some plausibility on the surface. He failed to mention, however, that Newton had recovered relatively quickly from this lapse. Moreover, as a later commentator on the Newton papers Abraham Yahuda put it, however odd they might first appear the theological writings nonetheless "bear the stamp of Newton's genius and will always have value."[49] In other words, despite their eclecticism these papers give no evidence of being written by a man who had lost his grip on reality. Though Biot's contentions could easily be mistaken for mere libel, it was in fact his unwavering admiration of Newton that caused this portrait of a purely rational man who descended into insanity to appear. For, surely, such a keen mind could not have fascinated upon something so obscure as religion, theology, and alchemy unless a fundamental brokenness had fixed itself upon his genius.

Newton's genius and his madness became pieces of the larger narrative puzzle of the period of time known as the Enlightenment. The spirit of the age involved a self-conscious rebellion from what had come before. Nicolas de Condorcet (1743–1794) in his *Sketch for a Historical Picture of the Progress of the Human Spirit* for example wrote that "the triumph of Christianity was the signal for the complete decadence of philosophy and the sciences."[50] And he was hardly alone in this opinion.[51] Escaping from such decadence was deemed an absolutely necessity, and the vehicle to do this was the newly charged *Encyclopedia* tradition, which attempted to gather together all of the new knowledge of the day and press it into a single comprehensive vision. As quoted in the last chapter, "The narrative of 'new science' progressively dismantling all remnants of superstition and Scholasticism in its way was . . . central to the self-perception of the *philosophes* [that is, French intellectuals in the eighteenth century],"

---

47. Force, "Newton's God of Dominion," 76.

48. On Biot and Laplace's attempts to discredit Newton's theology, especially as it related to his scientific endeavors see: Manuel, *Isaac Newton, Historian*, 5, 255–57n22.

49. Quoted in Dry, *The Newton Papers*, 162.

50. On Condorcet and others, see Brewer, *The Enlightenment Past*, 24–74.

51. Brewer, *The Enlightenment Past*, 24–48.

as Dan Edelstein puts it.[52] It was a story that obscured how, at root the Enlightenment was not anti-religious however much it was innovative, but was obsessed with religion, even Christianity.[53] Indeed, even the great polemicist Voltaire was "at his most religious or theological" precisely during the height of his attacks on orthodoxy.[54]

> If we examine the foundations [of the *philosophes*] we find that at every turn [they] betray their debt to medieval thought without being aware of it. They denounced Christian philosophy, but rather too much, after the manner of those who are but half emancipated from the "superstitions" they scorn. . . . They dismantled heaven somewhat prematurely it seems . . . In spite of their rationalism and their humane sympathies, in spite of their aversion to hocus-pocus and enthusiasm and dim perspectives, in spite of their eager skepticism, their engaging cynicism, their brave youthful blasphemies and talk of hanging the last king in the entrails of the last priest—in spite of all of it, there is more of Christian philosophy in the writings of the *Philosophes* than has yet been dreamt of in our histories.[55]

The anti-clericalism of Voltaire and others has ever since been "confused with a thoroughgoing secularism [they] never espoused."[56] We are, at bottom, creatures who thrive on stories however, and the story of warfare was an epic too good to be forgotten.[57] There was thus a double rewriting: the narrative of the *philosophes* was adopted, but it was itself rewritten as a wholesale secular and scientific war overcoming the belligerent religion that had the poor sense to stand opposed to it. The irony of a continued, covert religious affinity continues when we learn that the first uses of the story of scientific or natural philosophical knowledge displacing religious fantasy was actually born by Protestants critiquing what they perceived as superstitious Catholic excesses.[58] The cult of the saints, the veneration of relics, the doctrine of transubstantiation, and more were all Protestant targets, seeing the surge in natural knowledge as paralleling their own reformation of religion. Thus, while the Catholics themselves were deeply intertwined into the great number of Enlightenment movements of their own,[59] the eventual diminution of superstitious logic at the hands of science was in fact a common storytelling device in Protestant interpretations of history. In the hands of the *philosophes,* this origin was

---

52. Edelstein, *The Enlightenment,* 117.

53. On Voltaire's religion, for example, see Erdozain, *The Soul of Doubt,* 118–72. On the Enlightenment and religion generally, see Bulman and Ingram, eds., *God in the Enlightenment*; Firestone and Jacobs, *The Persistence of the Sacred in Modern Thought*; Dupré, *The Enlightenment*; Barnett, *The Enlightenment and Religion.*

54. Erdozain, *Soul of Doubt,* 118.

55. Becker, *The Heavenly City,* 30–31.

56. Erdozain, "A Heavenly Poise," 74.

57. Smith, *Moral, Believing Animals,* 73.

58. Harrison, "Science and Secularization," 47–70; Ungureanu, *Science, Religion, and the Protestant Tradition,* 133–44.

59. Lehner, *The Catholic Enlightenment.*

forgotten and a countermemory was put in place. "It was all too easy [for historians]," writes Daniel Brewer, "to tell the Enlightenment's story by adopting the adversarial stance the *philosophes* developed . . . phrasing things in terms of the opposition between the benighted and the enlightened, the old and the new."[60]

Despite the fact that Newton was a darling of the French Enlightenment, this was initially far from an organic fit. His baptism into the church scientific of the French Enlightenment was part of this broader process of propaganda and rewriting history. When Newton's *Principia* first came out, no one in France read it. A century later, however, this had changed entirely. A Newton emerged who was held up not just for his brilliance in science, but as the very emblem of humanity coming of age in a modernity whose signature move was to cast off its old superstitions. As J. B. Shank outlines in his fascinating book *The Newton Wars and the Beginning of the French Enlightenment*, much of this was due to the concerted campaigning and propaganda of no less than Voltaire. A lover of English science, Voltaire returned to France after a period of exile and immediately began disseminating his new gospel. "Voltaire finally appeared," exclaimed one reviewer, "and at once Newton is understood, or is in the process of being understood; all Paris resounds with Newton . . ." It was not so much the content of Newton's ideas, however, that made him appeal to Voltaire for such conscription. Rather, the height of their glamor came by the way they could be deployed to shape debates and reinforce the general dichotomy between enlightened and benighted, between the new and the old. Newtonianism was a convenient meeting place that entangled "all the hot button topics of the day."[61] This included things like science, nature, and experiment, to be sure, but also a litany of other pressing matters like radical religion, the philosophy of the controversial Baruch Spinoza, materialism, publicity, politics, and aesthetics. Voltaire, as it happens, was one of the first to reflect upon the notion of a philosophy of history, and "this inauguration of the philosophy of history" by Voltaire was represented as "an emancipation from the theological interpretations, and [was] antireligious in principle"[62] despite relying on any number of reworked theological ideas to make its points. It is little surprise, therefore, that "Newtonianism in France," as Shank concludes, "functioned less as a coherent and consistently defended set of scientific ideas, than as a political position that united savants with very different intellectual agendas."[63] As we saw above with Newton's biographer Biot, Newton's religion and his theology were easily pared off, overlooked, or forgotten for the sake of curating a specific historical image.[64]

60. Brewer, *The Enlightenment Past*, 16.

61. Shank, *The Newton Wars*, 163; cf. also Shank, "Between Isaac Newton and Enlightenment Newtonianism."

62. Löwith, *Meaning in History*, 104.

63. Shank, *The Newton Wars*, 408.

64. As McMahon, *Enemies of the Enlightenment*, argues, however, on the other side of things the "religion" often seen as specifically opposed to the Enlightenment is itself often a rhetorical creation not just by Voltaire and the philosophes, but by the groups specifically opposed to the Enlightenment

The same could be said of Descartes. While Newton's papers were hidden—and a scandalous portrait of his madness drawn—the philosopher René Descartes had suffered a fate much more macabre, but with similar results. Descartes (1596–1650) will no doubt be familiar to many philosophy students because of his immortalized saying "I think, therefore I am." Apart from this curious nugget of argument, however, Descartes was a vastly talented mathematician, physicist, and astronomer conversant with the findings of Copernicus and his contemporary, Galileo, even putting forward his own cosmological theories on the matter.[65] While any number of figures could be selected to illustrate how their theological and religious roots have been obscured in the historical record, to select Descartes at this juncture is no haphazard choice. If Newton represented the revolution in terms of his science, Descartes did so in terms of philosophy (though it should be noted that the two areas were hardly as distinct as we often see them today). His attitude of radical, systematic doubt, his approach to knowledge apparently independent of religion, his placement of the thinking subject at the center of thought, all were thought to characterize the start of truly modern philosophy. He sought to uproot everything previously taken for granted as secure; to doubt everything until he could uncover that last, recalcitrant kernel of knowledge that was so secure in its obstinance the possibility of doubt could not by definition even arise. Yet even with the notion that we should represent Descartes as primarily a philosopher we encounter not a neutral conceit, but a decision born from propaganda—both his own and that of posterity, in particular it became part of Voltaire's ever expanding historiographical arsenal. Descartes's physical theory and his mathematics, though surviving in a variety of idiosyncratic blends and hybrids, were rendered obsolete nearly immediately upon the arrival of Newtonian theory. Not wanting to lose one of their secular saints, however, in turn Descartes was "placed in the philosophical canon, so that although his research in mathematics" along with his physical theory and crude experimentation came to be viewed as defunct, he nonetheless "retained his heroic status" of genius by being reinterpreted primarily as a philosopher.[66] Indeed, instead of theology, metaphysics, or even epistemology the primary driver of this curated image of Descartes was "compatibility with [French] Newtonianism," and the image of modernity it supported above all else.[67]

In this manner Descartes was seen as the "Father of Modernity," the man who, in the words of twentieth-century philosopher Ernst Cassirer, single-handedly forged "the spiritual essence" of the new epoch, which would "permeate all fields of knowledge." Though different than Voltaire's portrait, perhaps no work solidified this

(sometimes referred to as the counter-Enlightenment). By creating "religion" as a sort of intentional hodgepodge of everything the philosophes were opposed to, an intentionally exaggerated irritant could be created to enrage austere Enlightenment sensibilities and set up an off-balanced strike primed for an immediate counter-blow. We thus have both "science" and "religion" fashioning and intentionally grooming themselves as opponents for each other for equal but opposite reasons.

65. Ariew, *Descartes Among the Scholastics*, 179–216.

66. Fara, *Newton*, 131.

67. Gaukroger, *Descartes*, 4–6.

sobriquet indicating Descartes's paternity of all things modern than the work of the German Kuno Fischer (1824–1907). In a gripping multi-volume narrative, writes Christia Mercer, Fischer waxed eloquent about Descartes' "groundbreaking ideas and their impact on subsequent thought."[68] In so doing, Fischer represented Descartes in the now familiar image as a man who has turned from the Christian scholastics building their metaphysical castles in the sky, to a focus obsessed with epistemological concerns about how we even know anything in the first place. These concerns were often understood both by posterity and by Descartes' less than friendly contemporaries as completely separate from religion, and was necessitated precisely by the failure of theologians to adequately answer such a pressing question. Doubting everything, so the story goes, he whittled down all previous knowledge until he arrived at the famous "I think, therefore I am," which he could not doubt, and which in turn provided the bedrock foundation from which to rebuild the entire system of knowledge. From Descartes's "epistemological turn," a story emerged about how Descartes invented a new philosophy in desperate search for new foundational certainties that broke radically with a benighted theological past no longer able to cope with the realities of human existence.[69] The picture of Descartes the modern epistemologist that Fischer completed from the earlier efforts of Hegel and others, in other words, "excised philosophy of its theological and religious underpinnings"[70] and reinforced the broader currents of opinion on what exactly constituted the modern period as opposed to the religious past from which it now should be separated. In this manner Descartes "became the 'father of modern subjectivity' (or epistemology, or science) only anachronistically, that is, in light of what his sons and daughters . . . went on to do."[71] These curated images of Newton and Descartes represent the twin pillars emblematic of the revolutionary atmosphere of the day, along with innumerable other figures like Galileo. With minor variations, this is exactly the image of history any number of philosophers, historians, and scientists were taught at university. As historian Stephen Toulmin proposes, this model of history can without exaggeration even to this day be called "the standard account."[72]

> This preoccupation with [a de-theologized] history [meant] Descartes announced positivism and Hume criticism, both were therefore important

68. Mercer, "Descartes Is Not Our Father."

69 However, the nature of modernity as searching for rationalistic answers, or having a less credulous and more skeptical attitude than the Reformation and Medieval periods has been thoroughly demonstrated to be false. See: Popkin, *The History of Skepticism*, 3–16; Schreiner, *Are You Alone Wise?* 3–37.

70. Mercer, "Descartes Is Not Our Father." Cf. Wilson, *Descartes*, 3–4 who notes that Descartes "has consistently been 'interpreted' by English speaking analytic philosophers as if [he] were [offering] self-standing arguments," despite the fact that Descartes "repeatedly insists on the great importance . . . of God."

71. Clayton, *The Problem of God*, 53–56.

72. Toulmin, *Cosmopolis*, 12–13.

philosophers [in the standard account]. . . . The history of philosophy, such as it was then taught, did not especially stress what had interested the philosophers themselves, but rather what was considered important in its own right in philosophy. . . . What they were looking for in Descartes toward 1900 was [for example] the forerunner of scientism. . . . Students were introduced to a Malebranche without theology, to a logicized Leibniz who had lost interest in his religious organization [and so on it went] . . . perhaps the most remarkable manifestation of all . . . was the treatment of [positivism itself, which was] . . . mutilated and reduced to the introductory lectures on the classification of the sciences. . . . At any rate, it was impossible to expect either a philosophy of religion or a metaphysics from a history that was [now] no more than the agony of religion and metaphysics.[73]

As was hinted, this rewriting of Descartes had a much more ghoulish element to it, however. The extent that the French Enlightenment wanted to rewrite history was hardly confined to texts. The government of France instituted a concerted "de-Christianization" program that often resulted in the intentional destruction, desecration, and repurposing of religious buildings and works in virtually every town within France's borders—from small villages to Notre Dame itself. Sculptures, graveyards, and cathedrals all became victims of this translation. The cries to tear down the old regime were, in these instances, quite literal. In particular, one notable target of the new administration was the church of St. Geneviève, the patron saint of Paris. Taken over by the state, it became rechristened as "the Pantheon." Harkening to the ancient Roman building dedicated to all of the gods, the new Pantheon was not dedicated to divinities but to the men of France, to their genius, their reason, and their fame. Though in a very real sense perpetuating the Christian cult of the saints, it was devoted specifically to secular reflection—a secular temple housing the spirit of the age. It was also, as it happened, the building where Descartes had previously been buried.[74]

For the concerted de-Christianizing program, Descartes's burial meant to explicitly memorialize his Christian faith would simply not do. The state of Descartes's bones quickly became an object of debate among the new order. Condorcet—who above we saw penned the idea that Christianity brought a complete halt to the advancement of learning—led the campaign and made a persuasive case for enshrining Descartes amongst the newly sanctified heroes of the French Enlightenment. It was Descartes, he argued, who brought philosophy back to reason from its decadence within the systems of the scholastics. Now the French state must likewise honor Descartes and bring his remains home from a Christian burial that only dishonored his memory. Ironically, soon after Condorcet himself was executed as a victim of this new spirit, Descartes's bones were exhumed and taken from the old ground of St. Geneviève to be rededicated in the newly christened Pantheon. Several pieces of the man, mysteriously,

73. Gilson, *The Philosopher and Theology*, 29–30.
74. Shorto, *Descartes' Bones*, 43–128.

did not make the short journey—including no less than Descartes's skull—and some even turned up as grotesque pieces of jewelry crafted by a man named Alexandre Lenoir, distributed like relics of the saints of old in the name of the new philosophy of the genius of men.[75] Even secular science, it seems, has its relics.

It was thus on the eve of the Reign of Terror that Descartes—in what one historian has called one of "modernity's most sharply honed acts of self-expression"[76]—received the dubious honor of being enshrined as a secular saint. Whereas Newton's papers were separated out, and indeed the man's personality was divided against itself by means of a rumored madness, Descartes's indebtedness to the Christian past was quite literally buried. Such was the ambiguity of the revolution, however, that not only was Descartes's tomb disturbed for ideological reasons, but atheists who were entombed elsewhere at the Church of Saint-Roch—men like Diderot and Baron D'Holbach—had their own long slumbers disturbed as the surrounding church felt the embrace of liberty, fraternity, equality. Nothing was safe in the frenzy, it seems.[77] It is fitting that Lenoir, the man who absconded with a few bits of Descartes, also created what may be the first ever museum. As with a typical complaint with how museums display information, so here Descartes is a particularly poignant example of an object dislocated from the contexts that originally gave it meaning, coherence, and purpose, situated now in an alien structure that lays a forgetfulness upon it, one that appears as memory.

Though many movements in the Enlightenment were initially responsible for these images we are still often taught today, it remained for the nineteenth century and the rise of a movement known as positivism to turn these images into an explicit method for the history of science that would both exemplify and reinforce Newton, Descartes, Galileo, and a host of other figures as men of science or philosophy who had little to nothing to do with the Christian culture, thought, and practice that in fact saturated them. So deep did positivism embed these portraits, that they became part of the implicit and explicit meaning of what we now refer to as "the Scientific Revolution," itself. It is, as we shall see, similar to what Thomas Kuhn famously observed mid-century: "Textbooks . . . have to be rewritten in whole or in part whenever the language, problem-structure, or standards of moral science change. In short, they have to be rewritten in the aftermath of each scientific revolution, and once rewritten they inevitably disguise not only the role but the very existence of the revolutions that produced them. . . . Textbooks thus begin by truncating the scientist's sense of [their] discipline's history, and they proceed to supply a substitute for what they have eliminated. . . . The textbook derived tradition in which scientists come to sense their participation is one that, in fact, never existed."[78]

75. Shorto, *Descartes' Bones*, 107.
76. Shorto, *Descartes' Bones*, 112.
77. Spencer, *Atheists*.
78. Kuhn, *The Structure of Scientific Revolutions*, 136.

—Chapter 3—

A Quiet Revolution

The New Historiography of Science and Religion

Embedded in the Enlightenment's retelling of the narrative of Western culture is a partial denial of . . . multi-disciplinary approaches, insofar as theology . . . has been excluded from the mix in most secularist readings . . . In short, the exclusion of the role played by theological truth in the unfolding of Western history assumes what the telling of this history is meant to establish, that the progression of Western history represents a kind of inevitable development from the religious to the secular. No consideration is given to the idea that all of modernity may rest on a flawed and socially distorting, sublimated theology . . .

—Larry Chapp, *God of Covenant and Creation*

AMONG THE MANY NOTABLE figures introducing the world to a period of history known now as "the Scientific Revolution," few have been so formative (and today so forgotten) as Auguste Comte (1798–1857). He was, as far as the relevant scholarship can tell, the first to pair the terms *science* and *revolution* together to refer to this era, its figures, and the new scientific methods they used. For quite some time "revolution" was in fact originally associated with God's providence affecting political change from behind the scenes. Sharing a similar fate as we have seen marking "the genius," in the history of ideas, however, revolution now came to pronounce monumental changes put into motion not by God but by a specifically godless humanity. Almost by definition the concept "scientific revolution" edged the Almighty out of the picture, and this under the guise of an austere "objectivity" of method.[1] Though the deletion of theology in this period was driven more by the "narratological" historical rewrites we have been witnessing, Comte's positivism "clothed Enlightenment attacks on traditional

---

1. Cohen, *Revolutions in Science*, 51–76.

religion in the garb of scientific neutrality and historical inevitability, spurred on the academic secularizers who sought to reduce religion's public influence, and emerged during the middle decades of the twentieth century as a commonplace of modern sociology."[2] In other words, it drove the polemics of the *philosophes* underground to appear as matter-of-fact textbook knowledge. A "conversion of the Enlightenment into positivism" occurred, as Theodore Adorno and Max Horkheimer write. "To the Enlightenment, that which does not reduce to numbers . . . becomes illusion; modern positivism writes it off as literature."[3] Speaking generally, positivism is a particular form of secularization theory inaugurated by Comte who used it to establish the discipline of sociology. Secularization is a broad church with many rooms, but in essence it represents the idea that through the march of time humanity would depart from religion, and as it becomes more technologically and scientifically adept, this decline of humanity's primitive inclinations toward superstition and belief would accelerate exponentially.[4] Comte's positivism thus gave specific contours to this general idea of how secularization would occur. In his *Course of Positive Philosophy*, written between 1830–1842 and spanning five volumes, Comte envisioned all of humanity inevitably progressing through three stages or "stadial" periods: the theological, the metaphysical, and then finally the scientific or "positive."[5] Although today Comte is not a name frequently mentioned even amongst sociologists, in his own time his was a name "in the mouth of every man."[6] He was bathed in tones both reverential and fearful, and placed alongside those hoary enemies of the faith like Friedrich Nietzsche, Ludwig Feuerbach, and later Karl Marx and Sigmund Freud (often referenced by the nefarious moniker "The Masters of Suspicion").[7]

It would be difficult to overemphasize the impact that positivism has had on intellectual history. When Jerimiah Lewis Diman (1831–1881)—himself a forgotten figure, but one of the most renowned American theologians of his day—was asked to speak upon the greatest trial of his age, it was not Darwin he named, but Comte. "The general body of doctrine that passes under the designation of positivism is the bugbear of the modern religious mind."[8] Diman's opinion was not unique. As forgotten as it might be today, positivism was widely believed to represent "the [nineteenth

---

2. Witmer, "Auguste Comte's Theory of History Crosses the Atlantic," 96.

3. Horkheimer and Adorno, *Dialectic*, 26.

4. A great summary is Harrison, "Religion, Innovation, and Secular Modernity"; see also Harrison, "Religion, Scientific Naturalism, and Historical Progress."

5. On this see Dawson, *Progress & Religion*, 15–29. "Periodizing" history in this way has had a very powerful secularizing current. See: Davis, *Periodization and Sovereignty*, 77–102.

6. Cashdollar, *The Transformation of Theology*, 142–81.

7. One of the best introductions to Comte remains de Lubac, *The Drama of Atheist Humanism*, 131–268; see also Cashdollar, *The Transformation of Theology*; Caldwell, *Beyond Positivism*; Milbank, *Theology and Social Theory*, 49–144.

8. Quoted in Cashdollar, *The Transformation of Theology*, 6.

century's] most fundamental challenge to religious belief."[9] In fact, Darwinism was often tarnished in the religious mind precisely because it came to be associated with positivism and interpreted through it, not the other way around.[10] Epistemologically, positivism questioned not just particular theological claims, but the very coherence of the human ability to do theology. Relegating belief to a passing stage of humanity's infancy, like the philosopher David Hume before him, Comte insisted that all we can claim to know by this radically "scientific" purification are the "coexistences and sequences of phenomena."[11] Ontology, metaphysics, and theology were rendered obsolete not so much by argument but by definition. In terms of God, by undermining our ability to coherently speak of causality and other metaphysical notions, it carved off at their very joints the resources many of the so-called proofs for God's existence relied upon by in a sense defining them out of existence.[12] But even beyond the proofs, it drained theology proper of all apparent meaning. Ironically in one and the same move, even atheism was denied, since it constituted a knowledge claim that traveled far beyond the barbed wire of positivism's newly narrow world.[13]

Positivism as such provided both a method and a powerful lens through which to view both the whole course of human history, and individual thinkers, movements, and discoveries. It became deeply embedded into the warfare narrative and its corresponding methods of writing scientific history. The family tree of positivism blossomed through time, and some particularly rambunctious albeit relatively distant cousins to Comte's version of positivism in the twentieth century embodied in a coterie of academics known as the Vienna Circle would weaponize this narrowed methodological and historical vision even further. This group, which named their methods Logical Positivism, included such major names as Rudolf Carnap, Moritz Schlick, Kurt Gödel, and Otto Neurath. Beyond the circle, they counted among their contacts and allies in the world such luminaries as the philosopher Ludwig Wittgenstein, the physicist Ernst Mach, and the philosopher and mathematician Bertrand Russell. Though there was a great deal of variety among them—certainly more than is typically recognized in textbook accounts—in essence the Logical Positivists argued that only meaningful statements could be given scientific consideration. And meaningful statements in this case meant those that were either verified analytically (that is, things that are true by definition, such as all bachelors being unmarried men), or synthetically, meaning factual statements verified by empirical evidence. Everything else was neither true nor false, but quite literally nonsense. "The denial of the existence of a transcendent

---

9. Cashdollar, *The Transformation of Theology*, 6.

10. Cf. Gillespie, *Darwin and the Problem of Creation*. More on this in later in this chapter and in chapter 11.

11. Quoted in Cashdollar, *The Transformation of Theology*, 11.

12. For examples, see Levering, *Proofs*; Feser, *Five Proofs*; Küng, *Does God Exist?*; Wolterstorff, "The Migration of the Theistic Arguments."

13. See the discussion in Knight, *Liberalism and Postliberalism*, 37–124.

external world would be just as metaphysical statement as its affirmation," wrote Schlick. "Hence the consistent empiricist does not deny the transcendent world, but shows that both its denial and its affirmation are meaningless."[14] This provided a particularly dense nettle of problems for theology in the twentieth century,[15] but an often untold part of the story is how deep positivism's influences ran into how much it was responsible for the byways carved by the new genre of the history of science and its relationship to religion. Part of the lasting power of positivism was the rhetoric of realism untainted by theology or religion that they so often used (despite, at the end of the day, having fairly complicated positions within the philosophy of science). As one of the main members of the Vienna Circle, Otto Neurath put it: "The representatives of the scientific world-conception stand on the ground of simple human experience. They confidently approach the task of removing the metaphysical and theological debris [from both present and historical accounts]. Or, as some have it: returning, after a metaphysical interlude, to a unified picture of this world . . . free from theology . . ."[16]

Such histories, written through the filter that positivism often ruthlessly applied, appeared in turn to reinforce the narrowly confined methods being proposed. And so, the circle bent upon itself and was complete. John William Draper's biographer puts it well: "On the surface," he says, the "Positivistic habit of mind was radically empirical; it swept on to the rubbish heap anything unconfirmed by direct experimental observation." And yet, ironically enough, "The Positivists took an empirical attitude toward everything but empiricism." Their empirical methods were in fact resting upon "a priori reasoning" which is to say, an adamantine set of presuppositions about what was allowed to be seen as empirical, and so be seen at all. Rather than a hard-nosed investigative program, positivism ironically "chang[ed] a provisional technique into an article of faith."[17] In essence, by this new faith's rampant pruning, the deletion and separation of theology that the *philosophes* often began was transmuted and systematized into a method thought to unify the sciences and their history into a single, comprehensive vision. The notion of conflict was thereby reinforced by the apparent general absence of religion, except in cases of explicit friction and overt conflict.[18] Duhem became very aware of this general situation as his own research was bearing fruit:

> They [the positivist historians] show us how all the sciences are born of the fertile Greek philosophy whose most brilliant exponents left to the vulgar the ridiculous concern of believing in religious dogma. They depict to us shockingly that night of the Middle Ages during which the schools, subservient to the agencies of Christianity and exclusively concerned with theological

14. Quoted in Caldwell, *Beyond Positivism*, 13.

15. For a summary, see Knight, *Liberalism and Postliberalism*, 125–226. For the prehistory, see Cashdollar, *The Transformation of Theology*.

16. Neurath as quoted in Critchley, *Continental Philosophy*, 95.

17. Fleming, *John William Draper*, 43; Zammito, *A Nice Derangement*, 8.

18. Asúa, "The 'Conflict Thesis' and Positivist History of Science."

discussions, did not know how to gather the smallest parcel of the scientific bequest of the Greeks. They make shine into our very eyes the glories of the Renaissance where minds, liberated at long last of the yoke of the Church, have found again the thread of scientific tradition at the same time as they found the secret of scientific and literary beauty. They delight in contrasting from the 16th century on the always ascending march of science, the ever-deeper decadence of religion. They believe themselves to be authorized to predict the imminent demise of religion and at the same time the universal and unchallenged triumph of science. This is what is being taught in a number of chairs, this is what is being written in a multitude of books.[19]

Duhem's contemporary and German counterpart, positivist and physicist Ernst Mach is perhaps the most outstanding example of historical rewriting under the flag of positivism. Indeed quite ironically Duhem's historical influence, where it was felt at all given the opposition he faced in Paris and beyond, "was only allowed to enter the scene because hints of it were swept in by the coattails of Mach and Schlick," and so especially in America was subject to "chronic misconstruction," that (humorously enough) placed Duhem's historical reconstructions into the positivist camp, and let Duhem's scholastics speak only insofar as they had mathematics or straightforward empirical observations to contribute.[20] Beyond this, Mach had written one of the first histories of mechanics,[21] and indeed embodied every point in Duhem's paragraph above. He represents the positivist influence on the history of science par excellence. "The main effect of Mach's book in the various editions it went through between 1883 and 1912," writes historian of science H. Floris Cohen, "was that it pinpointed [a certain image of] Galileo as the central figure in the birth of modern physics."[22] Where Voltaire spoke of Galileo as the first to end the night of reason, Mach described the method by which he supposedly did so. This then was a Galileo whose thought was shorn of its theological, metaphysical, and general religious contexts. Mach's focus generated a very distinct coloring to history, where theology and physics were heavily contrasted as either having nothing to say to one another or being purely antithetical—"a conception that has continued to fit in well with the prejudices of many philosophically and historically untrained scientists regarding the nature of their craft."[23]

The warfare thesis found a rich and nurturing environment in Mach's method, which as a matter of course created a stark line separating science from theology, indeed painting them in antitheses to each other. Curiously, Mach admitted openly that "conceptions which completely dominate modern physics all arose under the

19. Quoted in Jaki, *Uneasy Genius*, 399.
20. Jaki, "Introduction" to Duhem, *Prémices Philosophique*, ix.
21. Mach, *The Science of Mechanics*, esp. 39–45.
22. H. F. Cohen, *The Scientific Revolution*, 41.
23. H. F. Cohen, *The Scientific Revolution*, 41.

influence of theological ideas."[24] Theological questions "were excited by everything," he says, "and modified everything." But, in order to escape the gravity of this uncomfortable confession for the historical march of science, Mach attempted to dislodge the demonstrable relationship between theology and physics by utilizing a typical positivist defense. Declaring that theology was only an improper guise for a more fundamental source of inspiration, it could be written off as an accidental by-product of the innate impulse of the human spirit to gain a comprehensive view of the whole of things. Monotheism was merely a curious and temporary historical derivative of this more basic urge, he says, one that produced both the sun of science and its lackluster twin, the shadow of God.

For Mach, the history of science and religion is little else than a narrative outlining their merciful separation over time. More accurately stated, like any good underdog story, it is a breath-catching tale of science's emancipation and escape from the torments of its jealous captor, religion. That Mach refused to see otherwise is evidenced in the lack of revision to his *The Science of Mechanics* after the startling rediscovery of da Vinci's notebooks in 1881. "It is interesting to follow Mach through successive editions of his book," writes one historian, "as he tries to come to grips with that new information [of the medieval and theological precedents for da Vinci and Galileo discovered by Duhem] while at the same time maintaining substantially his original view."[25] Others concur: "Mach succeeded to his own satisfaction in accommodating views that soon were to lead to a very different picture of the 17th century Revolution in Science, as if they were nothing but welcome support for his original discontinuity view [that is, discontinuity between medieval and modern, the theological and the scientific], which he continued to uphold in the main body of the text of the *Science of Mechanics*."[26] Such revisions would clearly not have made it into his first edition, but as Duhem's biographer points out, despite Mach's acknowledged deep respect for Duhem, after the full publication of Duhem's three volume *Etudes sur Léonard de Vinci* [Studies on Leonardo da Vinci] demonstrating medieval precedent for modern mechanics—out seven full years before the *sixth* edition of Mach's book—Mach made no changes to his overall theory.[27]

Were it not for the immense influence of Mach—and positivism generally—on writing that followed, such infelicities would hardly be remarkable. "In the scholarly tradition stemming from nineteenth century positivist Ernst Mach," however, "historians have told a story that stresses the radical discontinuity of the Scientific Revolution from what came before," including religion, and locate this "point of rupture in the mind of Galileo. Subsequent historians of science—whatever their historiographical predilection—all tended to accept Mach's" peculiar portrait of Galileo and other figures "as the

24. Mach, *The Science of Mechanics*, 551.
25. Shapere, *Galileo*, 8.
26. H. F. Cohen, *The Scientific Revolution*, 45.
27. Jaki, *The Road of Science*, 158.

revolutionary moment . . . [and despite a few differences all have] generally accepted this story."[28] Perhaps the most explicit (and openly influential) work was the Positivist Hans Reichenbach's 1951 opus *The Rise of Scientific Philosophy* which is bold in its brazen preference for a positivist notion of history and makes no apologies in seeing most of philosophy as mawkish failures that sought comfort in the irrational whose time has passed. Nonetheless, it was often the subtle and implicit repetition of the form of positivistic history that had the most impact. As we saw above, though Westfall is an excellent historian and no positivist, the basic story line of the separation and overcoming of theology repeats itself. Other works, like E. A. Burtt's 1951 *The Metaphysical Foundations of Modern Science*, while in general attempting to counter the histories of positivism by demonstrating how deeply metaphysics had travelled in lock-step with scientific history, the major story beats of Mach's adamant fable of separation are repeated, only now in a tone of lament. For those less cautious or able than Westfall or Burtt, such stories can threaten to spiral out control into bizarre manifestations, such as religious prejudices making Christians unable to comprehend how to make good ladders and useable hammers (yes, that is a real charge made by a best-selling author, as we shall see).[29]

These wooden representations appear most often in introductions to the subject matters at hand, creating a fundamental, almost primal association. Several years back, for example, the philosopher and historian Dudley Shapere collected a large number of rather humorous physics textbook snapshots of Galileo that were, in essence, little more than a repetition of Mach.[30] In other textbooks—in particular introductions to sociology and anthropology—the bouquet of mythology surrounding those like Galileo is not only repeated, but the anti-religious implications are often emphasized with gusto, brought to the fore and made to appear almost as the very point of invoking these past figures at all. After surveying dozens of textbooks in sociology, anthropology, and a few in psychology, Thomas Aechtner came to the frustrating if not totally surprising conclusion that "contemporary postsecondary textbooks and reference materials of various disciplines still present the conflict model's narrative as *the* historical account of religion and science interactions. . . . The model persists not merely as a popular artifact, but also as a conspicuous historical narrative in modern university-level pedagogical and reference materials."[31] Not only is this misrepresentation disheartening in itself, we have to remember that it is most likely to occur in introductory works—meaning that the first impressions students will have to any given scientific discipline is one that has couched it, however much in passing, as an endeavor fundamentally at odds with Christianity or religion.

Beyond Mach or other particular figures writing histories of science, positivism's lingering grip affected historical method itself, broadly conceived. Nowhere is

---

28. Osler, "The Canonical Imperative," 10; cf. Osler, "The Changing Historiography."

29. More on this in chapters 8 and 9 as we briefly turn to the work of Catherine Nixey.

30. Shapere, *Galileo,* 3–8.

31. Aechtner, "Galileo Still Goes to Jail," 210. More on this in chapter 9.

this more evident or ironic than in what many take to be the legacy of Leopold von Ranke, the German father of the American academic search for absolute historical objectivity.[32] While Ranke was not directly involved in the creation of the warfare narrative, the ideals of his methodology claiming to capture history "as it really happened" (*wie es eigentlich gewesen* as his famous German phrase goes), certainly were. Indeed, "almost every major debate in German or American historical thought on the nature and method of historical research has centered around, or at least involved, the acceptance or rejection of Ranke's methodology and philosophy of history."[33] And yet, principally in America, Ranke's Christian Idealism was (and is) shorn off by the very academy that embraced him as an icon.[34] For Ranke, though he would hardly be mistaken for a theologian of orthodox persuasion, "history provides the locus where God is witnessed, and historians stand as the 'priests' who decipher its divinely guaranteed coherence."[35] Much as with the age-old theological dictum that both the rational order of the world and the human mind's ability to embody and discover that order in its own musings were the gift of a rational Law-Giver, so too was the objectivity of history vaunted by Ranke and his students justified by way of a very theological rationale: that of the support and guarantee of God's providence.[36] In this way too the human and interpretive character of history could be upheld, whereas—as many like Wilhelm Dilthey in Germany argued—a positivist approach whittled men and women down to less than ghosts, leaving behind the bleached bones of scattered and valueless facts. Indeed, the very notion of humanity awaking upon the shores of this world and venturing forth in history as a linear, unique, non-repeatable but still meaningful quest was a very distinctive contribution of Judeo-Christianity to the ancient world. As the philosopher Hans-Georg Gadamer put it, "the uniqueness of the redemptive event [of Christ] introduced the essence of history into Western thought."[37] To take this away was also surprisingly to evacuate history of its broader reason. It was also much more difficult in practice to exorcize than many initially thought. Nearly all

32. Novick, *That Noble Dream,* 21–47.

33. Iggers, "The Image of Ranke," 17.

34. Iggers, "The Image of Ranke," 18.

35. Clark, *History, Theory, Text,* 202n46; Iggers, "The Image of Ranke," 25.

36. Indeed, the "objective observer" model of history has been remarked to itself be a legacy of monotheism. See: Novick, *That Noble Dream,* 409. Cf. Gadamer, *Truth and Method,* 207: "The idea of infinite understanding . . . for which everything exists [for God] simultaneously . . . [was] transformed into the original image of historical impartiality. The historian who knows that all epochs and all historical phenomena are equally justified before God approximates that image [of God's omniscience]. . . . The more [the historian] is able to recognize the unique, indestructible value of every phenomenon, that is, to think historically, the more his thought is God-like. That is why Ranke compares the office of historian to that of priest. 'Immediacy to God' is for the Lutheran Ranke the real content of the Christian gospel. The re-establishment of the immediacy that existed before the fall does not take place through the church's means of grace alone. The historian has a share in it too, in that he makes mankind, which has fallen into history, the object of his study, and knows mankind in the immediacy to God which it has never entirely lost."

37. Gadamer, *Truth and Method,* 419; Howard, *Historicism.*

post-Enlightenment historical thinkers, their orientation to the future, their desire to find objective and immanent solutions within history are "still in the line of prophetic and messianic monotheism." Whatever their protests, historians are still "Jews and Christians, however little we may think ourselves in those terms" because philosophies of history are elaborated by "secularizing theological principles and applying them to an ever-increasing number of facts."[38]

The lingering presence of positivism not only hid theological and religious contributions at large, it also hid its own borrowings in these areas. The supreme irony of positivism in the end was that it could present itself as a universal and "scientific" discipline only "to the extent it conceal[ed] its own theological borrowings and its own quasi-religious status."[39] The positivistic concept of science "decapitates philosophy" and in turn itself, precisely because it is "historically speaking, a residual concept." While often still reliant upon pictures of the world inherited from Christianity, "It has dropped all the questions which had been considered under the now narrower, now broader concepts of metaphysics, including all [religious] questions vaguely termed 'ultimate and highest.' Examined closely, all the excluded questions derive their inseparable unity from the fact that they contain . . . the *problems of reason* in all its particular forms."[40] Despite his own strong opposition to positivist historiography, the reworking of Ranke made him an exemplary founding figure to many in America unfamiliar with Ranke's professional habits in Europe. And so, positivism appeared to gain support from yet another angle. A uniquely American image was produced of the historian obsessed with the impossible put supposedly "purified" scientific objectivity of documentary history,[41] the same image thought to have produced the notion of warfare through objective and dispassionate analysis. While this played out in many different ways across the landscape of philosophy and history, in regards to the warfare thesis in particular it affected Darwinism and the Victorian period as "the secular interpretation of Victorian and general nineteenth-century intellectual life very much reflected the concerns of mid-twentieth-century American university intellectuals" as Frank M. Turner writes.[42] Myth and method converged. That is to say, history became whittled down to a clash of technical science overcoming an overstuffed and thoroughly subjective theological view of nature.

The Fall of Positivism

To get rid of the positivist inside us is precisely what we have to learn in order to become viable historians of science.

38. Löwith, *Meaning in History*, 191–203.

39. Milbank, *Theology and Social Theory*, 52; Nisbet, *The Sociological Tradition*.

40. Edmund Husserl, quoted in Pfau, *Minding the Modern*, 20.

41. D. Ross, "On the Misunderstanding of Ranke."

42. Turner, *Contesting Cultural Authority*, 5.

—H. Floris Cohen, *The Scientific Revolution*

Despite how it has lingered at the popular level, the movement known as Logical Positivism died a mercifully decisive death in the mid-twentieth century. In fact, Logical Positivism is about "as dead as a philosophical movement ever becomes."[43] And this obituary was old news even in 1967, the year it was published.[44] To a large extent it was killed both by its own internal contradictions, as well as by new generations of students no longer satisfied with its pretense to rigor. While its death occurred because of a series of devastating refutations by thinkers like Willard van Orman Quine and Thomas Kuhn, its demise was embodied as well in a sequence of startling events. Most recently for example, was the shocking admission of one of the premier avatars of positivism, Antony Flew, that he had come to believe in God despite having written one of positivism's most recognizable manifesto documents in the essay "Theology and Falsification," as well as being one of the most infamous and outspoken atheists of the twentieth century.[45]

Perhaps no story embodies positivism's death more literally than the tragic murder of Moritz Schlick—the founding father of Logical Positivism and the Vienna Circle—by one of his students.[46] The student in question, Johann Nelböck, shot Schlick as he was descending the stairs after leaving his class lecture on June 22, 1936. When testifying in court, Nelböck both admitted to the murder and denied any sense of remorse. The strict rigor of Logical Positivism did not just abandon metaphysics as nonsense, a majority of philosophy, politics, ethics, aesthetics, and even economics were likewise rendered absurd under the strangulating pressures of the so-called verification principle. To reiterate, the principle stated that a scientific statement must either be true by definition, or empirically demonstrable—the major irony in large part responsible for positivism's downfall being that the verification principle was itself neither of these. Regardless of whether the principle was consistent or not, a majority of what we consider vital to human existence cannot, of course, live in the arid wasteland of such proof. Even something that presents itself as ostensibly data-driven as economic theory, for example, spent the better half of two decades working its way out of positivism in terms of both history and theory.[47] As far as ethics, positivism demanded that it be nothing more than emotivism—that is, ethical statements are merely the expression of preference and do not mirror objective reality. "Do not kill" means something like "I find killing unpleasant, kindly refrain." To speak of truth in ethics, therefore, made as little sense as reflecting upon the truth statement of someone shouting *ouch!* (or something less savory) after stubbing a toe on a rock.

43. Passmore, "Logical Positivism," 5:52–57.
44. This anecdote is taken from Garcia and King, "Introduction," 21n19.
45. Flew, *There Is A God*.
46. Edmonds, *The Murder of Professor Schlick*.
47. Caldwell, *Beyond Positivism*.

Such things are not even "truth conducive" as the philosophical phrase goes, meaning they could be neither true nor false. That was the case here, or so the very unstable Nelböck claimed, hiding his crime in the phantasmagoric state that was ethics in the wake of positivism.[48] While political events in Germany at the time played a major part, a love triangle had evolved between Schlick, Nelböck, and another student Sylvia Borowicka. Logical Positivism, said Nelböck, destroyed any sense of his own moral restraint, and so jealousy won out.

Anecdotes aside, positivism had been languishing for some time before its death. When the adherents of the Vienna circle largely transplanted to America after the Nazi Party took power in 1933, their roots were unsuited to the soil of this new world. In its original European context, Logical Positivism bloomed as a powerful revolutionary force, tightly interwoven with other local revolutionary flora spanning the arts, the sciences, politics, and society, while its own ambitions as we have seen aimed at nothing less than a total refashioning of humanity. Logical Positivism saw itself as no mere philosophy, but as, in the grandest sense of the term, generating a worldview (*weltaunschauung*) that would lead a scientific race of men and women into the future. As we have seen it also generated distinctly positivist renditions of the past as well. Once in America, however, much of this had to be culled for the transplant to be successful. Its revolutionary politics for example, especially its associations with Marxism, were quickly pruned off to make Logical Positivism look more like a variety of the prevalent commonsense empiricisms and pragmatisms on offer in its new home. This recasting of itself in an Americanized, scientific image not only allowed an eventual refuge from the paranoid gaze of rampant McCarthyism mid-century, but it also through this self-presentation gained a steady stream of funding from the National Science Foundation and several other grant bodies. In fact, however, this strategy ultimately backfired.

As Michael Friedman puts it, this pruning was *too* successful, as Logical Positivism began to be "identified with a rather simpleminded version of radical empiricism."[49] Just so, it was domesticated in the historical memory of the philosophical community as relatively innocuous, merely another sub-species of philosophy of science in the gardens of the American academy soon to be forgotten.[50] In other words, part of the problem for the history of science and religion was, ironically, that a domesticated positivism died so hard in the mid-twentieth century that the images of historical figures it had produced lived on without now being explicitly associated with the

48. On ethics in the wake of positivism, see: MacIntyre, *After Virtue*, 23–108.

49. Friedman, *Reconsidering Logical Positivism*, xiv. Cf. also McCumber, *Time in the Ditch*, who presents a clear (and somewhat disturbing) argument that in the wake of McCarthyism a majority of philosophy in America similarly pruned itself to be seen as little more than a handmaiden to the sciences. In this manner a newly castrated scientist philosophy was also seen as unilaterally anti-theological, and this post-McCarthy characteristic in turn reinforced the broad anti-theological attitudes of the 1960's and the 70's. That is, until post-positivist philosophy began to take hold and allowed for more theologian-friendly approaches.

50. Friedman, *Reconsidering Logical Positivism*, xiii–xiv.

castrated movement. As with so many phenomena in this chapter, "the unanimity of earlier historians, it seemed, had been the result of their borrowing from each other's narratives instead of returning to the original texts."[51] While it was lambasted as a crude joke that had taken its fair share of philosophers for a temporary joyride, the shadowy images of genius haunting the corridors of science past like Galileo, Newton, and Descartes still retained a stranglehold on textbooks and classrooms, especially, as we will see in chapter 4, with the work of the singular George Sarton, father of the American branch of the history of science.

Most surprising of all, this realization was not always left up to investigators to discover. Much as with Flew's conversion to theism mentioned above, other former positivists—if not undergoing conversion per se—indeed confessed at least a few past historiographical sins:

> And since confession is (said to be) good for the soul, I will admit that for a long time, I (along with quite a few other philosophers of science of recent times) have been a "sinner." Some of us have been satisfied with a "smattering of ignorance" in regard to the historical development of the sciences, their socioeconomic settings, the psychology of discovery, and of the theory of invention, etc. A few of us, though proud of our empiricism, for some time rather unashamedly "made up" some phases of the history of science in a quite "a priori" manner—at least in public lectures and classroom presentations, if not even in some of our publications. . . . Even if the sources were not always complete, and not always accurate, they were available, but we rarely consulted them. Most of us have come to repent of this inexcusable conduct.[52]

This rather revealing confession was from no less than Herbert Feigl, one of the original members of the Vienna Circle and the author of a prize-winning essay on Einstein and relativity theory that won the affection of the legendary physicist himself. No doubt Feigle (who noted that his father's atheism made his own early emancipation from Judaism at the age of eight "quite easy")[53] would deny that his confession was making room for the reintroduction of the truth of metaphysics and theology. Nonetheless, one can certainly read it as an oracle, for in the mid-twentieth century the pictures still haunting the academy would finally begin to change.

Back to Newton and Descartes

The problem as I slowly—very slowly—came to perceive it,
was indeed a historiographic one.

—B. J. T. Dobbs, "Newton as Final Cause and First Mover"

51. Toulmin, *Cosmopolis*, 13.
52. Feigl, "Beyond Peaceful Coexistence," 3.
53. Edmonds, *The Murder of Professor Schlick*, 109.

Turning again to Newton and Descartes as our example, as positivism began to lose its hold through the twentieth century previously forgotten details arose, new questions were asked. For example, scholars began wondering if there was "any firmly established, rather than generally plausible, connection between the Scientific Revolution and what the German sociologist Max Weber called, in an unforgettable phrase, 'the disenchantment of the world'?"[54] We might recall Westfall's quote above, where the "Scientific Revolution" was seen as essentially defined by its separation from things like theology. A small band of scholars who had "gone ahead to investigate detailed aspects of these historical processes," with new eyes, began to come to the conclusion that the connection of science with the decline of religion so firmly entrenched in our collective consciousness, as a matter of fact, "seem[ed] hardly to exist."[55] For example, with an extensive reliance upon Duhem, Lynn Thorndike began work on the almost equally massive eight-volume *A History of Magic and Experimental Science* in 1923 that included a deep engagement with Christian "magic" (here a broad category not neatly separated from theology, which we might for convenience gloss as "interest with the non-mechanical and secret ordering of the world"). Thorndike's line of inquiry became all the more fantastic (and renowned) with the work of Francis Yates. Yates discovered a previously neglected network of inquiry and investigation undertaken by a group of Christian mystics in the Renaissance who called themselves Rosicrucians. Their thought, along with Hermeticism, influenced figures such as Francis Bacon, Johannes Kepler, Rene Descartes, Robert Boyle, John Dee, Isaac Newton, and indeed nearly the entirety of the host of the Royal Society. The so-called "Rosicrucian turn" in the historiography of science composed a robust (but not always compatible) supplement to Duhem's more rationalistic line for the pedigree of the sciences.[56] Summarizing some of this new work in the early 1980s, following Thorndike and Yates the historian of science Charles Webster could write that suddenly it seemed to historians of science that "the new science was accompanied by a less radical epistemological shift [away from religion, magic, theology] than has hitherto been thought."[57]

As part of this shift, early on in the century some began to notice that the potpourri of different narrations of the life of Newton were such a mess of startlingly different portraits that the genre of "meta-biography" was essentially invented to take an inventory and sort through the jungle of sketches and caricature.[58] As time went on, it was increasingly realized that a "process of selectivity . . . tending to highlight modern elements in the thought of Newton's generation, while discretely allowing

---

54. H. F. Cohen, *The Scientific Revolution*, 177.

55. H. F. Cohen, *The Scientific Revolution*, 178.

56. See: Yates, *Giordano Bruno*; Yates, *The Rosicrucian Enlightenment*; cf. also Fleming, *The Dark Side of the Enlightenment*, 107–214.

57. Webster, *From Paracelsus to Newton*, 12.

58. Higgit, *Recreating Newton*.

anything of a contrary nature to fall into the background," was a prime culprit.[59] Far from separation between theology and science, the eventual opinion emerged that the Scientific Revolution and the Enlightenment were often seen as the perpetuation of a "divine science." Around the time Duhem's works were fighting to be published, we can at this point return to Keynes, who found himself bidding halfheartedly on the Newton lots at the Sotheby auction. After winning some of his initial bids, however, his appetite grew. He came back the next day, armed with a more aggressive intent and the not inconsiderable wealth he had accumulated through investments and the sales of his latest book. The second day of the auction was Keynes' to do with as he pleased, and he ended up purchasing thirty-eight of the over three hundred lots of Newton's work on offer that day. All in all, he spent today's equivalent of £25,000. The other major restorer of the theological Newton, Abraham Yahuda would, at this auction and others, end up purchasing the equivalent of £50,000 in total.

Beyond these two, there were thirty-seven total purchasers that day who left with pieces of Newton's hidden work. They were out in the world now, scattered across the continents and just waiting for posterity to realize that these pages spoke of a Newton who was much stranger, much more complicated, than the paeans to his science had initially allowed. Still, this information was slow to emerge as relevant to histories of science. It will come as no surprise that George Sarton, much as he did with Duhem's reworked Leonardo, responded only with cool indifference to the theological and alchemical papers of Newton. "He declared that as a scientist he was personally no more concerned with Newton's non-mathematical works than a medical man would be with the rabbinical books of Maimonides." It wasn't really until P. M. Rattansi and J. E. McGuire's seminal 1966 paper "Newton and the 'Pipes of Pan'" that Newton's deep concern with biblical interpretation, alchemy, and theology were no longer just that awkward hobby he would put away when polite company came over. They became seen rather as core features of his entire life's work and provided a natural theological structure that framed not just his physical and mathematical theory, but revealed erased and overlooked subtexts of the dawning age of science.[60] Indeed both theology and his alchemy were considered by Newton to be "a study of the modes of divine activity in the world."[61]

Keynes would in his newfound fascination lovingly call Newton not the first scientist, but rather, in line with Thorndike, "the last magician." This was no insult, and Newton was not therefore imagined as a mere conjuror of cheap tricks. Keynes, rather, painted the portrait of a man who was deeply invested in seeing the world as a vast and organic lattice of secrets begging to be uncovered by a variety of overlapping approaches. This was a complex and many-hued portrait, we might add, whose rich and often bewildering religious colors were exactly what the palette of "genius"

---

59. Webster, *From Paracelsus to Newton*, 1.

60. Rattansi and McGuire, "Newton and the 'Pipes of Pan.'"

61. Dobbs, *The Janus Faces of Genius*, 13; Iliffe, *Priest of Nature*.

was often meant to overcome.[62] So intricate is the network of religion and natural philosophy in Newton that it is useless to look at Newton in terms of how his religion affected his science, or vice-versa. They were not separate enterprises impacting one another in an interdisciplinary way. They were aspects of one single though internally differentiated enterprise, natural philosophy, which spoke of all things and their relationship to God.[63] This can be expanded at large to much of the work being done in the modern period.[64] A recent commentator, James Force, goes so far as to call Newton's thought "a seamless unity of theology, metaphysics, and natural science." This is because "Newton's God of dominion, i.e, the total supremacy of God's power and will over every aspect of creation, colors every aspect of his views about how matter (and the laws regulating the ordinary operation of matter) is created, preserved, reformed, and occasionally interdicted by a voluntary and direct act of God's sovereign will and power."[65]

These revelations about Newton were not alone. Elsewhere, in 1913—the same year Duhem landed his contract to produce the ten-volume *le System du Monde*—his fellow Frenchman Etienne Gilson defended his doctoral thesis at the University of Paris. In the thesis, entitled "Liberty in Descartes and Theology," Gilson—who would go on to become one of the most celebrated historians of medieval philosophy in the twentieth century[66]—argued that the philosopher, mathematician, and physicist René Descartes, despite his own endless assertions of originality and his representations by posterity as something of a singularity, was in fact deeply indebted to the scholasticism he renounced. This was a remarkable claim for a number of reasons, not least of which because at the time that Gilson was writing, Descartes was not seen as an inheritor of Christian philosophy precisely because the prevailing wisdom was that no such paradoxical entity even existed.[67]

In the seminal histories of philosophy written by G. W. F. Hegel and Jacob Brucker, for example, each argued that because theology dominated all other discourses in the Middle Ages, philosophy, understood as an autonomous discipline, did not exist.[68] Indeed, it was with Descartes that philosophy began to free itself from the grasp of meddlesome theologians, according to both men. A live debate about these historical opinions began to rage amongst high profile theologians, philosophers, and historians, especially in the first half of the 1930s in the wake of Gilson's revisions.[69] "Can

---

62. Fara, *Newton*, 155–91.

63. A. Cunningham, "How the *Principia* Got Its Name," 381–82.

64. Popkin, "The Religious Background of Seventeenth-Century Philosophy," 393–422.

65. Force, "Newton's God of Dominion," 84.

66. Murphy, *Art and Intellect*; Cantor, *Inventing the Middle Ages*, 326–36.

67. Sadler, *Reason Fulfilled by Revelation*; Sadler, "The 1930's Christian Philosophy Debates."

68. On Brucker and Hegel's narratives of the history of philosophy respectively, see Inglis, *Spheres of Philosophical Inquiry*, 30–40, 49–53.

69. Sadler, "The 1930's Christian Philosophy Debates."

it be seriously maintained," asked Gilson, "that modern philosophy from Descartes to Kant [and beyond] would have been just what in fact it was, had there been no 'Christian philosophers' between the end of the Hellenistic epoch and the beginning of modern times?" The answer, as many have increasingly affirmed since Gilson, is a resounding "no." Even a summary examination of the "philosophic output of the seventeenth, eighteenth, and even nineteenth centuries will at once reveal characteristics very difficult to explain unless we take into account" the Christian legacy.[70] "It is a curious fact often denied," he continues, "that if our contemporaries no longer appeal to the *City of God* [by St. Augustine] and the Gospel as Leibniz [for example] did not hesitate to do, it is not in the least because they have escaped their influence. Many of them live by what they choose to forget."[71] Indeed, as a more recent commentator has put it, contemporary philosophy "is very much the misbegotten child of theology . . . its longings and nostalgias, its rebellions and haunting memories . . . even its most strident rejections of faith are [often] determined by the Christian tradition, and by the Christian West's internal struggle against itself."[72] Even in the vein of logic and Anglo-American analytic theology the great twentieth-century philosopher Anthony Kenny (himself a former Christian and now agnostic) remarks that though it is still not very appreciated, nonetheless "medieval logicians had addressed questions that had fallen into oblivion after the Renaissance, and many of their insights had to be rediscovered during the twentieth-century rebirth of logic . . . inaugurating a new phase in the reception of medieval philosophy in the general, secular academic world."[73] It is no coincidence that the fall of positivism mid-century was also responsible for the

70. Gilson, *The Spirit of Medieval Philosophy*, 13. Cf. as well Gillespie, *Theological Origins of Modernity*; Farrell, *How Theology Shaped Twentieth-Century Philosophy*; Baring, *Converts to the Real*; Agamben, *Omnibus*. In some respects Gilson's claims parallel the independent work of M. B. Foster beginning with his publication in the issue of Mind in 1934. He is credited alongside Duhem with jump-starting the revised notion of Christianity's influence upon the sciences, also paralleling some of the earlier assertions of Alfred North Whitehead's lectures in 1925 on the medieval theological origins of modern science. For Foster, modern science is summarized by empiricism, an empiricism won ultimately by a theological vision of the world that closes off rationalistic speculation theorizing how the world "must be" prior to and apart from empirical inquiry and investigation. Rather "the voluntary activity of the Creator . . . terminates in the contingent being of the creature . . . But the contingent is knowable only by sense experience." See: Foster, "The Christian Doctrine of Creation"; Foster "Christian Theology and Modern Science." For a summary of Foster's work, see Wybrow, ed., *Creation, Nature, and Political Order*; Davis, "Christianity and Early Modern Science," 75–95; and for a general outworking of Foster's thesis, see Oakley and O'Connor, eds., *Creation: The Impact of an Idea*. Foster's work is in general a bit confused and confusing. Nonetheless it is useful as a broad supplement to Duhem, and has been continued indirectly by the work of Brague, Funkenstein, Henry, Hooykaas, McGrath, Oakley, Osler, Torrance, Taylor, and others who see value investigating the impact of voluntaristic theology on the sciences. For a rebuttal, however, see Harrison, "Voluntarism and Early Modern Science," who argues that other theological areas were more influential.

71. Gilson, *The Spirit of Medieval Philosophy*, 17.

72. Hart, *The Beauty of the Infinite*, 30; cf. Baring, *Converts to the Real*.

73. Kenny, *A New History of Western Philosophy*, 259. Cf. as well Farrell, *How Theology Shaped Twentieth-Century Philosophy*.

rise of Christian philosophy and its increasing gains of elbow room in the American academy.[74]

To reiterate, this is not suddenly to conjure a history where Christianity, philosophy, and science were the tightest of chums through history. The situation is one full of complexity—but it is precisely that complexity that allows us more breathing room than the strictures that are necessarily part of the ensemble of ideas ballasting the warfare thesis. What might seem initially to be a relatively narrow, focused inquiry, Gilson's thesis can rather be seen in some sense to represent a newborn intuition that sought to overturn one-sided historical pictures, that is, to lay Descartes's bones to rest again in their Christian memorial. This is hardly to baptize Descartes (or Newton) as orthodox Christians. It is rather to call our attention to a baseline demonstration of how deeply theologies of all sorts were (and still are) operating in thought typically represented as free from such things.[75] It is to remark quite simply that Christians populated all aspects of that adventure of human spirit we call the history of ideas. This demonstration, even at such an initial stage, deeply disrupts the self-fashioning narratives we have all inherited about who we are, where we have been, and where we are now going. Much as with the works reevaluating Newton, or Duhem's work with Leonardo, they were not just an investigation into one man, but were rather charged with the energy of reevaluating opinions on the nature of entire historical periods. As one scholar has recently reminded us: "There has been considerable debate over the last hundred years about Descartes' originality, almost all of it bound up with a debate about the origin and nature of modernity" itself.[76] This remains true even today. Though many of Gilson's conclusions and arguments have been questioned or refined, his claims of Descartes and his successors' continuing reliance upon the theological and philosophical work of their predecessors and their contemporaries has only been strengthened. The philosopher Christia Mercer, as but one example, has recently demonstrated Descartes's dependence upon his relative contemporary, the Carmelite mystic Teresa of Avila, having crafted his *Meditations* along the lines of Teresa's famous work *The Interior Castle,* which was a mystical treatise describing one's spiritual preparation and journey to God.[77] Descartes in fact received many of his most memorable images from her—the evil deceiver demon, the notion of radical doubt, subjective inwardness—which were part of a broader inheritance starting with Augustine, who anticipated in his own way Descartes's argument of "I think, therefore I am."[78] Others in a similar vein took up Gilson's challenge and have shown how

74. Wolterstorff, "How Philosophical Theology Became Possible Within the Analytic Tradition of Philosophy."

75. See, e.g. Wilson, *Descartes*, 3–4 who notes that Descartes "has consistently been 'interpreted' by English speaking analytic philosophers as if [he] were [offering] self-standing arguments," despite the fact that Descartes "repeatedly insists on the great importance . . . of God."

76. M. A. Gillespie, *The Theological Origins of Modernity*, 189.

77. Mercer, "Descartes' Debt to Teresa."

78. See for example Augustine, *On the Trinity,* 10.10.14; *City of God,* XI.26; *Enchiridion* 7.20.

deeply Descartes's work in natural philosophy was indebted to scholastic discussions of God and God's relation to the world.[79] A rich vein of theological anthropology and its assumptions regarding the nature the human and her knowledge from what was or was not lost by Adam in the fall also ran through Descartes, upon which he was deeply reliant.[80] Broadly put, scholarship since the 1980s has in fact largely come to the awareness that "Cartesian methodology and epistemology depend on Cartesian theology."[81] This was only the beginning of theology's comeback story in the history of ideas and practices. "What is . . . crucial [to realize], then," writes one commentator, "is that the religious resurgence [in historiography] had been taking place on the highest planes of philosophical reflection for well over a century." This includes "the radical religious turn of thinkers [reconstructing those] like [Soren] Kierkegaard and William James, the mystical turn of [Henri] Bergson and [Ludwig] Wittgenstein, and the religious murmurings of [Martin] Heidegger, as well as of recent postmodernists like Gianni Vattimo, Jacques Derrida, and Jean-Francois Lyotard." To which we must conclude that a stunning array of scholarship is pointing to a historiographical change where "religious revival would seem to be not merely a popular, sociological phenomenon, but a philosophical one, a shift in ... balance occurring at the profoundest levels of Western thought."[82]

A Quiet Revolution: Rise of the New Historiography

We . . . remain unaware of the full extent to which characteristic concepts and patterns of . . . philosophy, [science], and literature are displaced and reconstituted theology, or else a secularized form of devotional experience. . . . [We] readily mistake our hereditary ways of organizing experience for the conditions of reality and the universal forms of thought.

—M. H. Abrams, *Natural Supernaturalism*

As the twentieth century continued, the newly "re-theologized" portraits of figures and movements and even entire disciplinary branches continued to emerge. In 1988, for example, Boyd Hilton listed geology, astronomy, magnetism, physics, biology, and natural history as key areas where historians had come to a deeper appreciation of the impact of Christian theological convictions—among other things—upon the sciences. He himself added social theory and political economy to the list, which continued

---

79. Secada, *Cartesian Metaphysics*; Ariew, *Descartes Among the Scholastics*; Osler, *Divine Will and the Mechanical Philosophy*; Marion, *On Descartes' Metaphysical Prism*.

80. Harrison, *The Fall of Man*.

81. Clayton, *The Problem of God in Modern Thought*, 66.

82. Melzer, *Philosophy Between the Lines*, 328.

to grow even more rapidly through the 1990s and 2000s.[83] Figures such as Robert Boyle and Gottfried Leibniz, and indeed the entire experimental tradition, had their complex relations to religion restored as shades and hues and ambient light to their intellectual portraits and contributions.[84] Fringe religious figures like Thomas Hobbes, radicals like Benedict Spinoza, or outright atheists like Friedrich Nietzsche, the utilitarian John Stuart Mill, or even Sigmund Freud have had their own covert reliance and perpetuation of various strains of theology once again placed in the puzzle that is the broader intellectual and practical contexts informing them and their works.[85] Elsewhere, even a few historical enigmas were solved by reinserting theology into the picture, as when Patrick Riley discovered that the political notion of "general will" so central to Jean-Jacques Rousseau and which had provided no little mystification for historians, appeared to pop into history as if summoned from nothing because its origin was in fact a theological one that had migrated into the later compartmentalized domain of civic theory.[86] And such stories only multiply as our travels through the twentieth century continue. Despite raising eyebrows in academic departments if mentioned, nonetheless the fact today remains that "If the theological was marginalized in the age of Western secular modernity," or, at least, if our memories once told us that this was so, "theology has now returned with a vengeance." While it is a bit too much to say that "theology is reconfiguring the very makeup of the humanities in general," it is nonetheless true that "disciplines like philosophy, political science, literature, history, psychoanalysis, and critical theory in particular feel the impact of [theology's] return."[87]

Such was the storm of realizations regarding the theological and religious pedigree of what many had taken for granted as containing no such thing, a debate erupted mid-century over this whole issue. Quite famously in the so-called "quarrel of secularization" that occupied many academics mid-century, Hans Blumenberg powerfully defended the "legitimacy of the modern age" in his magnum opus by the same title. In it he was challenging the "secularization thesis"—but not in a way we might assume. Rather than question the thesis in the name of religion, Blumenberg wanted to question what he called the "expropriation model."[88] By this he means to question those who see historical processes as following the model of a transference of "goods" (in this case, ideas and practices, as we have been seeing) from the church to the world. Here in his famous spat with philosopher Karl Löwith, Blumenberg is specifically contesting

---

83. Hilton, *The Age of Atonement*, x.

84. Shapin and Schaeffer, *Leviathan and the Air Pump*; Vanzo and Anstey, eds., *Experiment, Speculation, and Religion*; Crombie, *Robert Grosseteste and the Origins of Experimental Science 1100–1700*; M. Hunter, *Boyle*; Mercer, *Leibniz' Metaphysics*; Backus, *Leibniz*; Webster, *The Great Instauration*.

85. Martinich, *The Two Gods*; Erdozain, "A Heavenly Poise"; M. A. Gillespie, *Nihilism Before Nietzsche*; Webster, *Why Freud Was Wrong*.

86. Riley, *The General Will*, 4–5.

87. Davis, "Introduction," in *The Monstrosity of Christ*, 3.

88. Blumenberg, *The Legitimacy of the Modern Age*, 21.

Löweth's thesis that in essence modernity was merely Christianity in secularized form. Löwith's work was a concerted demonstration regarding how "providence" became "progress."[89] The political theorist Carl Schmidt in the same vein and around the same time infamously argued that "All significant concepts of the modern theory of the state are secularized theological concepts not only because of their historical development—in which they were transferred from theology to the theory of the state, whereby for example, the omnipotent God became the omnipotent lawgiver—but also because of their systematic structure, the recognition of which is necessary for a sociological consideration of these concepts."[90] Thus Blumenberg—though hardly to be equated with the sophistry of the philosophes or the metaphysical naivety of the positivists—attempted in a profound and powerful but ultimately unsuccessful way to rally against the idea that secularization is "the final theologoumenon . . . which seeks to impose upon the heirs of theology a guilty conscience . . ." for forsaking or having forgotten "its true presuppositions" in theology.[91]

No simple proponent of an areligious or non-theological history, Blumenberg's complex arguments—which are still worth revisiting today—ultimately did nothing to slow the quiet revolution. The long neglected theological and religious subtexts still so firmly riveted to the Western world proved to be too fertile an avenue of historiographical research to be completely ignored. With the snowball steadily rolling down the hill, scholars not just in the history of science and religion but in regards to the West at large began to uncover just how far removed from a neutral stance or objective description the secularization theories that underlay the warfare narrative of science and religion, actually were.[92] Rather than factual descriptions, they often involved actively producing and reinforcing the secular pictures of reality they described by gerrymandering representations of the historical record, intentionally or not. José Casanova refers to theories of secularization in this sense as often being "self-fulfilling prophecies"[93] of sociologists, historians, and others. There is now a "growing consensus that the characterization of the Enlightenment as the Age of Reason, in which reason was diametrically opposed to religion, cannot be sustained."[94] Even the notion that this period signaled an exponential rise of unbelief, where identifiable parties of

---

89. Löwith, *Meaning in History.*

90. Schmidt, *Political Theology,* 36. Cf. as well Brague, *The Law of God*; Kantorowicz, *The King's Two Bodies*; Oakley, *Natural Law, Laws of Nature, Natural Rights*. Similar claims were made of ethics as well, as in MacIntyre, *After Virtue,* 60, who noted that today confusion arises precisely because "moral judgments are linguistic survivals from classical theism which have [today] lost the context provided by these [religious] practices."

91. Blumenberg, *Legitimacy,* 72–73.

92. For a wonderful overview of some of the scholarship, see Brooke, "Religious Belief and the Content of the Sciences."

93. Casanova, "Beyond European and American Exceptionalisms," 24.

94. Barnett, *The Enlightenment and Religion,* 26.

deists, skeptics, and atheists first emerged, has had something of an (all-too-quiet) "death-knell sounded for it."[95]

Not only is there "no corpus of evidence to suggest that the use of reason constituted the motor of changing attitudes toward the church,"[96] in fact the elevation of "reason" as among the primary rhetorical and material tools used in the polemics of the time was a product of intra-Christian conflicts over orthodoxy and knowledge claims generally.[97] Even the history of "Western atheism, then, is not a story of external assault on Christianity. It is a story of Christians and post-Christians attacking from within," as one historian puts it. "The ferocity of such [atheist critique] remains traceable to religious values," even those supposedly under critique, and is "characterized by the internalization of religious ideas, not their disintegration."[98] This perhaps seems a minor point, for many the dissolution of Christianity no doubt remains just that, regardless of cause. But not so. Even the "death of God," we might remind ourselves, was not just originally a theological theme but in some sense constituted the very core of the gospel itself.[99] The broader point drew scholarly attention to the fact that "critical

95. Barnett, *The Enlightenment and Religion*, 31.

96. Barnett, *The Enlightenment and Religion*, 38.

97. Schreiner, *Are You Alone Wise?*, esp. 3–36 for a good summary. One could make the argument that this is a conclusion that follows from Jonathan Israel's monumental work, *Radical Enlightenment* as well. The figure that occupies him as he narrates the emergence of secularity and naturalism is neither Descartes nor Newton per se, but rather Benedict Spinoza. Israel seems to exaggerate the novelty of Spinoza to enhance the rationalist break from Christianity. Much of the same observations that have here been used on Leonardo, Descartes, and Newton (and will be used later on Copernicus, Galileo, and Darwin) also apply to Israel's Spinoza. It seems that Israel exaggerates the difference between Spinoza and Christianity by primarily using creedal orthodoxy as the comparison. Nonetheless, "when one uses the Jewish, Christian, and Islamic philosophers as the standard . . . the lines of continuity and influence emerge rather more vividly," writes Philip Clayton, who concludes "the distance between Spinoza and traditional theism has frequently been overstated. . . . To recognize Spinoza's location within this tradition is to recognize how misleading is the mantle of atheism and irreligion that is often used to define his life's project" See: Clayton, "The Religious Spinoza," 81, 83. Cf. Erdozain, *The Soul of Doubt*, 69–117: Spinoza's "unbelief" is "heterodoxy, not atheism."

98. Erdozain, *The Soul of Doubt*, 5–6; Ryrie, *Unbelievers*, 5. Indeed, it has been a basic and increasingly confirmed suspicion of historians that atheism as a historical phenomenon is primarily a negative image of the theism it is negating. This leads to Protestant, Catholic, and Eastern Orthodox variations of atheism, for example. See: Buckley, *At the Origins of Modern Atheism*; Hyman, *A Short History of Atheism*; Kors, *Atheism in France* vol. 1; and Leech, *The Hammer of the Cartesians*, who each in their own way elaborate this dialectical thesis. Nonetheless, "negation" here is not primarily a negative concept, but a creative one. That is, far from being able to shake the religious impulses they are overcoming, atheism continues them, albeit in new form. Atheism in this sense is better understood, not as non-religious and anti-theological, but as "an irreducibly complex, dynamic and emergent mode of making meaning in a world . . . an (often unrecognized) aesthetic . . . that, at its very core, is theologically significant" (Callaway and Taylor, *The Aesthetics of Atheism*, 4, 7).

99. Cf. Mulhall, *Philosophical Myths of the Fall*, 29: "It may seem obvious that the madman's address to the theists [in Nietzsche's *The Joyful Science*] when he forces his way into their churches and sings his requiem to God, constitutes a straightforward act of blasphemy, the turning of a liturgical form of appeal to God into an insult. But in fact, every element of the madman's proclamation to the atheists . . . can be seen as internal to orthodox Christian belief; more precisely, it amounts to a call . . .

examination of the faith has been carried out and discussed not only from outside, but above all within theology." Theology quite often does not need to make itself relevant, precisely because "where [secular] concerns are in fact justified, they have been perceived with far greater sophistication within the theological discussion in the form of a self-criticism of the tradition of faith."[100] That is to say, many avenues supposedly free of theology are not in need of theological application but of a recollection that many of their own components are latently theological. Such points of contact already exist, but were waiting for someone to rouse them from their slumbers.

The reinsertion of religion and theology into the course of ideas and practices reveals how far theologically informed reasoning is still operating even in areas of explicit rebellion and critique. It has become evident to many scholars that shifts and transformations within theology need to be accounted for when speaking of its defeat or its triumph. Far from being conquered by "outside forces" so to speak, various strands of theology often go through makeovers and even self-marginalization, where they explore an avenue of inquiry, fail, and peter out or run underground. This even shows up on occasion in the historical record as something of a confession. Pierre-Joseph Proudhon (1809–1865) for example, advocated for a ruthless "humanitarian Atheism" but nonetheless confessed his helplessness to do so without repeating the very religious formulae he sought to escape and undo. He is, he says, "forced to proceed as a materialist . . . and to conclude in the language of a believer, because there is no other way; not knowing whether my formulas, theological despite myself, ought to be taken as literal or as figurative . . ."[101]

On the other hand, the (quite apocryphal) tale recounting the great mathematician Pierre Simon de la Place informing Napoleon Bonaparte he had "no need of that hypothesis [God]" for his nebular theory of solar system formation, for example, is a helpful set piece to note how changes in theology affected the notion of science relating to that theology.[102] For it was the God of Isaac Newton that la Place was rejecting, and to which his system had no need for reference.[103] This God occasionally intervened to tinker and so maintain the proper orbits of the planets that otherwise threatened to tumble into the sun because of the relentless tug of gravity, among other myriad and blue-collar workaday checks on the list of cosmic to-do's. Thomas Aquinas or Augustine of Hippo could have only nodded in agreement with la Place, but to a much different effect. To them, the thought that God was somehow in competition or working

---

to remember that Good Friday and Easter Sunday are conjoined by Holy Saturday . . . that the cross and resurrection are held together by the grave."

100. Pannenberg, *Christianity in a Secularized World*, 47.

101. Quoted in Abrams, *Natural Supernaturalism*, 66.

102. See especially Numbers, *Creation by Natural Law*, 77–87. And for how la Place was used in the Darwin debates, 105–19.

103. Laplace as such played a major part of secularizing Newton, and separating the new physics from theology. See: Hahn, "Laplace and the Vanishing Role of God in the Physical Universe," 85–95; Hahn, "Laplace and the Mechanistic Universe," 256–276.

alongside nature on the same playing field would have been a devastating theological blunder, even a blasphemous one.[104] The distinction between primary and secondary causality meant that God—as prime cause—created everything and continues to work through everything, understood as secondary or created causes. Saying God caused a thunderstorm was not thereby to put it under threat by a meteorological explanation, but was rather a description at a theological or even metaphysical level of what meteorology was explaining naturally. The notion of the quasi-physical nature of God doing very particular workmanlike jobs within the universe to keep it running, or to fill explanatory "gaps" represents a very peculiar mutation in the modern theological tradition, one sometimes referred to rather broadly as "physico-theology" that was occurring often among "lay" theologians like Newton or Robert Boyle.[105] Ignoring this leads to grand scary stories, historical claims illustrating science pushing "God" or "Christianity" out full stop. Yet, the fact that natural explanations and God are even colliding as potential explanatory adversaries often indicates theology has shifted in a particular period into an entirely different register.[106] As the theologian William Placher once famously summarized the issue: "some of the features contemporary critics find most objectionable in so-called traditional Christian theology in fact come to prominence only in the seventeenth century. Some of our current protests, it turns out, should not be directed against the Christian tradition, but against what modernity did to it."[107] When theology was deleted at large from the historical record, such nuances were immediately lost as well. In its place, the "theology at large lost to reason and science" trumpeters came marching in.

Everything we now know about the crisis of faith, the rise of atheism, and the decline of traditional religious practice where it did occur, however, tells us that "the smoking gun of science never fired." Even the "'luminous fire' of German atheism was rooted in fierce Christian conviction."[108] Religion in the modern period was, in fact, all around, and the infinite was in the air. The sacred was roaming about as the minds of

104. Tanner, *God and Creation*; cf. also Dodds, *Unlocking Divine Action*, 11–44.

105. See Funkenstein, *Theology and the Scientific Imagination*.

106. For a tidy summary, see Placher, *The Domestication of Transcendence*, 111–28. To be clear what is meant is not that theology was traditionally seen to have no natural consequences. Rather, what is meant is that God's transcendence was viewed "non-contrastively" or "non-competitively" with creation, and as such God's agency was not in a zero-sum system with natural agencies. So to speak, to mention that "God did it" and to provide a scientific analysis of an event would be two complementary descriptions, one at the physical and one at the metaphysical level. As the modern period progressed, theological descriptions increasingly obtained "physical" rather than metaphysical characterizations as impatience with metaphysics became more prevalent. In consequence, theology was suddenly in competition with other types of description, including the scientific, having been reduced and seen to provide the same sorts of explanations as scientific analysis.

107. Placher, *The Domestication of Transcendence*, 2. Cf. Leech, *The Hammer of the Cartesians*; Grant, *The Foundations of Modern Science*, 125–26; Surin, *Theology and the Problem of Evil*, 38–11; Tanner, *God and Creation*, 81–119.

108. Erdozain, *The Soul of Doubt*, 263.

men and women with it polished their new methods, their new findings, indeed even the new world until they all shone—or so it was thought—with hints of God. This God often looked very different amongst the various projects (some admittedly were a bit hard to distinguish from a sinister twin or a less-talkative anemic cousin)—of this there is no denying. It nonetheless took very concerted efforts of redescription to hide the fact that God was still the fuel of many systems. To be sure, one can tell the story of the rise of atheism with the arguments it put forward over time. But to do so is to miss the true picture, and the deeper question: what had changed, or what was changing in how the world appeared to give the questions and arguments a newly vitalized razor's edge? For there were few new arguments that had really ever been offered. Something else, the very "feeling"[109] of the world, how it presented itself and was received by people, had shifted, was being constructed in such a way that instead of internal critique they terminated in a horizonless view.[110] As the philosopher and sociologist Charles Taylor puts it, setting the agenda for his own enormous, nearly 900-page project, "To put the point in different terms, belief in God isn't quite the same thing in 1500 and 2000." That is, "the whole background framework in which one believes or refuses to believe in God," shifted, warped, altered.[111] A major part of this came not by the logic of the arguments, but as I have been arguing in this book, by their broader placement within story. There was "an urge of secularism to build a narrative of objective, scientific reason triumphing over ignorance and superstition."[112] Some of this was driven by professionalization, some by economics, some by personal vendetta. But as this particular framing progressed the notion that many of its most effective tools were crafted within Christianity lost their luster or were simply deleted as inconvenient. Moreover, as the philosopher Bruce Kuklick notes in his *Churchmen and Philosophers: From Jonathan Edwards to John Dewey*, reminiscent of Casanova's remark above about "self-fulling prophecies" regarding secularism there is a reoccuring "presentist bias" in historical scholarship that has made it route to recount the transition away from our theological past to our very particular secular present as inevitable when it was anything but that.[113] This makes it more pernicious sounding than it really was, of

109. A history of atheism from the perspective of a more emotions- or feeling-first view has expertly been done by Ryrie, *Unbelievers,* and finds a good deal of resonance with our own work here. "The emotional history of Western atheism, then, is not a story of an external assault on Christianity. It is a story of Christians and post-Christians attacking from within, and doing so from the moral high ground [of Christianity itself]" (6).

110. On what can be called the "phenomenology of atheism" then, see the excellent work of Minich, *Bulwarks of Belief.*

111. Taylor, *A Secular Age,* 13.

112. Erdozain, *The Soul of Doubt,* 262.

113. Kuklick, *Churchmen and Philosophers,* xix–xx. Cf. 252–53: "Dewey's naturalization of the self-realization ethic was congruent not only with neo-orthodoxy, but with the vision of Jonathan Edwards. Both men were engaged in a dialogue that in the eighteenth and nineteenth centuries was widely recognized as religious. Both saw that salvation was contingent on relegating the self to its appropriate place in the scheme of things. But in another sense Dewey and Edwards were at odds.

course. Many historians were well-intentioned, and having little to no knowledge of theology had no reason to include it in their stories of science. Others, of course, were perhaps less well intentioned. Ultimately, the effects were the same.

On the other hand, Christians eager to find cause for their aggravation in clearly marked enemies "outside" the faith, reinforced the neat separation with that special glee reserved for armchair prophets shouting at what they very often did not understand. Just so, resources generated within the histories of Christian reflection and exploration and conflict were erased or overlooked, even as they continued to be utilized:

> [All the various subtraction narratives of secularization] make a crucial move which they present as a "discovery," something we "come to see" when certain conditions [like the Scientific Revolution, or Enlightenment, or the Death of God] are met. In all cases, this move only looks like a discovery within the frame of a newly constructed understanding of ourselves, our predicament and our identity. The element of "discovery" seems unchallengeable, because the underlying construction is pushed out of sight and forgotten. . . . All these accounts "naturalize" the features of [a current supposedly secular situation]. They cannot see it as one, historically constructed [and for our purposes, theologically constructed reality] . . . among others.[114]

A good story with easily identifiable villains and heroes gripped the imagination. Complexity, as so often, remained rejected by the antibodies the reigning historical methodologies had produced against it. "The forging of such links [between the rise of reason, atheism, and the decline of Christianity, e.g.] was related to the desire of modern historians to find the "modern" in past periods like the Renaissance, the Enlightenment, or the Scientific Revolution. Yet, in turn "the 'modern' [in these periods] had to have its own roots, so historians then sought the proto-modernity in earlier periods."[115] The red carpet welcoming the celebrities of the "modern" stretched backward further and further through time as a consequence. A gilded but retroactive pathway of reason which, as it pressed on, circumvented religion precisely because "intellectual historians . . . assigned a low priority to the history of theology and religious ideas."[116] And, precisely because "once an ideological construction becomes dominant, [it] accrues the collective force of a respective and thus powerful layer of professional historians."[117] As this academic commonplace swelled outward—not just

---

For Edwards only supernatural grace could overcome the natural and achieve the proper integration of the individual and the cosmos. Dewey succeeded in infusing the ostensibly natural instrument of science with this supernatural power." This is a theme we will continue to see in our next chapter with Thomas Huxley, and again later on with Andrew Dickson White and John William Draper as well.

114. Taylor, *A Secular Age*, 571.

115. Barnett, *The Enlightenment and Religion*, 23.

116. Coffey and Chapman, "Intellectual History and the Return of Religion," 3.

117. Barnett, *The Enlightenment and Religion*, 24.

through generations of students, but even into pop culture—it became an enormous, almost intractable problem to overcome.

Uncovering to what extent theology was removed from historiographical reconstructions of scientific and philosophical history remains an ongoing project, to say the least. And understanding its significance one way or another is a daunting interdisciplinary goal. Over time the continual unspooling of the thread that Duhem and others had used to plumb the depths of this previously unexplored theological labyrinth connecting us to scientific history more and more led historians of science "away from the image of scientific knowledge as completely autonomous, gradually accumulating and floating above the sites in which they took shape." The broader sociological and contextual conditions led directly into a "greater interest in religious parameters, because they, after all, have been constitutive elements in many of the contexts in which science has been pursued. . . . An approach that historians have found useful here is to ask what *function* the theology may be playing within the science and vice-versa. In either case attention must also be paid to historical context."[118] Moreover, the very fluidity of our contemporary categories through time led scholars to realize that the only way to approach the truth of the historical path of science was actually "to underline the artificiality of abstracting the 'science' and the 'religion' from past (and present!) contexts with a view to establish some notional, unmediated, relations between them . . ." By focusing on the historical location of ideas and practices, historians began to show how the sacred can be present in what might otherwise pass as the secular, and vice-versa.

As different avenues of research advanced and the dust of different archives were brushed off by these new methods, a grand tapestry began to be restored, unfurling before the eyes of often astonished historians—many of whom had no particular bent toward being religious themselves. It was as if an entirely alternate history had suddenly stood as an open door to a different world, as if from nowhere. Only this was not some science fiction or fantasy conspiracy—for what was produced was not a shadowy cabal of unknowable figures, but the theatrical display of the human adventure (and debacles) of exploration, discovery, and knowledge that had been quashed under the unyielding reductionism of positivism and scientism applied to history. To the shock and dismay of many, and the delight of others, the fruit of these meticulous studies have over time begun to reveal an epoch of modernity—and indeed, a postmodernity[119]—completely unintelligible without the continuing permeation of theology and religion in all spheres of thought and practice.[120] In 1991, for example, John

118. Brooke and Cantor, *Reconstructing Nature*, 23–26.

119. On the theological inheritance of some postmodern currents, see, e.g., Baring, *Converts to the Real*; Holsinger, *The Premodern Condition*; McGrath, *The Early Heidegger and Medieval Philosophy*; Tyler, *The Return to the Mystical*; Jordan, *Convulsing Bodies*; Coyne, *Heidegger's Confessions*; Falque, *The Loving Struggle*.

120. As Howard, *Religion and the Rise of Historicism*, 22 puts it: "Modernity, in other words, did not spring into existence from outside history. The cultural legacy of Christian-theological ways of

Hedley Brooke released his somewhat unassumingly titled *Science and Religion: Some Historical Perspectives,* which arguably became the most important book written on the history of science and religion up to that point. The uniqueness of Brooke's work is that it encapsulated, summarized, and expanded upon—through a mesmerizing command of the sources—the massive sea changes indicated here and beyond. The historian of science Noah Efron recollected that his first encounter as a PhD student with Brooke's newly minted work was so profound, that when he picked it off the fresh arrivals shelf in the library, he sat down next to the shelf on the carpet (the book not having yet been catalogued for borrowing), and did not move until that evening when his wife came to pick him up. "The person she found crumpled on the carpet was different from the one she had kissed goodbye ten hours earlier," he recalls.[121]

The ramifications of all of this are no doubt legion, and certainly not all positive for Christianity. For our story, however, the takeaway is simple: the clean juxtapositions needed to create a narrative of science on the one side and religion on the other so as to speak of their unending war appeared more and more elusive as the twentieth century and its historical scholarship advanced. Again and again, regardless of discipline, or thinker, or era, the new face of these sources clearly told the curious tale of a war that wasn't. To be sure, conflicts existed—and in abundance. But the battle lines were never between science and religion because these things occurred in curious intermixtures on all of our previously clear-cut "sides." In both triumph and tragedy, extensive theological and religious judgments appeared everywhere. Duhem's thread led deeper and deeper as adventurers with new tools and resources dared plumb further and discover the menagerie of exotic religious-scientific and scientific-religious creatures roaming the labyrinth, as hybridized as any minotaur. Between Newton's papers and Descartes's bones, then, history as we thought we knew it was beginning to change, revealing religious hues, deep springs of theology, a lost or forgotten God slumbering beneath our typical stories of religion's humiliation at the hands of that young upstart, science. But a question in the last third of the twentieth century had also began to arise: what would happen if they turned their new methods upon many of the most famous purveyors of idea that science and Christianity had been in continuous conflict through history? What would one find? Could historians discover the bricks and mortar used to build the historical diorama of science vs. religion? To that part of the story we now turn.

---

thinking and organizing experience must be regarded as an enduring existent in evaluating the putatively novel claims of modernity."

121. Efron, "Sciences and Religions," 245.

Part Two

The Lords of Time

We write the history we want to continue. . . . So contemporary historians described earlier philosopher's projects in terms they wished to share.

—Susan Neiman, *The Problem of Evil in Modern Thought*

Unbelief for great numbers of contemporary unbelievers is understood as an achievement of rationality. It cannot have this without a continuing historical awareness. It is a condition that cannot be only described in the present tense, but which also needs the perfect tense: a condition of 'having overcome' the irrationality of belief.

—Charles Taylor, *A Secular Age*

Since the sixteenth century . . . historiography has ceased to be the representation of a providential time, that is, of a history decided by an inaccessible Subject who can be deciphered only in the signs that he gives of his wishes. Historiography takes the position of the subject of action—of the prince, whose objective is to 'make history.'

—Michel de Certeau, *The Writing of History*

—Chapter 4—

What the Bulldog Saw

Huxley, The X-Club, and the (Re-)Writing of Scientific History

Science versus Religion — the antithesis conjures two . . . entities of the later nineteenth century: [Thomas] Huxley [Darwin's Bulldog as] St. George slaying [Bishop Samuel] Wilberforce smoothest of dragons; a mysterious undefined ghost called Science against a mysterious indefinable ghost called Religion; until by 1900 schoolboys decided not to have faith because Science, whatever that was, disproved Religion, whatever that was.

—Owen Chadwick, *The Secularization of the European Mind in the 19th Century*

TURNING THEIR ARSENAL OF new questions and new ways of looking at things upon the warfare thesis itself, what historians discovered was that the creation of the warfare myth did not grow like a seed, sprouting from small beginnings as it stretched its budding branches upward and outward. Rather, the myth was constantly invented, plunging from heaven suddenly here and there like a deluge, but then also tumbling backward to soak history, framing it in its own image and connecting with other previous bursts of the myth like droplets of water pooling into one body. Just so, the illusion was fostered of a long and continuous struggle through time. The war between religion and science is not often present in historical events per se, but in the memories of them, in the constant retellings molding them like continuously reworked clay. If one were to travel back in time, in other words, we would not find what we expected—a war between two well-defined sides. One of science. One of religion. Rather, we would discover that our textbooks have often pressed all too simple shapes upon history, much like the phantom glamor of nostalgia that ambers the memory of youth in pure, golden hues. To reiterate the major point of the last chapter: "historians concerned with the secularizing novelties of modernity have been reluctant to recognize theology's resilience in the face of criticism. A host of avowedly secular thinkers have been singled out as defining the discourses of modernity—Comte, Marx, Engels, . . . [Nietzsche, Freud], Darwin . . . and theologians and other religiously inclined thinkers normally do not 'make the cut.'"[1] Not only do they not make the cut, the continuing religious and theological components of the secular pantheon themselves are over-looked, or forgotten. The discovery of such deficiencies led to the general conclusion explored at length in the last chapter:

> [These grand stories of *the* relation between science and religion throughout history] are vulnerable because they are selective in their use of evidence. They gloss over the diversity and the complexity of positions taken in the past. Each tends to assume that "science" and "religion" can be given timeless definitions and that there is some inherent, some essential, relationship between them. . . . Many such attempts have been made in the past to construct an ideal model. The study of history is humbling because it shows how ephemeral most have been. [Thus there is] value in a historical approach if it alerts us to the way in which prior interests, political, metaphysical, and religious, have shaped the models that have been sought. . . . The point is that there is no single story one can tell about this.[2]

Despite the increasing recognition that the conflict of faith and science has been historically constructed, nonetheless it remains true that "sometime in the middle third of the nineteenth century some observers began to suspect that every new conquest achieved by science involved the loss of a domain to religion."[3] It turned out this was itself a major key to unlocking the secret origins of the warfare thesis. Many will not be surprised that the story of warfare between science and religion—even if only for the convenience of retelling—starts in relation to Darwin. His name, most assume, shoots shivers down parsons' spines and sends priests to fumble at their collars out of worry. And to be sure, there were many heated debates over the theological meaning and scientific legitimacy of Darwin's theory of natural selection and common ancestry. But warfare finds a different sort of birth here—not as a war of something called science battling something called religion, but amongst the historians and public persona of the time, who wanted, so to speak, to tell the story of their victory as a newly instituted professional class in advance. Or, in less polemical terms: they wanted to tell the stories of the meaning of their identity as professional scientists and historians in advance of those identities being completely real, or their recognized authority publicly secured.

Much as the "Scientific Revolution" carried with it a series of anti-theological connotations, the similarly titled "Darwinian Revolution" was meant to affect notions

1. Howard, *Religion and the Rise of Historicism*, 4.
2. Brooke and Cantor, *Reconstructing Nature*, 21.
3. Numbers, "Aggressors, Victims, and Peacemakers," 20.

of comparable transition. This "myth of the Darwinian revolution," as Darwin historian Peter Bowler writes, entails that "Darwin is often seen as having single-handedly introduced and popularized the essentially materialist view of evolution still accepted by modern biologists."[4] Moreover, the Darwinian Revolution is seen "as a watershed separating modern culture from the traditional roots of Western thought."[5] As we saw with Newton and Descartes, however, the revolution was in fact thoroughly theological. So too, historians have come to discover that "Darwinism promoted a revolution within, not against, natural theology."[6] Slowly, realizations such as this started with works like the 1959 book by Charles Gillespie, *Genesis and Geology*. Later works added to this process, including the likes of Neil Gillespie's 1979 *Darwin and the Problem of Creation,* or Dov Ospovat's path-setting work *The Development of Darwin's Theory* published in 1982 and Robert M. Young's *Darwin's Metaphor* in 1985. However, none of these works tackled the "warfare thesis" head on, however relevant in passing that may have been. This was left to James Moore's earth-shaking 1979 *The Post-Darwinian Controversies,* which concentrated all of the revisionary work up to that point into a laser focus upon the thesis of conflict and found it, unsurprisingly, to shatter as Moore's work revealed how deeply history had been reshaped into its image. As peculiar as it might seem, from there essays with titles like "Charles Darwin's Use of Theology in the *Origin of Species,*" or historians claiming that the *Origin* is the "last example of [the tradition of] Victorian natural theology," have begun to invade the literature. Darwin inherited a picture of the world common to the natural theologians of his day like William Paley, many have argued, asked the questions they asked, and right up to the *Origin* operated with many of their assumptions—often to flip them on their heads, yes, but never to remove them from theology or religion as a whole.[7] This is to such an extent that it can even be claimed that "To ignore or attempt to explain away Darwin's theism is to cut oneself off from understanding much of Darwin's science."[8]

Yet these reconsiderations were (and are) fighting an uphill battle, with a particularly materialist and godless picture of Darwinism deeply entrenched in our collective memories, one where evolution and Darwinism are also cast in terms that set it apart as its own religion in competition to Christianity and others.[9] When positivism cast its shadow over Darwinism, remarkably widespread interpretations of evolution that saw it as a grand form of God's providence (even by the younger Darwin himself) likewise were dissolved.[10] Not, it should be stated, because the theory demanded such

4. Bowler, *The Non-Darwinian Revolution*, 2.

5. Bowler, *The Non-Darwinian Revolution*, 174.

6. Bowler, *The Non-Darwinian Revolution*, 198; cf. 174–95.

7. Ospovat, *The Development of Darwin's Theory*.

8. Ospovat, "'Darwin's Theology,'" 520; cf. Ospovat, "God and Natural Selection."

9. Ruse, *Darwinism as Religion*. It is notable that the historian of science Ruse, a card-carrying atheist and evolutionist, is the one making such a claim.

10. Cf. Gillespie, *Darwin and the Problem of Creation*, 41–67, 146–156. E.g. 156: "The interest in

be the case, but more often because of the positivist philosophical and historical cast placed upon it. As Robert Richards notes of *The Origin of Species*, for example, "in this work of twenty-years maturation [of Darwin's research, Darwin] continued to suggest that the laws of evolution, those secondary laws, ought best be conceived as God's commands."[11] As such when we turn to the debates over Darwinism and creationism, for example, positivist interpretations of Darwin—and often not Darwinism per se—were the core fires driving conflict with belief and theology.[12] This continued on into the twentieth century, with this materialist version of Darwinian historiography finding an anchor point in the Darwin centennial mid-century. "Certainly, if one message tended to come through much of the work surrounding the [1959] centenary of *On the Origin of Species*," as the historian Frank Turner reminisces, "it was the generally positivistic [and explicitly materialist, anti-theological] character of Darwinian science."[13] Such readings constituted a tactic of ignoring the theological and other non-positivistic reasoning saturating Darwin, his contemporaries, and their successors, and "citing only the positivistic passages from Darwin and other scientists."[14] A theme now familiar to us. Even by 1981 as the scholarship revealing the previously ignored theological currents was increasing considerably, James Moore complained that it nonetheless still repeated the broad story of theology's dissolution "that typifies and consummates the positivist historiography of Darwinian scholarship" that has made up "a century's [worth of] writing on Darwin-and-religion."[15] This is not, we should add, necessarily to condemn the science on offer. It is rather to call into question the broader meanings and implications attached to it. Such seminal works coming out of the centenary like Gertrude Himmelfarbe's *Darwin and the Darwinian Revolution*, or John C. Greene's *The Death of Adam* in tones of both lament and praise, summarized the anti-Christian and materialist effects of Darwinism for a public increasingly eager and ready to receive such images as the 1960s commenced. It also set the tone for what would seem to count for many as the points of Darwinism in need of rebuttal. It is no accident that two years after the Darwin centenary the reaction of creationists commenced, with the 1961 publication of Whitcomb and Morris's *The Genesis Flood*, which set the tone for creationism for the latter half of the twentieth century, and beyond. The open question is just how much these "theological" responses were

---

The Origin of Species to the historian is intensified by the fact that while it was [taken as] a harbinger of a new positive biology, it was paradoxically also one of the last major theoretical works of science to be significantly dependent on theology for the force of . . . part of its argument."; Sloan, "Darwin on Nature and Divinity," 251, who notes far too many see Darwinism in terms of "nonmetaphysical positivism rather than as a theory deeply imbued with metaphysical commitments of a constitutive character."

11. Richards, "Theological Foundations," 65.

12. Cashdollar, *The Transformation of Theology*, 182–208.

13. F. Turner, *Contesting Cultural Authority*, 18.

14. F. Turner, *Contesting Cultural Authority*, 24.

15. Moore, "Review: Creation and the Problem of Charles Darwin," 189.

narrow overreactions to an image of evolution itself rendered artificially atheological and atheistic. And so, it was thus that all the straw men set out to war.

One place to start the strange tale of how the conflict of science and Christianity was written into history is by starting in the middle of things. Let us travel to Oxford to witness a debate between a bishop and a scientist over the new theory known as Darwinism. The plot twist here—which in some sense our first chapters already spoiled—is that Christians and non-Christians densely populated all aspects, sides, and flavors of this debate, though this tends to be totally obscured by the narrative of warfare, and the deletion of theology. As we peer beneath the veil of our urban legends, we find a complex labyrinth where allegiances were not clear cut, and the boundaries of things like "science" and "religion" we often unreflectingly consider today to be solid, shift and shimmer as parties negotiated identities—both of others and of their own.

What we are about to see, then, is a story of self-fashioning, where a group of friends and scientists began to tell stories about themselves to themselves and to others. To let themselves be known, yes, but also to aid in becoming the very things they said they were. To be sure, they were hardly the first to see the historical relationship between religion and science in terms of antagonism and conflict. But they universalized and institutionalized it in a way without precedent. Conflicts that bore all the marks of their local and contextually sensitive origins shed these skins and expanded to fill our memories with the grand historical struggles that we are so familiar with today. From a group of well-positioned friends—who called themselves the X-Club—the story of warfare would wend its way into European Universities and find sympathetic generals on the ground in the X-Club's similarly well-placed American friends. This deeply affected the notion of "science" that even to this day carries connotations of war with religion with it. In fact, far from war originating in history itself, we can put it boldly with many historians today that the origins of warfare be relocated to the foundations of the history of science as a university discipline at the turn of the twentieth century.[16] Indeed, without noting this, or paying attention to the "'silent practices' of the naturalists," that is, their reworking of history in positivist and conflictual terms, "it would not be possible to comprehend how the conflict . . . arose or why positivism triumphed over its rivals."[17] It was here that the findings of science became seen as at the same time the retreat of religion. And so, the notion of a perennial war and its set pieces like the Galileo affair, the flat earth, or the Oxford debate (to which will turning at the end of our book) all came and were marshalled together to produce a vision of an (ironically ahistorical) historical vision of warfare funding the discipline of the history of science. This vision even to this day often hangs over science and religion dialogue like

---

16. The best recent debunking the war of Christianity and evolution from this historiographic angle is Kemp, *The War That Never Was.*

17. Gillespie, *Charles Darwin and the Problem of Creation,* 4.

distant storm clouds, dark-eyed sentinels guarding travelers from passing the horizon on to new frontiers.[18] But we are getting ahead of ourselves.

Apes and Angels

The year was 1860. On a hot summer's day in June, amidst riotous applause and—in a language of praise now lost to us—amongst a field of fluttering white handkerchiefs waving like battlement flags, Bishop Samuel Wilberforce (1805–1873) sat down within Oxford's new university museum looking satisfied with himself. Gazing aristocratically upon a room of intellectuals who would—surely—declare his cause victor before the night was over, he peered imperiously once more at the man he believed he had just made look the fool. That man was Thomas Huxley (1825–1895),[19] or, as he was sometimes called for his ferocious defense of evolution, "Darwin's Bulldog." The insult that Wilberforce had just uttered before the breathless room was to ask Huxley: "tell me, was it through your grandfather, or your grandmother that you claim descent from a monkey?" By our standards, this is, perhaps, tame; in polite Victorian culture such a breach of etiquette showed how high the tensions were. As one account noted—"in [another] idiom lost to us"—a woman, at this remark, "showed her intellectual crisis by fainting."[20] Calls came now for Huxley's rebuttal.

The clapping no doubt muffled it from reaching Wilberforce's ears, but unperturbed, even delighted at the bishop's public impropriety, whispered to the man sitting next to him before he stood saying, "the Lord hath delivered him into my hands." This was ironic, for Huxley—who would several years later coin the term "agnostic" for himself and others like him—was not sure he even believed in the Lord.[21] But, as the story goes, there was truth in his saying nonetheless. Just one year before, Charles Darwin's epoch-making *The Origin of Species* was published. Having spent from 1831 to 1836 traveling the globe on the *HMS Beagle* as a gentleman naturalist, the observations Darwin made of finches and other local flora and fauna on the Galapagos islands—as the story goes—revealed to Darwin an entirely new map of how life came to be. When he left England at the age of twenty-seven, Darwin was a firm believer in the fixity of species (that is, dogs stay dogs and birds stay birds, whatever differences might accrue among different dogs and different birds over time). By the time he returned, the notion of transmutation had slowly, reluctantly, taken its hold in Darwin's head and his heart, though to say this out loud to his friend Joseph Hooker felt, he said, "like confessing a murder."[22]

18. Kemp, *The War That Never Was*, 96–188./
19. The best biography on Huxley is Desmond, *Huxley.*
20. Irvine, *Apes, Angels, and Victorians*, 5–6.
21. Lightman, *The Origins of Agnosticism.*
22. Desmond and Moore, *Darwin*, xviii.

Already tinged with the romantic air of a youth traveling the world to discover himself and follow his passion, the mythos of Darwin's quest for origins was increased as he carried with him John Milton's *Paradise Lost,* itself describing an epic reimagining of the fall of Adam and Eve, and their temptation by Satan. With his discovery of evolution—or, more properly speaking, the principle of natural selection (since evolution had arrived in Darwin's day through many contributors)—Darwin now peered behind the flaming swords between us and Eden, but saw that there was no Adam, and no Eve; Eden itself was not a garden but a "warm little pond," as Darwin described it, brewing primordial life. Was there, perhaps, also no God? Darwin for most of his life considered himself a theist, and then an agnostic (borrowing his friend Huxley's term). His loss of faith later in life came not from his theory—as many have misleadingly stated—but from the death of his father and his beloved daughter, Annie.[23] This was a loss from which his faith never recovered, not even on his deathbed as another myth has proposed.[24] Nonetheless, many of the most vocal and influential promoters of his theory, scientists or otherwise, couched the theory in terms of a historical narrative, one where it brought about—inevitably, inexorably—the historical decline of religion. "Although atheism may have been logically tenable before Darwin," writes Oxford biologist and provocateur Richard Dawkins, "Darwin made it possible to be an intellectually fulfilled atheist."[25] As an episode within this victory march of science, Huxley overcoming Wilberforce carries with it, in the opinion of many, the glorious banner of fate.

As Huxley stood, then, the uncomfortably packed space of (at least) 700 people fell silent. The stage had been set for his theater, and in a voice both solemn and grave Huxley began down the path providence had apparently set for him. He retorted with all the gravity he could muster in the wake of Wilberforce's unseemly question and answered that he would not be ashamed to have a monkey for an ancestor; he would, however, be ashamed to be connected to a man who used his great gifts to slander and obscure those "wearing out their lives" in search of truth. This had done it; more riotous applause equal, if not greater, than that given to Wilberforce thundered from the seats. Handkerchiefs again sprouted in approval. Flags of a new war. The evening, it seemed, had slipped from Wilberforce's fingers, as he "suffered a sudden and involuntary martyrdom."[26]

Far from an incidental skirmish, the encounter between Huxley and Wilberforce, ornamented now by the drama of memory, is one of many episodes of supposed conflict between faith and science held up like a Byzantine icon, a picture summarizing an episode of a larger narrative considered canonical. Only in this story the good news of the evangel is the light of science warring to overcome the ignorance

23. Moore, "Of Love and Death."

24. See: Moore, "Telling Tales," 220–234.

25. Dawkins, *The Blind Watchmaker,* 10.

26. Irvine, *Apes, Angels, and Victorians,* 6.

of dogmatic superstition and theological repression. Huxley besting Wilberforce is as such represented not just as one man topping another. "In these scenarios, Huxley and Wilberforce are not so much personalities as the warring embodiments of rival moralities," writes Sheridan Gilley. "Huxley [is portrayed as] the archangel Michael of enlightenment knowledge, and the disinterested pursuit of truth; Wilberforce, [as] the dark defender of the failing forces of authority, bigotry, and superstition."[27] Set in a Victorian backdrop, "only the stock conventions of melodrama can do it justice" continues Gilley, "and so it lives on in the popular mind as the best known symbol of the nineteenth century conflict of science and religion."[28] Others have not failed to find similar ways to describe the event. As John Lienhard put it, it was "the first major battle in a long war [against Christian fundamentalists]."[29] It is, as an image, one of the cornerstones of the historical idea of the "perennial warfare of science and Christianity." Indeed, it is the rock upon which that peculiar church is built. Or at least, it is one of them. The historian Ian Hesketh thus concludes: "If it appears that there was a campaign to rid science of all theological remnants and to historicize the relationship between science and religion as a great battle, that is because there was in fact such a campaign."[30] And it paid off.

Repeated so often in stories about the inevitable flight of dogmatic religion before the march of science, a recent scholarly exposé of this somewhat mindless repetition in both popular and academic publications humorously refers to it as "1859 and All That."[31] In a particularly apt example the New Atheist Christopher Hitchens writes without any sense of irony or apparent awareness of the cliché he just wandered into that the Huxley-Wilberforce debate was the "tipping point" that turned the tides in the fight of evolution and Christianity. "In front of a large audience," he says, "Huxley cleaned Wilberforce's clock, ate his lunch, used him as a mop for the floor, *and all that* [emphasis added]."[32] Hitchens is loath to leave this event without friends, and so in addition places it between the flat earth on the one side, and the Scopes Trial on the other, as links in the chain of religion's humiliation at the hands of science. Aggregating these mythic episodes together, as we shall see, is a frequent literary strategy.

Despite the pomp and circumstance, however, the Oxford debate—like so many episodes in the supposed history of warfare between science and religion—is in fact a myth. "Myth" as we are using it here does not indicate that the story is simply untrue.

27. Gilley, "The Huxley-Wilberforce Debate," 325–40, quote at 325.

28. Gilley, "The Huxley-Wilberforce Debate," 325.

29. Quoted in Livingstone, "Myth 17," 152.

30. Hesketh, *Of Apes and Ancestors*, 100.

31. James R. Moore, "1859 And All That." A special word of thanks to Multnomah University and Seminary Librarian Suzanne Smith, who helped me track down this article, which proved especially hard to obtain.

32. Hitchens, "Equal Time."

Rather by myth we mean "ideology in narrative form," to borrow a phrase.[33] Myth is an arrangement of facts just so—though the facts themselves will not often be left without smudge—so as to provide a lens to see the world. And it is these mythic histories, both perceived and imagined, that mark and inscribe day-to-day judgments about the general course of the relationship between Christianity and the sciences, and really the entire course of the fortunes and failures of Christianity in Western history. The problem is that the conceptual heavy lifting these grand stories like the Huxley-Wilberforce debate often do, and the way they predispose us to be inclined toward certain types of judgments regarding science and religion, has seldom been recognized and—especially at a general level—has rarely been examined critically. Until recently, as we have already begun to see.

The memo that the war is not only over but perhaps never even began, however, has failed to be transmitted very far outside of a few select circles of academia, and the typical account of the rise of Darwinism still in essence follows the impression set by one of the main initiators of the warfare thesis whom we will meet in the next chapter, Andrew Dickson White. White wrote that "The *Origin of Species* came into the theological world like a plough into an ant-hill. Everywhere those rudely awakened from their old comfort and repose swarmed forth angry and confused."[34] No doubt Darwin's work caused a great stir, and both infuriated and stimulated a good deal of people—not just the theologians. However, this account of White's represents what has been termed a narrative of self-fashioning in which—not the victor, but those longing for victory—had begun to write the histories of their desired triumph in advance. In fact, as we shall see some in this chapter and later on, "the impact usually associated with Darwin, [Herbert] Spencer, [Alfred Russell] Wallace, Huxley, *Essays and Reviews*, and John Tyndall, was part of a larger movement embracing a number of naturalistic approaches to the earth, life, and man—in utilitarianism, in population theory, in geology, phrenology, psychology, and in theology itself."[35] That last one—theology—will no doubt startle some. But this is the paradox that lay at the heart of stories of the triumph of naturalism over theology: naturalism was achieved initially on theological premises, and then—quite literally as we will see in a moment—its theological aspects were written out of history as a pure triumph of naturalism over an inferior religious rival.[36]

As Jessica Riskin puts it in her book-length study on just such an issue: "In short, a contradiction sits at the heart" of modern narratives of the rise of science as a triumph over theology and religion. "The central principle responsible for defining scientific explanations as distinct from religious and mystical ones was the

33. Lincoln, *Theorizing Myth*, 207.

34. A. D. White, *Warfare*, 1:70.

35. R. Young, *Darwin's Metaphor*, 4.

36. Cf. Ruse, "Removing God from Biology," 141: "For Darwin, the problem of final cause was a Christian . . . problem. And his solution, natural selection, [he saw as] a Christian solution!"

prohibition on appeals to agency and will. This principle itself [historically] relied for its establishment upon a theological notion," that is, God as the creator of the laws of nature. "To put it another way," she says, "when the inventors of modern science banished mysterious agencies from nature to the province of a transcendent God, they predicated their rigorously naturalist approach on a supernatural power." Ironically, in terms of the history of ideas, "a material world lacking agency assumed, indeed required, a supernatural [Designer] god."[37] What else, it was reasoned, could account for the regularity instead of chaos, the wondrous interconnection instead of meaning-lessness? Who else but the grand Artisan could make such a machine? Nor was this solely pictured in terms of direct intervention, but "Providence was also, in turn, a shelter under which the ideas of self-organization could grow."[38] The choice between the mechanism of law or direct creation was not one between science and theology per se, but between two visions of science *within* differing theologies. No less a com-mentator than the great Benjamin Warfield—defender of the inerrancy of scripture and conservative Christianity at old Princeton University—argued that John Calvin himself had laid down a theological path to a "naturalistic explanation of nature," and even more provocatively that though obviously Calvin was not privy to many subjects then under debate like the age of the earth, his theology could nonetheless be viewed as "a precursor of modern evolutionary theorists."[39]

James Moore, who set the precedent for attacking the warfare thesis as mentioned in the chapter intro above, can therefore record in his seminal study that orthodox Calvinism was, surprisingly enough, a historically important factor in priming posi-tive reception of Darwin. One must always caution generalization, but nonetheless he argues that in many areas in the nineteenth century Darwinism was "the legitimate offspring of an orthodox theology of nature."[40] Famously, the Marxist historian Jo-seph Needham even claimed that the lack of belief in a supernatural designer God and its entailments for a view of nature's regularity and intelligibility was one of the main reasons scientific naturalism did not take hold in China and the East as it did in Europe.[41] Today, of course, says Needham, laws no longer carry the connotations of a lawgiver or of a command to be followed. Rather, they are statistical regularities, de-scriptions rather than prescriptions. Nonetheless, he muses, whether the recognition of these statistical regularities and their theoretical formulation into law would have

---

37. Riskin, *The Restless Clock,* 5.

38. Sheehan and Wahrman, *Invisible Hands,* 44; cf. Funkenstein, *Theology and the Scientific Imagi-nation,* 202–289 for a related history regarding divine accommodation and providence and their transformation into human power and natural self-organization.

39. Warfield, *Evolution, Science, and Scripture,* 308–9.

40. J. Moore, *The Post-Darwinian Controversies,* 16. Cf. England, "Natural Selection, Teleology, and Logos," 271.

41. Needham, "Human Laws and the Laws of Nature."

been reached "by any road [other] than that which Western science has traveled," and which required a very particular theological view of the world.[42]

The point of these observations for our purposes is not necessarily a recommendation of such theological positions, nor that the bewildering array of religious responses to Darwin were restricted to this happier strain. It is rather to say, once again, that history is far more complex and interesting than anything that can be reduced to the formula "science vs. religion." Even if we limit ourselves to viewing, say, historical Calvinist responses to Darwinism—and so contain our inquiries to people who ostensible hold similar or even identical theologies—there are a variety of responses that seem to be conditioned by geographical and so local-contextual notions shaping the available rhetorical and conceptual space in which Darwinism entered. As the historian David Livingstone has investigated in detail, in Edinburgh there was an almost casual acceptance of evolution, while in Belfast Darwinism met with violent rebuffs.[43] The variety merely increases as other Calvinist homesteads are surveyed. What this suggests is that what is typically narrated as the rise of naturalism and secularism *against* theology was actually quite often a debate amongst a constellation of positions *internal* to theology itself—in other words historians have recently begun to discover "how largely the crisis [of Darwinism] arose and was resolved within the framework of established religious beliefs."[44] These debates were also colored by local circumstances, and one or two points of a constellation of different positions that were either justified by or saturated with theological considerations—like a mechanical world—were often carried forward at a popular level of reception but eventually renarrated as the triumph of a godless or disenchanted universe.

Whether or not our increasing awareness of theology's role in these historical transitions and mutations means that theology shot itself in the foot, I will leave up to the reader. What is clear, however, is that theology and religion need to be reinserted into their proper places in these historical narratives to truly understand what was going on. As Charles Taylor writes of these events leading to the rise of naturalism and the idea that God now had no job, materialist and atheistic interpretations are, in fact, merely one way that the story could have gone. "That [methodological naturalism in Darwin and others] came to serve as grist to the mill of exclusive [atheist] humanism is clearly true. That establishing [methodological naturalism] was already a step in that direction [toward atheism] is profoundly false." Indeed, says Taylor, "This move had a quite different [theological] meaning at the time, and in other circumstances might never have come to have the meaning that it bears for unbelievers today."[45] For in undoing special creation Darwin—especially early on in his notebooks, but even for some time after the *Origin* was published—noted that his theory evidenced the

---

42. Oakley, *Natural Law*, 37.

43. Livingstone, *Dealing with Darwin*.

44. Moore, *The Post-Darwinian Controversies*, 13.

45. Taylor, *A Secular Age*, 95.

immediacy of God's wisdom working in the world to produce and maintain the laws that led to such "endless forms most beautiful." In this manner several scholars have even gone so far as to suggest (somewhat audaciously to our contemporary ears) that "the *Origin of Species* is the last great work of Victorian natural theology."[46]

Ironically, if the naturalism so often equated with the triumph of evolution was borrowed from theology, so too were many of the polemics against religion borrowed from amongst the religionists themselves. In this vein, many "historians popularized the warfare between Darwinism and dogma by drawing on the military metaphor [contemporary] to [nineteenth century] culture, and by exploiting liberal and prot-estant hatred of a [then] conservative papacy."[47] James Ungureanu has recently dem-onstrated at length that many of the most vicious attacks that we now associate with atheists against religion were actually directly cribbed from Protestant critiques of Catholic "superstition" and "anti-scientific dogmatism."[48] The complexities of history do not allow us to describe this ongoing conflict in the simple terms of "science vs. religion"—even in seemingly clear-cut cases like the 1925 Scopes Monkey Trial, as we shall see. Indeed, as two historians commenting upon the evolution of the univer-sity in America write, "in retrospect, the latitude that colleges and universities gave natural and social scientists to pursue their investigations seems far more pronounced than theological opposition to such efforts."[49]

All of the preceding is to say that motivations, allegiances, and the very catego-ries used like "science" and "religion" are not stable across time and cannot be forced to stand still for the sake of taking a neat and tidy picture.[50] In the Victorian conflicts rather than science vs. religion, Darwinism rather "intruded on a complex debate *within* Christendom itself between liberal and conservative theologies"[51] as well as a debate—as often among Christians as against them—regarding professional power being located amidst the Anglican clergy or the emerging class of professional scien-tists. Moreover, Darwinism itself was hardly a homogenous theory or the property of some homogenous group we can label "scientists." Darwinism arrived in the midst of methodological and theoretical debates "raging among geologists, naturalists, biologists, paleontologists, morphologists, biblical scholars, various religious sects, politicians, clergymen, historians, industrialists, merchants, aristocrats, socialists, and Chartists."[52] Instead of an argument against religion from a group termed "scien-tists" many of these arguments occurred across a myriad of disciplines, and were part of the struggle to form a coherent notion of what an identity as "scientist" (among

46. Durant, "Darwinism and Divinity," 16.
47. Gilley, "The Huxley-Wilberforce Debate," 327.
48. Ungureanu, *Science, Religion, and the Protestant Tradition*.
49. Roberts and Turner, *The Sacred & The Secular University*, 63.
50. Lightman, "Does the History of Science and Religion Change?"
51. Gilley, "The Huxley-Wilberforce Debate," 327. Emphasis added.
52. Hesketh, *Of Apes and Ancestors*, 8.

whom were innumerable theologians) even meant. In other words, neither "science" nor "evolution" arrived as a stranger outside and alien to Christianity, nor something native and internal to science, but was born within and continuously folded into the ongoing internal reflections and self-criticism of Christianity and, slightly different, Christendom. And all of this amidst broader flurries of various disagreements amidst a rapidly changing society.

Unfortunately, though these assertions of complexity are in some sense representative of an emerging consensus of historians working in the field of the history of science and religion over the last half century, these revisions have been slow to trickle down into the public imagination at large. What is being claimed by historians, in brief, is that the primary conflict turned not on science vs. religion, but upon an explicit act of the *writing and rewriting of identity*. The "warfare of science and religion" needs to be relocated not as a natural event between two defined entities, but at the origins of *historical writing* whose early methods deemed this, for innumerable reasons, the best way to tell the story. This brings us right back to Wilberforce and Huxley to help all of this seem much less abstract and theoretical. To be sure, the confrontation between the two men did occur in June of 1860 at Oxford, during the annual British Association for the Advancement of Science (BAAS) meeting. And, as Brooke puts it, "whatever construction we place on the event there was clearly a commotion of a kind."[53] But of what sort, exactly?

Back to the Bulldog

Images of a valiant Huxley leading science to a lopsided victory against the blowhard bishop are prime examples of the sort of historical rewriting that litters the genesis of the warfare narrative. These accounts of the flabbergasted bishop undone by an indignant yet eloquent Huxley are stitched together from mostly contradictory reports "created 20 years later and . . . [which are] still trotted out uncritically to this day," as historian Frank James points out.[54] By the mid-twentieth century, however, these variation of events were so ingrained as a bad historical habit that even when Lynn A. Phelps and Edwin Cohen set out one of the first attempts at "reconstructing" the event in a short 1973 article, their "reconstruction" turned out to restate what everyone thought they already knew—that Huxley's response proved "forensically fatal" and that he "destroyed" both Wilberforce's arguments and his credibility.[55] Five years later in 1978 the BBC broadcast their *The Voyage of Charles Darwin,* which included for posterity a highly dramatized version of Wilberforce's drubbing that used as their sources only Huxley's own letters recalling the event. This spread the already deeply embedded mythology much as we will see the movie *Inherit the Wind* framed public

53. Brooke, "Wilberforce-Huxley," 131.
54. James, "On Wilberforce and Huxley."
55. Phelps and Cohen, "The Wilberforce–Huxley Debate."

opinion about the 1925 Scopes Monkey Trial over creation and evolution even more than the trial itself, or how Berthold Brecht's play *Galileo* burned yet another "Christianity vs. science" set piece into the public imagination.[56]

On the second level of the Natural History Museum today a plaque stands watch next to a nondescript door stating "A meeting of the British Association held 30 June 1860, within this door was the scene of the memorable debate on evolution between Samuel Wilberforce Bishop of Oxford and Thomas Henry Huxley." Ironically, despite the plaque's assertion of its memorability, the evidence we do have is somewhat scant. Even the great hall that housed this verbal bout of fisticuffs—a room that could hold up to a thousand people, and indeed may very well have on the legendary day in question—has been transformed into a narrow hallway-like storage room. Not even the physical memory of the architecture, it seems, can now summon the once-upon-a-time it housed in a bygone era. Current passersby are no doubt left perplexed that a debate could have even taken place in this spot—to say nothing of the particulars of the debate itself. Ian Hesketh has recently compiled the most thorough investigation of the event to date in a book-length sleuthing, *Of Apes and Ancestors: Evolution, Christianity, and the Oxford Debate*. "For a historian," he says, "it is a humbling experience to see that the room where it took place no longer exists . . . that [the debate] must be imagined by considering the sparse documents written at the time, documents that are themselves plagued with conflicting recollections and blatant exaggerations." Even further, "compounding the problem is that the debate was initially memorialized by one side, becoming in hindsight a key battle in the rhetorical war between science and religion, a Galileo affair for the nineteenth century."[57] The main contours of the story handed down until recently are indeed those provided by Leonard Huxley and Francis Darwin—the sons of Thomas Huxley and Charles Darwin—as these had been told to them by their fathers and their fathers' friends.[58] Such pedigree, of course, does not automatically disqualify the information given, but when combined with the fact that there are numerous other contrary reports and circumstantial evidence, our ability to clearly discern fact from fabrication demands some effort. As far as we can tell, for example, both men—Huxley and Wilberforce—left the clamor that evening feeling they had each emerged as the clear victor. And this was not just their opinions. As the paper *Athenaeum* put it, the bishop and the Bulldog "have each found [in the other] foemen worthy of their steel, and made their charges and countercharges very much to their own satisfaction and the delight of their respective beliefs."[59] Others reported that Huxley's voice was too weak to carry to the whole crowd and was heard by, at best, perhaps half of those in attendance. Relayed by Huxley's friend and botanist Joseph

56. Larson, *Summer for the Gods,* 225–46.

57. Hesketh, *Of Apes and Ancestors,* 88–89. As we shall see soon, the Galileo affair is not what many make of it, either.

58. Hesketh, *Of Apes and Ancestors,* 97.

59. Quoted in Brooke, "The Wilberforce-Huxley Debate," 128.

Hooker, this information can hardly be suspected of forgery by one of Wilberforce's allies—though perhaps it was exaggerated slightly to support Hooker's opinion that it was *his own* speech to the crowd that won the day. Indeed, Hooker painted the scene as if it was a literal boxing match. "I was cocked up with Sam at my right elbow," wrote Hooker to Darwin, "and then I smashed him amid rounds of applause—I hit him in the wind at the first shot in 10 words taken from his own ugly mouth." And for the *coup de grâce* Hooker "proceeded to demonstrate in as few more [words, one] that [Wilberforce] could never have read your book and [two] that he was absolutely ignorant of the rudiments of Botanical Science."[60]

Amongst all the differing reports, one of the more peculiarly damning evidences against the Huxley-esque narrative of how the event at Oxford went comes quite surprisingly by way of one John William Draper. The forgotten keynote speaker of the BAAS gathering that day, and as we shall see in coming chapters was—along with Andrew Dickson White—one of the foremost initiators of the idea that institutional religion has historically and inevitably conflicted with the pursuit of science. It is all the more perplexing therefore that Draper does not mention Huxley's supposedly devastating victory in his most famous (and very frankly) titled book, *The History of the Conflict Between Religion and Science*. Some have remarked in defense of the orthodox Huxley-Wilberforce narrative that Draper's reasons for leaving his eyewitness account of the supposed victory of Huxley out of his recollections probably involves the fact that those who heard Draper's talk before the exchange noted that it was by all accounts "a crashing bore"[61] and really "flatulent stuff."[62] And these quotes come—not from an ardent opponent of Darwin or Huxley—but from Darwin's close friend Hooker. Not one to fall on his own sword for a cause, the reasoning seems to go, Draper ever so briefly mentions his paper in the *Conflict* and then quickly moves on with nary a mention detailing the evening. Letters recently rediscovered by James C. Ungureanu, however, tell quite a different story.[63]

In one, John expresses his excitement over his upcoming lecture at the BAAS meeting, saying it is a high honor that "they have given me the best hour." In another, dated about a week after the Oxford debate, Draper's eldest daughter Virginia writes of the experience at her father's lecture. It was held in a beautiful room "crowded to suffocation." Here we find that the woman who fainted did so because she was not getting enough air due to the heat and the crowd—and not, as it happens, because of anything either Wilberforce or Huxley may have said about grandmothers. Virginia continues and expresses her delight at the after-party she and her father attended, hobnobbing with many intellectuals who were also at the presentation earlier that

60. Hesketh, *Of Apes and Ancestors*, 65.

61. Lightman, "The Victorians," 67.

62. Lightman, "The Victorians," 67.

63. Ungureanu, "A Yankee At Oxford." All quotations from the Draper family letters in the following paragraph are reliant upon the excellent research of Ungureanu.

day. There is no hint of embarrassment on the Drapers' part. Most interesting of all is John's reflections later on after his presentation. "I have accomplished very thoroughly all that I came here for," he says in one letter. In another, the sense of accomplishment reaches a fever pitch: "I may truly say that I [have] never undertaken anything before which so thoroughly succeeded."

These are certainly not the words of a man ashamed of his presentation, perhaps later wanting to forget by omitting his lecture from later accounts. Nor indeed does he seem to have any inkling that his presentation would later be related to posterity as ill-conceived. Despite his remarks' length, Draper recalls, "I was listened to with the profoundest attention." Wilberforce makes an appearance in a letter as well, as he arose to orate a response to Draper's paper. This point fits with the usual story, as far as that goes. Far from an irritation, or a blotch on the day however, Draper recounts that the bishop was "a very fine speaker" who was, Draper recalls, extremely congenial despite his criticisms. What is more, as Ungureanu notes, perhaps the oddest feature of Draper's letters "is the conspicuous *absence* of Huxley" from the day.[64] While it is perhaps impossible to accurately speculate on why Draper failed to mention Huxley, regardless of the reason "what is abundantly clear [from these letters] is that by excluding Huxley the Drapers tell a story that conflicts" with the orthodox warfare story line, when they had every reason not to.[65]

To complexify things further, Huxley's writings indicate he was himself *against* those who claimed a necessary contradiction between religion and science. Quite surprisingly, Huxley writes that "the antagonism between science and religion about which we hear so much, appears to me to be purely factitious." Indeed, it has been "fabricated on the one hand, by short-sighted religious people who confound a certain branch of science, theology, with religion; and, on the other, by equally short-sighted scientific people who forget that science takes for its province only that which is susceptible of clear intellectual comprehension."[66] It is dogmatic theology and the theologians in power—not religion—that is the target of Huxley's general ire. While this might seem little more than an interesting—perhaps even disingenuous—historical anecdote, this attitude differentiating theology from religion will become a cornerstone to help us understand the origins of the warfare myth in the next chapter. On the other hand, those who did hear Huxley at the Oxford debate and gave their approval were not merely "godless" scientists, but among Huxley's endorsers were also orthodox churchmen. Above all, no less an individual than Hooker came from a fervently evangelical family, and there is no evidence he dissented from this religiosity. Quite the opposite. Letters from Hooker's father—who was the Regius Professor of Botany at Glasgow—indicate that the whole family saw faith and science as twin passions to worship God. Others with religious sensibilities who supported Huxley were

64. Ungureanu, "A Yankee at Oxford," 142.

65. Ungureanu, "A Yankee at Oxford," 144.

66. Huxley, *Science and the Hebrew Tradition*, 160–61.

full-throated theological and scientific endorsers of evolution,[67] and some were merely aghast at Wilberforce's abandoning propriety and "good Victorian manners" with his ape-ancestry joke.

We do well to note that the alleged joke would have been no mere breach of etiquette either, if it did in fact happen. Gowan Dawson humorously recounts just how vulgar Victorian culture perceived the rising use of ape-like cartoon representations of humans to be. In one sketch in particular, representing Huxley and Wilberforce as it happens, a conversation in some cigar-laden gentleman's club goes like this: "So long as *I* am a man, Sorr [sir], what does it matther [*sic*] to me whether me *Great-Grandfather* was an anthropoid ape or not, Sorr!" Which receives the (even today) rather shocking response, written in a lisp: "Haw! Wather disagweeable for your *Gwate Gwandmother,* wasn't it!"[68] Evolution—that is, true evolutionary theory—isn't really even on offer in the image of the ape-human traded in Victorian culture. A cruder, more direct (and I hope I do not need to add, much less scientific) human and simian relation is imagined. While it would be too much to say that this episode of warfare was distilled from one bishop's particularly cringe-inducing inability to read a room, there is some force behind this assertion. We do not know whether Wilberforce meant to invoke this bizarre sexual connotation with his attack if it happened,[69] but whatever aura of gentlemanly dignity his position as Victorian bishop afforded him on this stage would certainly have been forfeit to more than a few in attendance. Nonetheless, however much Huxley may have kept his composure that evening, it should be noted out of fairness that he was otherwise not above using the elderly to score a few rhetorical points, as when he gibes "old ladies of both sexes, [who] consider [*On the Origin of Species*] a decidedly dangerous book."[70]

In the end, Wilberforce's impropriety real or otherwise did not decide how the debate went. We have records of persons present who defected from being initial defenders of Huxley and Darwin that day, to being persuaded by Wilberforce's more genteel and less grandmother-directed arguments against evolution. Among the defectors was no less prestigious an individual than Henry Baker Tristram, who was one of the first to apply Darwin's principle of natural selection to animal camouflage. Apart from the largely anachronistic (and, largely American) twentieth-century image of the Creationist trying to pummel Darwinists with the Bible (perhaps literally),[71] Wilberforce's critiques certainly had theological aspects, but were on the whole based upon scientific assessments of the theory of evolution contemporary to his day, not on religious arguments. Whatever Darwin's eloquence and meticulousness may have

---

67. Livingstone, *Darwin's Forgotten Defenders.*

68. G. Dawson, *Darwin, Literature, and Victorian Respectability,* 58–59.

69. Hesketh, *Of Apes and Ancestors,* 97.

70. Quoted in Noll and Livingstone, "Introduction," 2.

71. As we shall have opportunity to see, even where largely critical, theological interactions with Darwin were often highly sophisticated and thorough. See Gundlach, *Process and Providence.*

been in the *Origin*, it needs to be pointed out that his version of the theory of evolution had hardly achieved any type of consensus at the time. As Huxley himself wrote in his essay "On the Reception of 'The Origin of Species,'" had a hypothetical "general council of the Church scientific" been convened, "we should have been condemned by the overwhelming majority [of scientists]."[72] This was in fact what fueled much of Huxley's rhetorical strategies. It was yet to be established as a theory, and indeed what exactly evolution even meant was heartily debated and endlessly nuanced among those who generally supported the theory, to say nothing of its detractors. Darwin (not himself present at the debate) had responded to Wilberforce's review of the *Origin* not with disdain but by remarking that many of Wilberforce's critiques were "uncommonly clever" and that he had picked out "all the most conjectural parts."[73]

As it turns out, one of Darwin and Huxley's chief scientific opponents, Sir Richard Owen (1804–1892), had been coaching Wilberforce behind the scenes for an encounter such as this. Wilberforce may have been a bishop in the Anglican communion, but the church he was explicitly representing was actually a rival school to Huxley, consisting of Owen's scientific interpretation vying for orthodoxy. Conversely, while Huxley—who had a deep falling out with his former mentor, Owen, and was deeply angered to detect his fingerprints all over the debate as he was working behind the scenes—was ostensibly defending Darwin, he only reluctantly and never fully accepted the only truly unique aspect of Darwinian evolution, its holiest of altars—that is, the mechanism of natural selection. Natural selection was not viewed by most as the key aspect of Darwinism until the 1940s, when Mendelian genetics provided a stable mechanism for trait inheritance and the centennial in 1959 rewrote natural selection as the singular element that Darwin injected into the debates of the time. And again between the centennial of 1959 through the 1980s another flurry of perspectival shifts on what constituted true "Darwinism" emerged.[74]

By the 1970s and accompanying these broader shifts in the notion of Darwinism, the traditional form of the Oxford debate narrative was slowly beginning to be questioned in earnest by historians of science.[75] Increasingly, they have shown how the Oxford debate is the product "of historical spin doctoring," and, ultimately, it was a rivalry that involved "personal struggles and jealousies" expressing itself in and supercharging the languages of science and religion surrounding the early reception of Darwin to vent themselves in public.[76] While Huxley hardly advanced the conversation in terms of the fine details of evolution, the press that he brought to bear on Darwin's theory—for good or ill—was invaluable. This was recognized by Darwin above all others, who was pleased that the Bulldog had been loosed among

72. Quoted in Gilley, "The Huxley-Wilberforce Debate," 327.

73. Taken from Livingstone, "Myth 17."

74. Hull, "Darwinism as a Historical Entity."

75. See the helpful summary in Ungureanu, "A Yankee at Oxford," 137–38.

76. Hesketh, *Of Apes and Ancestors*, 89.

the pigeons. "Facts alone would not be enough for evolution to make the jump from a heterodox and heretical theory of species development to one of science," observes Hesketh. "History itself would need to be on the Darwinians' side."[77] This is hardly to make the absurd claim that evolution is nothing but rhetorical trickery, or to dismiss its scientific explanatory power. It is, however, to point to the fact that the way many *use* "evolution" often implies much more than a mere scientific theory. The term drags with it a messy web of intrigue, worldviews, and even the almighty dollar. Indeed, it is significant to note that in terms of scientific theory as it is carried on today, neither "Darwinism" nor "neo-Darwinism" represent an accurate description of the content of the rapidly evolving fields of inquiry. Such terms are kept, despite this, precisely because they carry what are for many useful boundary-defining markers that make easy "us vs. them" shorthand references. Both in history and today, therefore, we have to be very attentive to the fact that "Darwinism" is ultimately a codeword that meant (and means) different things to different people. To truly understand the theory and its reception, we have to be attentive to how the term is being controlled and then received, even across different geographical spaces.[78]

Suddenly, the supposed war of science and religion many want to distill from the Oxford debate changes, and the "party lines" are hardly neatly mapped out in such terms. As we will see in our final chapter on the Scopes trial, and the events surrounding it, messiness can be broadened without fear of generalization to the debates over the reception of Darwin as a whole. Indeed, "contemporary readers who associate creationism with the teaching of so-called scientific creationists will no doubt be surprised by the small number of nineteenth-century creationists writers who subscribed to a recent creation in six literal days, and even greater the rarity of those who attributed the fossil record to the Noachic flood." Creationists in the nineteenth century on the whole were not only open to the geologic and evolutionary findings, but were often themselves the ones researching and driving those interpretations.[79]

Let us unburden ourselves with these many details by recalling the major point to be had: the reception of evolution, just as its iconic representation in the Oxford Debate—whatever it originally was—lives not just in itself but perhaps even more so in its retelling, its insertions into our collective cultural memory. And so our perception of just how much historians have often—sometimes innocently, sometimes intentionally—framed and shaped what we "know" has become more apparent, as we realize that "the concerns of the appropriators determine which past figures are useful, seminal, or important, rather than the intrinsic or timeless merit of their ideas."[80] As it stands, then, "whether or not Huxley actually won the debate is perhaps beside the point [for the creation of the warfare narrative]. He was certainly shaping the memory

77. Hesketh, *Of Apes and Ancestors*, 3.

78. Livingstone, "Re-Placing Darwinism and Christianity."

79. Numbers, *The Creationists*, 16.

80. Osler, "The Canonical Imperative," 8

of the debate [in the years to follow], and this mattered much more than anything that may or may not have been said at Oxford."[81] The primary sources tell us very little of what actually happened, concurs J. R. Lucas, "but about currents of thought in the latter part of the [nineteenth] century, it tells us a lot."[82] It turns out this is not merely true for the Oxford debate, but for the secret strategy that Huxley was slowly moving into position to execute at large.

The X-Club, Scientific Naturalism, and the (Re-)Writing of History

The eminent historian of the Victorian period Frank M. Turner has pointed out that what many want to narrate as "science vs. religion" is in fact better explained when we take into account, as one example, the fact that the younger professional class which included Huxley and his friends were attempting to establish themselves a professional identity as "scientists" (which we will recall was new, and yet to carve out a stable social space). To do so they had to oust those who held the jobs and social authority they wanted—the clergy.[83] Huxley and others like him found their weapon for social leverage in the conflict metaphor used to describe the historical relationship between science and theology and so, by extension, their institutional relation in Huxley's day.

Beneath the charges of religion vs. science (or theology vs. science), one can certainly detect more than simple hints of Turner's observation about the social standing of scientific naturalism and the scientists who championed it. For a great deal of time, natural history (parts of which we would later designate "biology") was done by learned gentleman naturalists, many of whom were churchmen. These were highly dedicated, well-read individuals who formed loose communities of recognition and so followed a rough outline of what was considered accepted procedure. The production of the *Bridgewater Treatises*—a set of works commissioned by the eighth Earl of Bridgewater to comment on knowledge of nature in terms of natural theology and design argumentation revealing a Designer—and the work of William Paley (also famous for his work on design arguments) fall into this category. Nonetheless, such practices could also lead to highly sensational works claiming that even human consciousness and will are purely natural phenomena dictated by God's imposition and constant preservation of natural laws on creation, as in the case of the initially anonymous *Vestiges of the Natural History of Creation*, later discovered to have been written by the Scotsman Robert Chambers.

We will have opportunity to speak more on design arguments later on in this book. The point for now is that to consider these men (though as we are rediscovering there were also women, who were in the meantime forgotten and suppressed—as is

81. Hesketh, *Of Apes and Ancestors*, 93.
82. Lucas, "Wilberforce and Huxley."
83. F. Turner, *Contesting Cultural Authority*, 171–200.

sadly too often the case)[84] "amateurs" despite not being "scientists," and because they were doing theology, or were theists, would be anachronistic, and likewise shows the great lengths of success the professionalization campaign of Huxley and his compatriots to transfer social prestige to "scientists" garnered. Perhaps the most amusing example of this is Francis Galton's *English Men of Science: Their Nature and Nurture* (1874). Galton—who was Darwin's half-cousin and the inventor of the notion of eugenics—like Huxley, wanted a war so as to split clean the line between professional scientists and amateur theologians. For his book, he sent out questionnaires to a great number of colleagues across the country. A great sinking feeling must have come upon him when a majority of the scientists he questioned reported back that religious beliefs in no way were a hindrance to scientific work. Perhaps taking a cue from his pet theory of eugenics, Galton decided to cull this herd of unworthy responses and produce his own version for the good of society. In his book, therefore, he simply asserts that religious convictions were "uncongenial" to the pursuit of science, in spite of the fact that his data not only did not support this conclusion, but practically howled the opposite at him.[85]

The new scientific communities that were developing did indeed rein in the speculation of the gentlemen naturalists, and created a more controlled and successful network of methodological criteria for knowledge of the natural world to succeed. But to see this as in discontinuity with patterns of thought, practice, and even societal structures among the natural philosophers and theologians is for the most part an immensely successful rhetorical ploy on the part of Huxley's campaigning, as we will see in a moment. To harp on our theme, the Darwin historian Robert Young emphasizes that to see these encounters as *between* science and religion is to miss the profound continuities that survived between the theological and scientific belief systems.[86] As has been mentioned, too often we treat things like "science," "theology," "religion," and the like as known quantities with well-defined borders, whose presence can then be traced through the ebb and flow of historical time. Young claims quite provocatively that the emergence of "scientists" in the evolution debate "was merely a demarcation dispute *within* natural theology" rather than seeing it as natural theology opposed to something else, say "secular" or anti-theistic science.[87] Historian of science Peter Harrison agrees, and argues that "perhaps these skirmishes should be thought less in terms of conflict between science and religion, and more as theological controversies waged by means of science."[88] Everyone in this sense was trying to follow the advice of Francis Bacon, often named the "Father of Modern Science," in his *Advancement of*

84. Schiebinger, *The Mind Has No Sex?*

85. Turner, *Contesting Cultural Authority*, 185.

86. Young, *Darwin's Metaphor*, 1–22.

87. Young, *Darwin's Metaphor*, 12, emphasis added. See also Bowler, *The Non-Darwinian Revolution*, 174–204.

88. Harrison, *Territories*, 197.

Learning—and before him thinkers like St. Augustine and St. Thomas—to "let no man . . . think or maintain that a man can search too far, or be too well studied in the book of God's word, or the book of God's works, divinity or philosophy; . . . and again that thy do not unwisely mingle, or confound these learnings together." The trick here, says Young, is to see that "everyone thought that he was following" Bacon's advice. That includes major thinkers like Cambridge geologist Adam Sedgwick, another renowned geologist Charles Lyell, indeed even Darwin himself, who quotes Bacon in the epitaph to *On the Origin of Species*. Everyone wanted to separate theology and science as distinct—even if closely related—disciplines. "The question was," Young notes, "where to draw the line"?[89]

In other words, many of those participating in the professionalization debates over science were both Doctors in Divinity and members of the Royal Society, and so to them "the problem was not whether God governed the universe" in an intimate act of providential care and ordering. The question was how. Regardless of where one stands on the legitimacy of their theology or not, when our historical inquiry looks at it from this perspective it was not a war of science and religion, but rather that many "of the evolutionists were explicitly looking for a grander view of God" than they found in the specific design arguments that Paley and the Bridgewater Treatises of their day could muster.[90] Thus the specialization and professionalization in the sciences, as a new and separate enterprise from things like the Bridgewater treatises, did not necessarily indicate that they were anti-theological, anti-religious, or anti-theistic. Nor did the rise in acceptance of evolution indicate a corresponding drop in religious belief.

Nonetheless, this theological finagling with the science of evolution struck Huxley and Tyndall especially as a rank half measure. It did not matter that these theologies were held by scientists in support of evolution, and even lent the still lingering social glamor of theology to the sciences in pilgrimage for such clout. That it was *theology* at all, however friendly, meant that it remained a point of access for ecclesial authorities to meddle. The issue for Huxley was never theology per se, even if that was often what appeared to be Huxley's target, but rather the authority and social ordering that such theology symbolized and to which it gave entrance. Huxley criticized what he called the *Pax Baconia* or the peace of Francis Bacon, where "Men were called on to be citizens of two states, in which mutually unintelligible languages were spoken and mutually incompatible laws were enforced: and they were to be equally loyal to both."[91] That his concern was primarily the location of authority in order to secure scientific freedom, and not theology per se, is especially clear in the case of Huxley's

---

89. Young, *Darwin's Metaphor*, 13.
90. Young, *Darwin's Metaphor*, 14.
91. Quoted in Barton, "Evolution," 262.

biology student, the Catholic St. George Mivart (who was also a student of Huxley's mentor-turned-rival, Richard Owen).[92]

Initially, Mivart came to Huxley "after many painful days and much meditation," when Mivart decided as a student he did not and could not believe in natural selection. He came to this conclusion on scientific grounds (albeit through what appears to be a misunderstanding of Darwin on several key points). He felt obliged to report this to Huxley. If he hoped it would be well received by his mentor, he was mistaken. "Never before or since have I had a more painful experience than fell to my lot in [Huxley's] room at the School of Mines on that 15th of June, 1869," Mivart lamented. "As soon as I made my meaning clear, his countenance became transformed as I had never seen it." Huxley was still kind and gentle, Mivart recalls, but in a deeply saddened tone "[he said] most firmly, that nothing so united or severed men as questions such as those I had spoken of."[93] Darwin, too, who felt kindly toward Mivart after a series of initial correspondences, was saddened—indeed shocked—when Mivart wrote what Darwin took to be a scathing review of his work. "He makes me the most arrogant, odious beast that ever lived," said Darwin, in a moment of indignation. He supposed that it must have been Mivart's Catholicism that was creating an "accursed religious bigotry."[94] In general, given the increasingly anti-modernist bent of the Catholic Church at the time, this could be seen as a reasonable conjecture. In terms of evolution specifically, however, it grows less warranted. Catholics—Magisterium included—still stinging from the historical fallout of how the Galileo affair was perceived at large, had been unusually tame when it came to pronouncing upon evolution. What is more, the Church never declared an official position on the matter.[95] In targeting modernism, the Catholic Church intended to combat not science per se, but rather the philosophical worldviews they perceived as increasingly wedded to such scientific projects.

Such remarks against Mivart by Darwin were therefore perhaps the speculation of a wounded man trying to make sense of what appeared to be a friendship broken off in its infancy. For, despite the charge of religious bigotry by Darwin, Mivart's criticisms had enough bite that he later devoted a new chapter to account for them in the sixth edition of *The Origin of Species*, as was Darwin's way of continual revision to account for his critics. More surprising for us today, Mivart was still an evolutionist. His negative position on natural selection was still completely within the pale of the evolutionary consensus at the time, such as it was. No less a figure than Alfred Russell Wallace—the co-discoverer of evolution via the principle of natural selection with Darwin—just six years after his co-discovery began to doubt the sufficiency of natural selection as a mechanism.[96] And as we mentioned, Huxley himself was never fully

92. On Mivart, see Artigas et al., *Negotiating Darwin,* 236–69.

93. Quotes taken from Artigas et al., *Negotiating Darwin,* 237.

94. Artigas et al., *Negotiating Darwin,* 238.

95. Artigas et. al., *Negotiating Darwin.*

96. Young, *Darwin's Metaphor,* 19; Riskin, *The Restless Clock,* 214 49.

comfortable with the idea, seeing it as too random to neatly fit into his absolutely uniform and determined view of nature. It is really only by wearing the spectacles of the twentieth century, where natural selection again became ascendant up through the 1970's as the central principle of Darwinism, that we can demand evolutionists of the nineteenth century must have also been adherents of natural selection. But this was not the case.[97]

As David Hull observes, "Mivart could easily have become a Darwinian. His views about evolution differed in no important respect from several key Darwinians. Like Huxley, he thought evolution was more saltative [that is, proceeding by major evolutionary jumps] than Darwin did. Like Gray, he thought it was directed. And like so many Darwinians, he did not think natural selection could do all that Darwin claimed of it."[98] Mivart would in fact go on to write a book published in 1871 titled *The Genesis of Species*, in which he promoted evolution but sanitized it of what he thought was the spurious mechanism of natural selection. And this for scientific reasons. In the course of his book Mivart claimed that Christians are completely free to accept the theory of evolution (as Mivart has presented it), and even cites St. Augustine, St. Thomas Aquinas, and the influential late sixteenth-century Spanish theologian Francisco Suarez, seeking to show that proto-evolutionary or, at least, compatible theology had been on offer. Mivart in a bold gesture even name drops Huxley, who had recognized that evolution—despite many claims—had not fully removed notions of purpose or directedness from creation as a whole. Indignant would not be a strong enough word for Huxley's reaction. He proceeded to spend his days rummaging through old Catholic encyclicals, and even cracking open the notoriously dense Suarez to garner weapons for his exclamation that evolution was in "complete and irreconcilable antagonism to that vigorous and consistent enemy of the highest intellectual, moral, and social life of mankind—the Catholic Church." Huxley then hammered the point home: Mivart could not by any stretch of imagination be "both a true son of the Church and a loyal soldier of science."[99]

To put it rather bluntly, it wasn't simply that Huxley thought evolution contradicted establishment Christianity; he *needed* it to contradict establishment Christianity. Huxley could deal with the Wilberforces of the world because, however irksome, they were an easier piece to fit in the narrative puzzle of anti-clericalism he was building. What vexed Huxley more than opposition to the theory of evolution were theologians who thought evolution and Christianity could be harmonized. We can see this in the fact that though critical of Paley, Huxley could even insist that because Paley, despite being the most well-known advocate for Design arguments, saw things as products of law. In this sense even Paley could be claimed by Huxley as a proto-evolutionist![100]

97. See Bowler, *The Non-Darwinian Revolution*.
98. Hull, "Darwinism as a Historical Entity," 797.
99. Quoted in Brooke, *Science and Religion*, 308.
100. Gilley and Loades, "Thomas Henry Huxley"—see esp. 290.

On the other hand, Huxley's savagery was at a high point when attacking the *Vestiges of the Natural History of Creation*, precisely because Chambers had provided a vision in which transmutation could become part of the natural theology tradition.[101] We saw the same rejection with Mivart's proposal. If harmony or—even worse!—mutual benefit between evolution and Christian theology was obtained, it would erase the fundamental distinctiveness that Huxley and his friends needed to differentiate their emerging profession of "scientist" from the still considerable social power of clergy everywhere. Moreover, though Huxley himself wrote that "true science and true religion are twin sisters"[102] any alliance with ecclesiastical authority and theology would in principle compromise Huxley's vision of a society ordered around free scientific inquiry. And this despite the fact that among the scientists—indeed among Huxley's friends—were theists aplenty.

As such Huxley along with a few like-minded and well-positioned friends formed the so-called "X-Club"[103] in 1864 precisely to expunge these theological half-measures. Ironically enough, Mivart was a member in this group until his attempts to reconcile evolution with Catholicism based in part on the very principles so dear to the X-Club got him exiled. As a sort of "Masonic Darwinian Lodge,"[104] they were a group of friends who dined together regularly at the St. George's hotel. Far from an innocent gathering,[105] they were, in the words of James Moore, "the most powerful scientific coterie" in Victorian England, and perhaps the world at the time.[106] Though some historians have followed Huxley's innocent dismissal of any conspiratorial intent, Ruth Barton has recently demonstrated how evident Huxley's intentions for the group were from the beginning, and how extensive their joint actions and planning were. Beginning around 1860 before they officially formed, through the 1880s, their concerted efforts were bent toward colonization of the universities and the elimination of theology from the natural sciences (and perhaps, altogether). Regardless of Huxley's hesitance regarding natural selection, as Barton has phrased it evolution became "The Gun in Huxley's War for the Liberation of Science from Theology," to take a piece from the title of one of her published essays.[107] And so the X-Club went on a campaign trail, similarly armed, to pull control of the universities from out of the clutches of the clerics. "It may be too strong to say that the scientific naturalists engaged in a conspiracy to take over science," writes Matthew Stanley, "but it is close."[108]

---

101. This does find a few counterexamples, as Huxley seemed to be more or less cordial with the Christian socialist and evolutionist Charles Kingsley.

102. Huxley, "Science and Religion," 35.

103. On the X-Club, see Barton, "'An Influential Set of Chaps'" and Macleod, "The X-Club."

104. Desmond and Moore, *Darwin*, 526.

105. Barton, "Huxley, Lubbock, and a Half Dozen."

106. Moore, "Theodicy and Society," 172.

107. Barton, "Evolution."

108. Stanley, *Huxley's Church*, 30.

They had multiple strategies, one of which was getting like-minded folks hired and placed in a variety of locations. Huxley "had already in a short amount of time," writes Stanley, "managed to place his students, allies, and demonstrators at a dizzying array of universities."[109] Another was utilizing the exam system to reshape the landscape of science by controlling the information being briefed through such testing. It was "a powerful way to ensure epistemological conformity."[110] A third strategy was the writing of textbooks—Huxley writing the *Introductory Science Primer* that was the first in a series laying out the principles of naturalistic science to children at school. As Lightman writes, "every school child that read this introduction to science would be trained to reject the very premises of the theologies of nature."[111]

Except, that was not really the case. The theologies of nature, as we have seen, were not only potentially compatible with evolution, they often set the tone in which the evolutionary debates were handled. So, what does Lightman mean? This leads to the main strategy of interest for our purposes: rewriting scientific history. As we have seen briefly, the above strategies—except the explicit attack on theology—could have in theory been accepted without much if any protestation by most of the X-Club's theological friends. We have to constantly remind ourselves that scientific naturalism was a live option for so many because a particular school of theological reason had initially put it on the table for discussion as the most proper understanding of how nature reflected the glory of the Creator. As we also saw, this caused Huxley and company to gnash their teeth over what they perceived as a fragile armistice that would end only in the church once again poking its nose where it did not belong. As such, as Lightman has shown elsewhere, Huxley actually co-opted the language and literary strategies of the natural theologians to turn them to his own ends of a repurposed naturalism.[112] Huxley in essence took over the notion of providence, transferring it exclusively to nature, while Herbert Spencer took the logic of theodicy justifying the ways of God to men in the face of evil, and argued that this is the best of all possible worlds, but now understood in an evolutionary context.[113]

In turn, such rewriting and cooptation became one of the most widespread strategies of Huxley and the X-Club, who reframed the rise of science in history in purely anti-theistic and non-theistic terms through (quite ironically) borrowing from those very theologies. Huxley thereby erased the strongly pervasive theological past that had contributed to the formation of science, in particular the history of naturalism:

> The naturalist's strategy was to rewrite the history of their discipline to erase the long tradition of theistic science. Scientists frequently reimagine their past in order to support their vision for the future, and the wave of scientific

109. Stanley, *Huxley's Church*, 244.

110. Stanley, *Huxley's Church*, 245.

111. Quoted in Stanley, *Huxley's Church*, 246.

112. Lightman, *Victorian Popularizers*, 372–37.

113. Lightman, "The Theology of Victorian Scientific Naturalists."

naturalists at the end of the nineteenth century did so to establish a particular way of thinking about science and religion. That is, that science as an enterprise only made sense in an areligious context. Concepts like uniformity, which were both theistic and naturalistic in practice, became recast as *only* naturalistic.[114]

Thus, the notion of warfare combined with the strategy of reframing the sciences purely areligiously that we met in the last chapter. While Huxley and the members of the X-Club had positivist tendencies, the deletion of theology and the rewriting of history was certainly not the exclusive property of positivism. The X-Club nonetheless did certainly dovetail with the broader trends started with Comte, even if Huxley could sneer at Comte's overall project that it was merely "Catholicism minus Christianity." In 1866 at another BAAS meeting in Nottingham, the botanist and X-Club member Joseph Hooker used the evening in 1866 as an occasion to recount the hostile environment that evolutionists had to endure through great acts of bravery such as Huxley's against a powerful and heavy-handed theological establishment. Thus, if rewriting was not the exclusive property of positivists, neither was it exclusively a non-Christian tendency. Many Christians as frustrated with the heavy hand of the establishment as Huxley could use the warfare thesis to gain some breathing room apart from ecclesiastical oversight in the sciences.

Eight years later, in the summer of 1874 in Belfast, Ireland, again at a BAAS meeting, John Tyndall—another X-Clubber and close friend of Huxley, Hooker, Darwin, and Draper—crafted a now-familiar historical narrative to frame the triumph of scientific materialism spanning the ages: he opens with the Greek "atomists" as examples of early scientific naturalism, brave men who warred against the "mob of gods and demons" that held back discovery of true knowledge. This was lost in the Middle Ages, says Tyndall, which was a period ravaged by the Church into "scientific drought." Then came Copernicus, Giordano Bruno, and Galileo, likewise persecuted by the Church but trying to bravely endure their hardships in the name of truth. Tyndall ended by lauding the achievements of Charles Darwin, and infamously claimed that he and his fellow scientists "shall wrest from theology the entire domain of cosmological theory."[115] The scandal this caused, as one might imagine, was immediate. Perhaps even more catalytic than the memory of the Oxford debate, "no single incident in the [Victorian] conflict of religion and science raised so much furor."[116] As we are beginning to see, however, all of these events are quite interconnected. Yet, though Tyndall's address is often held up as one of the chief examples of scientific materialism in the nineteenth century, a deeper look by historians has revealed that this is not the case, and is another casualty of framing things in terms of "religion vs.

---

114. Stanley, *Huxley's Church*, 248.

115. This account is indebted to Lightman, "The Victorians."

116. F. Turner, *Contesting Cultural Authority*, 196.

science." Tyndall was a pantheist (meaning that in some sense God or the divine is identified with the cosmos itself) wanting to ground materialism in this wider spiritual context.[117] His materialism was not just associated with scientific method, but with wonder, sexual passion, religious awe, and artistic creativity. In what has been called "natural supernaturalism,"[118] Tyndall echoed the position of German Romantics who grounded their materialism not on supposed "brute facts" like so many among today's atheists, but rather saw in nature the prerogatives and attributes previously reserved for the Christian God.

As is often the case, so too here: the supposedly hard-nosed, no-nonsense empiricism of scientific materialism and naturalism are in fact haunted by the metaphysics that initially accompanied their births. Tyndall's materialism jockeying against orthodox Christianity is not as such science vs. religion, as it is the clash of two titanic metaphysical visions attempting to organize the world in different ways. This gives truth to the historian of science Georgio de Santillana's observation that "the real conflict over these recent years of ours is not between 'science and religion': it is between romantic naturalism and a philosophy of order and design."[119] While it is far too much to merely swap these conflicts, Santillana's ultimate point—that at the heart of what is presumed to be science and religion in conflict, a great deal of other factors are actually at play—must be kept to hand at all times. Tyndall's Protestant and Catholic opponents bear a large portion of the blame for the historiographical rewrite of Tyndall as proto-New-Atheist along the lines of Richard Dawkins. Incensed by his presentation, these Protestant and Catholic scouts had no problem reporting back to home base by writing Tyndall into their own narratives of modernist and materialist attacks upon the ecclesial body of Christ. This in spite of the fact that Tyndall insisted on the importance of Scripture and the power of the Holy Spirit.[120] This image caught on in the press and has been an image hard to expunge ever since.[121] "What could be *said* about evolution and what could be *heard* about it were shaped by the memory" of Tyndall's bombshell for well over a generation.[122] The legacy of "what could be heard and said" shaped into terms of science vs. religion, but this must be deconstructed. The point is not to baptize Tyndall as orthodox in his theology—he certainly was not. But it is to insist yet again that "religion vs. science" does not describe what he was on about, however fractious his language and disposition was to the churchmen of his day.

There is, moreover, a substantial irony in how the Belfast Address eventually occurred. Despite the fact of the X-Club's explicit goals, this did not mean that their

117. Barton, "John Tyndall."
118. Abrams, *Natural Supernaturalism.*
119. Quoted in Torrance, *Theological Science,* xvii.
120. Cantor, "John Tyndall's Religion: A Fragment," 421.
121. Lightman, "Scientists as Materialists in the Periodical Press."
122. Livingstone, *Dealing with Darwin,* 60.

strategies were always so straightforwardly antagonistic. Indeed, what they truly wanted was a bloodless coup, where the theists discovered the castle had already been taken by the time any klaxons were blaring. Tyndall, in his preparations for the Belfast Address, was indeed initially indecisive about his presentation. He did want to present a history of materialism but having just come home from a very successful American tour where even the clergy were incredibly kind and welcoming, Tyndall was of a softer mind. Receiving letters thanking him for his visit, and letting him know that they were continuing to pray for his health and safety, Tyndall wrote to his close friend Thomas Hirst: "Some of the letters [from the Americans are] so full of indescribable sweetness it is a pain to give such people pain by differing from them."[123] That they told Tyndall they were praying for him was particularly interesting given that Tyndall had garnered an infamous reputation for engaging in the so-called "Prayer Gauge Debate"[124] attempting to determine empirically whether or not petitionary prayer was actually efficacious. Ever since 1867, when Tyndall's mentor and friend the Christian scientist Michael Faraday had passed, Tyndall became increasingly less cautious about his religious radicalism. Faraday had been a keen adviser to the younger Tyndall, and that was gone now. Ironically, because of his new boldness that manifested in things like the debate over prayer, the mayor of Belfast publicly protested Tyndall's appointment as a speaker. Whatever inhibitions Tyndall might have been preserving for the sake of decorum were now forfeit. Tyndall wrote to Huxley, "I wish to Heaven you had not persuaded me to accept that Belfast duty. They do not want me. Well, in return I may be less tender in talking to them than I otherwise should have been. . . . So I suppose I am in for it—and so are you, you know, and Hooker too,"[125] as both men would also be in attendance. Such was the growth in Tyndall's disgruntlement that a few months before the talk Huxley—Huxley!—was growing worried Tyndall would not be able to suppress his hostility toward his critics. The rest was history, with Tyndall booming that the explicit goal of the scientist was to take all of the theologian's toys away.

If Faraday would have been stung by Tyndall's choice of style and tone at the Belfast presentation, Tyndall's similar method to honor Faraday through writing his posthumous biography would no doubt have perplexed the man deeply. Faraday, a member of the highly doctrinal Christian sect known as the Sandemanians, saw his religion as a favorable and extensive influence upon his science. In telling the story of Faraday's life as a scientist, however, Tyndall not only did not mention his membership in the Sandemanians, "he was very successful at embracing Faraday's contributions to science in a way that made the theistic aspects of his science seem only naturalistic."[126] Indeed, though theistic scientists contemporary to Tyndall, Huxley,

---

123. Quoted in Lightman, "The Victorians," 72.

124. Mullin, "Science, Miracles, and the Prayer-Gauge Debate."

125. Quoted in Lightman, "The Victorians," 74.

126. Stanley, *Huxley's Church*, 251.

and the X-Club like Faraday, James Clerk Maxwell, Richard Owen, and others shared nearly identical notions of scientific method because—and not in spite of—their theism, this was summarily expunged from the records. "Opportunities to recast theistic science as naturalistic often appeared in the form of memoirs and memorials, which Huxley and friends were happy to take."[127] Or, in the case of those like Richard Owen who were seen as opponents, their opposition via preference for differing scientific explanations was rewritten as *nothing but* a by-product of antiquated superstition that an amateur allowed to contaminate the scientific process.

And these strategies stuck. Tyndall's portrait of Faraday ignoring his religious commitments was repeated up through the twentieth century. While the general strategy most particularly embodied by Huxley of rewriting the past, and indeed "placing his arguments in the mouths of historical figures," such as "[René] Descartes and Joseph Priestly," created a compelling and widespread opinion of history as the continuous, linear expansion of naturalism as anti-theism, occasionally stopped by orthodoxy, but whose victory was assured by the inevitable march of history and reason. Huxley as such did not have to remove the number of theistic scientists from his story—figures like Newton, Descartes, Maxwell, and Priestly could simply have their historical images tweaked slightly to support Huxley's vision instead of their own. In turn, he could paint their theological beliefs—which in fact had been variously and intimately part of their work in the natural world—as metaphysical nonsense uprooted by the axe of science: "theism and atheism; the doctrine of the soul and its mortality or immortality—appear in the history of philosophy like the shades of Scandinavian heroes, eternally slaying one another and eternally coming to life again in a metaphysical 'Nifelheim.'" Huxley desired to give such perennial philosophies a true and final death:

> Extinguished theologians lie about the cradle of every science as the strangled snakes beside that of Hercules: and history records that whenever science and orthodoxy have been fairly opposed, the latter has been forced to retire from the lists, bleeding and crushed if not annihilated, scotched, if not slain.[128]

When the later Fundamentalist controversy arose in America, evolution as a scientific theory had been primed as a historical foe of God and traditional morality by being embedded in an intricate historical and metaphysical story not inherent to the theory as *scientific*. Making matters worse, the specific Victorian contexts for this embedding were often forgotten, and conflict between science and religion writ large was, despite the efforts of many, seen as merely the natural state of things. "In their bitter battle for scientific hegemony the Victorian scientific naturalists fought largely in vain. But in establishing their myth of an enduring battle between religion and science, they were

127. Stanley, *Huxley's Church*, 254–55.
128. Huxley, quoted in Stanley, *Huxley's Church*, 352.

successful beyond their wildest dreams."[129] While we have seen that even Tyndall and Huxley ultimately viewed their rhetoric as for "true religion," and merely sloughing off the encrustations of dogmatic theology and ecclesiastical control, much as in the tale of the Sorcerer's Apprentice, so too in real life our creations often escape the intent originally envisioned for them. And so, the drums of war were now beating, and in the distance more drums were beginning to sound their responses. They only grew louder as Huxley, Tyndall, and the X-Club found like-minded allies in their American counterparts. To them we now turn.

129. Russell, "The Conflict Metaphor and its Social Origins," 26.

—Chapter Five—

The Armchair at the Center of the World

Andrew Dickson White, Religion, and Warfare in the American University

One who actually follows historians' footnotes back to their sources, accordingly, taking the time to trace the deep, twisted roots of the blasted tree of scholarly polemic, may well discover much more of human interest than one would expect in the acid subsoil.

—Anthony Grafton, *The Footnote: A Curious History*

Now mythless man stands there, surrounded by every past there has ever been, eternally hungry, scraping and digging in search for roots, even if he has to dig for them in the most distant antiquities. The enormous need of dissatisfied modern culture, the accumulation of countless other cultures, the consuming desire for knowledge, what does all of this point to if not the loss of myth?

—Friedrich Nietzsche, *The Birth of Tragedy*

AS THE WARFARE THESIS spread through the work of Huxley and the X-Club, the wheels of its travels were greased by an entire era that echoed in gunfire. Ideas of conflict, strife, and opposition were ready to hand, and warlike metaphors not only abounded on the page, but threatened daily to spring to life and spill quite literal blood through American houses, towns, and cities.[1] Anxious tendrils webbed indiscriminately between topics, placing all aspects of life in the bloom of an electric thrill akin to what one gets standing on the edge of a cliff, peering downward, fighting the pull of the drop. Memories of the French and American Revolutions burned in the back of minds everywhere. Even closer to home, the American Civil War was turning

1. Moore, *Post-Darwinian Controversies*, 50–100.

brothers and sisters against each other. The first shots would begin ringing in earnest the same year as the Oxford debate between Wilberforce and Huxley. Slavery and its horrors gnashed in the souls of southerners and northerners alike, affecting everyone, including historians. John William Draper—whom we met briefly in the last chapter and will be the main topic of the next—was so affected that his three-volume work on the history of the Civil War infamously represented it as two idealized, warring forces. This easy division of black and white, good and evil, primed his use of a similar abstraction when he turned his pen to writing his *History of the Conflict Between Religion and Science*. The history of religion and science was, wrote Draper recalling the exact same phrasing and imagery previously describing the Civil War, an "irrepressible conflict between [these perennial] opposing and enduring forces."[2]

At this stage of unrest America became a garden growing every type of new and exotic belief: Mormonism was flourishing, Mary Baker Eddy's Christian Science gained a foothold, the Jehovah's Witnesses were beginning to knock on doors, American transcendentalists, Blavatskian theosophists, and Freemasonry were suddenly to be found everywhere. And they were growing bolder by the moment. To make matters worse, while many abolitionist movements were being spearheaded by theologians, southerners and northerners alike had crippled the public respectability of scriptural reasoning by using it to promote and justify slavery, while others equally well used it to justify full-scale abolition.[3] A confused public began ever so quietly to put aside the authority of Scripture for public debates, precisely because as a matter of practice its most learned public exponents had apparently come to quite opposite conclusions using the same text. Above all, rumors of Darwinism had begun by the 1870s to take university campuses by storm. Christians went on high alert, and as the Church Father Gregory of Nyssa described the debates over the Trinity long ago, it was as a "battle at night," where the parties involved traded blows without quite knowing with whom they were feuding.[4] In part this was because party lines we have been describing were not always as clear as the participants themselves thought they were.

It was in this environment that lent its warlike energies to everything that the conflict narrative in the form we know it today first emerged. Amidst this tumult, war was quick to find itself upon the tongues of men. "Quite understandably," writes the historian James Moore, "the effect of [all of] these developments on the Christian public was to produce widespread confusion and strife. Condemnations tended to be categorical, novelty was frequently mistaken for infidelity, and opinion turned to prejudice on every hand."[5] These high-stake contexts gave Andrew Dickson White's "characterization of the relationship between science and religion as an epic and immortal battle"—and White's explanation of how he came to this opinion in the

2. Cantor, "What Shall We Do With The 'Conflict Thesis'?," 292–93.

3. Noll, *Civil War*, 157–62.

4. Quote in Robertson, *Christ as Mediator*, 1.

5. Moore, *The Post-Darwinian Controversies*, 30.

struggle with Protestant control of education—more plausibility than it ever should have had.[6] Like Huxley, both White and Draper were in the business of using history as one of their primary weapons. It is not too much to claim that White and Draper do not so much critique the history of the conflict of science and Christianity, as invent it. Both of their works are so full of sensation and outright error—both in method and fact—that no historian uses them any longer as serious sources. John Dillenberger's terse judgment reflects the common consensus of historians today: "there is no reason to regard [White's work] as a scholarly book today, much less as an adequate interpretation [of the history of science and religion]."[7] More damagingly than these errors, however, the metaphor of warfare itself pressed hard upon history, reducing a series of complex events into clear-cut silhouettes of their reality. The major point driving such criticism is the increasing realization amongst historians how much the elements of war "are more realistically to be located within [White and Draper's] own circumstance than within the histor[ies] they sought to record."[8] Conflicts specific not only to their own day, but very often to the local, personal, societal, and institutional contexts in which White and Draper moved expanded outward, and were projected backward, to remake history in their own images of struggle. This has been missed far too often, whether by historians operating within the terms Draper and White set, or Christian apologists who seek to counter their claims without questioning the nature and layout of the playing field itself. As Draper's biographer astutely observes, "[Draper's opponents] were beating him over the head with his own problem. They allowed him to set the terms of the debate."[9]

The same could be said for White. "Religion," and "theology" were often not just critiqued in his warfare narrative, but in many ways their essences and their histories were reconstructed to produce ideal opponents. The story of White nearly single-handedly creating the warfare narrative could easily be told from the perspective of the continued professionalization of "science" that we saw in the last chapter. For the sake of emphasizing how "religion" was similarly shaped at this time, it helps now to view White's historical crafting from that half of the "science vs. religion" binary. Yet, even from this perspective, professionalization is ever present not just in analogy but often still in direct association (if not membership) with the X-Club. Those like James Frazer, whom we shall meet later in this chapter, deeply contributed to the "religion" side of the shaping of the warfare narrative that White was crafting, and thought it was only anthropologists like himself—"strategically located at or near the metropole where the full range of comparative [religious] materials gravitated from every corner of the empire"—who were the ones allowed to formulate and control theories about religion. As historian Timothy Larsen summarizes the matter, Frazer's was "the

6. Principe, "The Warfare Thesis," 10.

7. Dillenberger, *Protestant Thought*, 14.

8. Russell, "The Conflict Metaphor and its Social Origins," 7.

9. Fleming, *John William Draper*, 131.

armchair at the center of the world."[10] The father of anthropology Edward Burnett Tylor was similarly "an armchair anthropologist,"[11] dealing with texts he received—mostly from missionaries!—as he created his theories of religion and secularization. One pictures them both sitting imperious and in the visage of the many gods and spirits they spoke of in their works, deciding the fate of information presented before them. Secularization theories describing the inevitable passing of religion that drove and were in turn driven by the warfare narrative were never about "*value-free armchair scholars*" impassively observing the world to describe it, but were active theoretical mechanisms "doing all in their power to guarantee [secularism] and make it a reality."[12]

Just as Huxley and the X-Club were rewriting scientific history often by relying in part on resources from the theologies they were excluding, so too Frazer, E. B. Tylor (who while not a member, was good friends with many in the X-Club), and others were providing resources for Andrew Dickson White's warfare by drawing from deep theological currents that helped them theorize upon secularization, religion, superstition, and the death of God. White—who considered himself a pious Christian—attempted to use these tools to purify the Christian religion, not destroy it. Like the sorcerer's apprentice, though, White quickly lost control of what he had summoned. While he did himself no favors in how he presented the warfare narrative, which left it ripe for such misinterpretation, and despite his own ultimate intentions for it, "White's narrative became . . . one of the most effective and influential weapons of unbelief."[13] Almost immediately picked up by atheists, freethinkers, and even more radical positivists, White's narrative (whatever its own defects, errors, and reductions) was reduced again into a tale of the absurdity of religion full stop, its war with science through time, and its eventual, inevitable defeat.

The Board Is Set, The Pieces Are Moving

Despite the emerging spread of methods and assumptions driving the warfare thesis, the X-Club and their associates were fighting the inertia of a deeply established Christian system. But what if one could start fresh? In America, to be sure, Christianity had in some ways an even stronger hold than its somewhat dilapidated form in Europe. And yet, just as much, America still represented something of a blank map upon which to draw new territory.[14] Though many during the time when Cornell University was established wanted to paint the old regime as a wilderness littered with denominationally driven, fortress-like Christian universities, this was hardly the

10. Larsen, *The Slain God*, 68.

11. Logan, *Victorian Fetishism*, 98.

12. Hadden, "Toward Desacralizing Secularization Theory," 590.

13. Ungureanu, *Science, Religion, and the Protestant Tradition*, 102.

14. Howard, *God and the Atlantic*.

case. Pre-Civil War education has been revealed by recent historians to have been a much more flexible, open network.[15] Ironically, the representation of pre-Civil War universities as a denominational wasteland was itself more often than not propaganda of the new post-Civil War regimes in one of its many moments of self-justification. Regardless, the Protestant establishment had began to fragment along the cracks of all the issues that would eventually culminate in the Civil War—regional, racial, ethnic, methodological, social class, and on it went. When the skeptical and positivist gales came, the ships of the prior universities were very often mid-scrap as parts were being refurnished for new vessels and opportunities, and many stood little chance of weathering the storm. Precisely for these reasons for Andrew Dickson White, Ezra Cornell, and their American confreres, "influence[d] by the English debates [over the nature of professionalization and the history of science and Christianity] that had surrounded the advent of Darwinism,"[16] the uproar in the wake of the Civil War meant the time was ripe for a new approach separate from denominationally driven Protestantism and its reigning methods of Baconian philosophy. The warfare thesis was the rhetorical wedge for such a campaign, and the great epic of secularization was the driving hammer. The antique impulses of humanity now organized together and called religion, it was said, were slowly, surely sinking, perhaps to disappear forever. It was time to get on the right side of history.

The major catalyst for White's war was set off in 1862 when the Morrill Land-Grant Act went into effect. The act granted vast acreages of land to each state for resale. The money acquired from these sales was envisioned to go in turn to providing funding for colleges and new educational programs. The entire fund had been placed in the hands of a single, second-rate institution that couldn't manage it, and so the suggestion arose that it be divided amongst several smaller institutions so as to maximize the potential use of funds. White, however, grew ill at this suggestion, and was hellbent on keeping the lump sum together as a single amount. To this end he approached the wealthy Ezra Cornell to persuade him in this endeavor and create a single elite institution modelling itself on the German University system. Free from the partisan interests of Protestant controlled higher education, the soon-to-be created Cornell University embodied White's utopian dream for education.

A man of many talents, White was serving as the senator of New York, and would later be appointed as a diplomat to Russia and to Germany. Cornell, in the meantime, had earned himself a fortune by entering into the telegraph business in partnership with Samuel Morse, having innovated the idea of using glass insulators at the junction of telegraph wires and the support poles to help avoid them shorting out. Finding in each other kindred spirits promoting what they determined to be freedom of inquiry from sectarian sensibilities, both White and Cornell saw an opportunity in the tumultuous atmosphere of the time to make a real impact upon American higher education.

15. Roberts and Turner, *The Sacred and Secular University*.
16. Marsden, *The Soul of the American University*, 117.

As we will see, this reform was not meant to destroy religion (or Christianity), but save the true nature of religion from theology and dogma. During his theological training White became particularly disgusted by the immorality of many of his zealously doctrinaire classmates, for example. Dogmatism, thought White, did not produce morality or true religion—which he saw as the inner moral impulses of the human heart—quite the opposite.

While the founding of Cornell University allowed White to enact some of his ideals, the notion of anti-sectarianism and anti-dogmatism had been something nurtured in White from his youth. This ethos was carried with him as White, along with nearly ten thousand other American students over the course of the nineteenth century, found himself relocating his education to Germany, in particular to the University of Berlin. Suitably enough for White, Berlin was the first German University to sever the centuries-old tie between confessional Christianity and its own education system. Paradoxically, however, because of the concerted efforts of Friedrich Schleiermacher, theology retained pride of place (albeit a precarious one) amongst the faculty there despite theology's fortunes drastically falling all over Europe.[17] With what has been called the "Berlin Paradigm" of theological education, the new standard of theology as *wissenschaft* (translated usually as "science," but having a considerably broader meaning than the English term) meant that "theology could only be a university subject by being transformed into a [purely] historical discipline."[18] Despite many seeing this as theology's death-knell, it proved not only resilient in these circumstances, but even thrived. For White, the stress upon theology as a historical discipline eventually allowed him to separate theology and dogma from religious morality, dispensing with the former in the name of purifying the latter. This purification would occur in the guise of a historical narrative of warfare, and the defeat of the absurdities of theology in the name of true religion and true science. The new frontiers of knowledge that science was opening were thus the perfect opportunity to cleanse what he saw as an orgy of traditional stubbornness and hypocrisy back in the states.

White and Cornell through much struggle to gain the land grant, pulled something of a coup. When the bill passed and the money from the grant allocated the money for Cornell University in 1868, a scandal-hungry press descended like vultures to accentuate the rumors of collegiate conflict in any way they could. The rigor and high standards of the German system could not be denied, but this enhanced rather than diminished German theology and educational theory as a threat in many American eyes. The Protestant establishment felt insulted by White and Cornell twice over. White, though hardly proposing an anti-religious or even areligious education model, dared to propose a German system whose primary goal was other than educating seminarians for the pastorate. Chief among the accusations was that Cornell sought to become a "godless institution," and "a garden for growing atheists." That sting was

17. Howard, *Protestant Theology*, 130–211; Zachhuber, *Theology as Science*, 12–24.

18. Hauerwas, *The State of the University*, 4; Frei, "Appendix A."

particularly sharp for White, who considered himself a devout Christian. In fact, he had made it his business to ensure Cornell had a strong interdenominational chapel, and encouraged the presence of a very active Young Men's Christian Association on campus. Moreover, beyond the fact that sour feelings are inevitable when large sums of money are involved, because the German system emphasized the lecture format and so enhanced the prestige of the innovative professor as lecturer, many felt even more upset at what they perceived as a cynical cash-grab going to fund a campaign of self-aggrandizement and narcissism on White's part.

White grew tired of these criticisms very early, as one might imagine. He found them cruel and thoroughly uninformed. Ironically, in the act of attempting to defend the old denominationally driven university system by critiquing White, in some sense what was achieved was the final seal of war upon White's images of history. His description of the struggle over the university system foreshadowed the nearly identical language he would eventually use for the relation of science and Christianity: "the introduction of this new bill into the legislature was a signal for a new war,"[19] he reflected at one point. Relatively soon after the founding of Cornell, White lost the stomach for coping with the onslaught, and decided to deal with his feelings on the matter by smashing his critics against the tides of history. This was particularly ironic as such historical deconstruction was precisely the calling card of the German system that landed White in hot water in the first place.[20] What smarted most for his targets was hard to tell: was it the laundry list of alleged historical sins against science and reason, or the fact that the whole thing was done, so to speak, in a German accent? As White wrote to Cornell, he was going to teach his religious opponents "a lesson which they will remember"[21] and he certainly lived up to that promise. Thus on a December evening in 1869—just nine years after the Oxford debate between Wilberforce and Huxley—when White ascended the hall at Cooper Union in New York City to a packed audience to speak on "The Battlefields of Science," this was not just ideological saber-rattling, but as with the mythologized Huxley-Wilberforce debate that White mimicked (or perhaps just as analogous, Tyndall's Belfast address), it was a manifesto for White's vision of science, reason, and education that was already underway.

His speech—much again like Tyndall's—caused nothing less than a sensation, and to his delight was published the next day in *The New York Daily Tribune*. There, readers of the paper were treated to the grandiose and nigh-eternal terms to which the warfare metaphor had now been set: a conflict "with battles fiercer, with sieges more persistent, with strategy more vigorous than in any of the comparatively petty warfares of Alexander, or Caesar, or Napoleon."[22] His opponents possessed the same nar-

19. White, *Autobiography of Andrew Dickson White*, 1:300.

20. Marsden, *The Soul of the American University*, 104.

21. Andrew Dickson White to Ezra Cornell, 3 August 1869, quoted in Moore, *Post-Darwinian Controversies*, 35.

22. White, "Battlefields of Science," 4.

row-minded dogmatism that caused the persecution of Galileo (this no doubt stung extra as it was accusing Protestants of acting like Catholics, another time-honored Protestant strategy of chastisement). Ultimately, the parade of examples was meant to establish without doubt that all "interference with Science in the supposed interest of Religion—no matter how conscientious such interference may have been—has resulted in the direst evils both to Religion and Science, and *invariably*."[23]

American Bacon

Penning his major work on the history of warfare, amidst many curiosities White had his thoughts turn to the topics of cheese and peacocks.[24] An unusual pairing, but both birds and curds were on White's mind because White believed he found in this bizarre juxtaposition of topics in the writings of St. Augustine (354–430) yet another delicious absurdity to ornament his historical narrative detailing the church's long list of scientific buffoonery.[25] In so doing, he was also directly damning the Baconianism of his elite Protestant opponents as the latest variety of such a superstitious conflation of science and theology. More in a moment. Though named after the "Father of Modern Science," Francis Bacon, Baconianism did not really reflect Bacon's own thought so much as an image of him produced by a Scottish Enlightenment school of thought termed Common-Sense Realism, which in turn became one of the most influential forces on Protestantism leading up to the debates over Darwin.[26] In 1823 for example, the editor of the widely read *North American Review* went so far as to proclaim "at the present day, as is well known . . . the Baconian philosophy has become synonymous with the true philosophy."[27] To an important extent this Christian fascination with Baconian method "may be understood as a counterthrust against the widespread effort in the eighteenth century [by the *philosophes* and others] to portray the scientific movement as innately hostile to traditional Christianity"[28] as well as an alternative to the "airy metaphysics" of the scholastics. Through Thomas Reid (1710–1796) and Dugald Stewart (1753–1828) this "Baconianism's" primary goal was to accumulate facts through a refined observation—and these facts equally included those culled from Scripture as well as nature. "[Reid and Stewart] argued straightforwardly that nature constituted one set of facts and the biblical scriptures constituted another, and that scientists and theologians could apply the very same

23. White, "Battlefields of Science," 4.

24. This anecdote from White has also been used to great effect by Hutchings, "Demonic Cheese–Donkeys and Immortal Peacocks."

25. White, *A History of the Warfare of Science*, 231.

26. Bozeman, *Protestants in an Age of Science*; Noll, *America's God*, 227–52.

27. Garroutte, *Language and Cultural Authority*, 25.

28. Bozeman, *Protestants in an Age of Science*, 44.

scientific method to the study of both."[29] But "facts," as with any theory, were defined in their own way. In the version popular at Yale, for example, empirical facts each had an idealistic correlate—so that "the mind acted creatively to make theistic sense out of the world that was the visible part of a divine universe."[30] While understanding God from the world was still an induction made by accumulating facts, the facts that these Baconians were dealing with were understood as intrinsically God-producing. This is because the mind's way of organizing and understanding those facts was also intrinsically God-revealing. While they can obviously be accused of some circularity here, it need not be a vicious one. For what was occurring was not a straightforwardly foundationalist argument that led one from the facts to God—but a sort of cumulative case; the discovered facts were part of the proof and demonstration, in other words, but so was the act and ability of humanity to make such discovery and organization. Our ability to recognize and understand was itself part of the argumentative case. As an epistemology, in contrast to the emerging elitism of professionalization that we saw in the last chapter, as well as the similar currents in positivism, Baconianism had the advantage in the States in that it reflected the deeply democratic ethos of nineteenth century American culture, since facts spoke to every human of God, and every human had the capacity to read the language of God.[31] Science was, within these stipulations, merely common sense rigorously applied. As one Baconian wrote in *Popular Science Monthly*, "the fundamental truths of morality and religion are [as self-evident] . . . as those of geometry."[32] Baconianism not only lionized the common man, but in effect placed science and theology onto the same playing field and under the umbrella of the same methodology and criteria, accessible to all not just in principle but in practice.[33]

The examples of cheese and peacocks gave White yet another example of these supposed populist "facts" with which to harangue his hapless foes. As legends contained in allegorical manuals on animals called Bestiaries went, there were stories that told of the divine favor that peacocks enjoyed, such that God preserved their flesh even after death and even though their beautiful feathers decayed. This was to serve, so said the Bestiaries, as a reminder of the enduring wisdom of God over against the vanity of lust. Even more outlandish, tales had been circulating that local innkeepers would mischievously put a mysterious alchemical tincture into their cheeses, transforming hapless travelers into donkeys after a long night of revelry, to be conscripted for use in free manual labor. The joke, of course, is that many innkeepers no doubt thought their patrons were already jackasses. But as strange as these tales were—they would

29. Garroutte, "Positivist Attack," 199.

30. Stevenson, *Scholarly Means*, 69.

31. Noll, *America's God*, 233.

32. Quoted in Garroutte, "The Positivist Attack," 198–99.

33. Noll, *America's God*, 227–52. As Noll remarks, this began to change mid-century as Charles Finney and other revivalists abandoned Baconianism for more "self-consciously romantic accounts of Christian belief" (263).

be laughed at "even by a schoolboy" of his day and age, says White—even stranger is White's selective reading of Augustine. Had his eyes strayed but a few sentences further in the passages of *The City of God*[34] that he himself cites with much amusement, he would have discovered that Augustine speaks about these tales to mention that he finds them ridiculous, not that he believes them. Regarding the peacock story, Augustine actually engaged in an experiment to put the matter to rest. "This [antiseptic] property [of the peacock's flesh] seemed to me incredible," he says about his first encounter with the legend. To investigate he ordered a slice of its meat be taken and kept for observation. Oddly enough, even after a year Augustine records that aside from shriveling up, the peacock meat actually did not decay, and emitted no rotten odors. Regardless of whether there were other factors at play, Augustine can hardly be accused of blind naivety on this point.

But, of course, in some sense this hardly matters. Many of White's positivist sensibilities caused him to expect such reasoning, and when he thought he found it, it hardly needed to be argued against so much as held up as the nonsense positivism already knew it was by definition. One can find any number of beliefs from the merely strange to the borderline insane, and then use those to represent the essence of what your opponents are doing. But what justification do we have for clumping such a misfit collection of discarded beliefs together under the heading "religion"? White made ready use of the new disciplines of positivist science to rewrite the reigning philosophy of Baconianism utilized by many in the Protestant academic establishment entrenched against both he and Cornell. Within the oculus of positivism any apparently supernatural fact in the world could be slotted together, from the benign to hysterical. Given the strong reactions Baconianism evoked, it is important to note if only in passing how much Baconianism was its own idiosyncratic variation within the broader Christian tradition. However influential at the time, it was atypical even if we isolate a relatively short span of time in America itself. "A theological Rip Van Winkle falling asleep in the early 1740's" writes historian Mark Noll, "waking up a century later would have found Americans speaking his language with such a decidedly strange inflection as to constitute a new dialect. . . . In turn, theistic common sense would exert a tremendous influence on theology in the nineteenth century." Baconianism was such a momentous change that it transformed "how Americans thought about human character, the nature of salvation, and the relationship of God to the world," to the extent many forgot that things had ever been done any other way.[35] At this point we might recall what was mentioned in passing in chapter 2: "How ironic it is to read in popular histories of the 'antagonisms of religion and the rising science.' That was precisely what the problem was not!" writes historian Michael Buckley. "These sciences did not oppose religious convictions, they supported them."[36] But this fusion be-

34. St. Augustine, *City of God*, XXI, 4.

35. Noll, *America's God*, 93.

36. Buckley, *At The Origins of Modern Atheism*, 347.

tween science and theology was unstable to the extent that when the winds of science changed, as they do, the religion wedded to it appears to become obsolete as well. As one scholar has put the problem: with theology and science newly flattened to occur on the same plane of meaning, God's newfound explanatory "transparency [meant] he was all the easier to identify and so kill."[37] Even before White's time in the early decades of the nineteenth century there were "vague murmurings of discontent" that began to be heard even from Christian "scientific men who were colliding against the hard walls" of Baconian methodology.[38] Yet the implications for Baconian Christianity in these murmurings were not yet grasped.

The many movements of positivism (some related to Comte, others not) that White was drawing together in his *magnum opus* were thus not alone in their growing outrage against the Baconian establishment. But, unlike other arguments against Baconianism, "positivism defeated Baconians primarily by a process of discursive colonization—that is, a process by which one discourse [positivism] compromises the symbols and assumptions of another discourse [by] explaining them in foreign terms."[39] Equating Baconianism with the earlier thought that was suspicious of the donkey-making properties of cheese, or imperishable peacock flesh, was but one tiny example of this broader strategy. For example, the use of analogical reason was defined into irrelevance, so that systems which appeared as analogies to a machine or a well-crafted piece of art were deemed flights of fancy. The internal relationship of parts (say of an organism) were atomized and treated as "parts" only because of their physical proximity to one another or relations of efficient cause and effect, not because they represented an essence or a nature, or in any way that the whole organism had any "top-down" type of feedback regulating the parts themselves via teleology. Nature herself became a completely depersonalized and non-metaphysical bundle of forces. Likewise, any reason involving teleology or tendency-like behavior was summarily dismissed as, at best, a useful fiction. Intuitions, which played a large part in Baconianism, were also deemed purely subjective and found themselves a particularly beleaguered casualty of the new psychological disciplines. The detailed mental sciences of Scottish Enlightenment figures like Stewart were precisely the targets many new psychologists aimed to replace by assuming much of their terminology and concepts but discarding their metaphysical arguments. Just so transposing and reducing Baconian vocabularies into purely materialistic and behaviorist sensibilities where they either no longer made sense in terms of Baconian argument, or appeared now to bolster the materialist framework of the behaviorists.[40] In this manner the entire storehouse of Baconian facts was picked through at once, and mutated. Each fact as it was filtered into the grammar of its new positivist masters was dissolved

37. Funkenstein, *Theology and the Scientific Imagination*, 116.

38. Daniels, *American Science*, 119.

39. Garroutte, "The Positivist Attack," 202; Garroutte, *Language and Cultural Authority*.

40. Dixon, *From Passions to Emotions*, 135–79.

into a myriad of contradictions, impossibilities, and subjective whimsies.[41] Of course, straightforward arguments were also used. But, in tune with the broader arguments of this book, the rewriting of history and the re-encoding of the entire Baconian lexicon so to speak stands out as of a piece with many of the broader strategies we have been seeing at large.

From this perspective Baconian design arguments easily became seen as of a piece with magical thinking like immortal peacock flesh precisely because positivism handily categorized all such arguments under the master category of "superstition." This labeling both allowed the attendant contexts of any given example to be ignored for such lumping, while also indirectly reinforcing the apparent unity of positivism as a systematic response in the face of such oddities. The positivist "sciences gain part of their respective notion of coherence in contradistinction to religion as an irrational belief system, while religion became a 'kind of negative image of science, and this contrast has become important for the integrity of the boundaries of science'" as a newly forming disciplinary division policing its territory.[42] Indeed, from the viewpoint of positivism *any* use of Christianity as part of a broader worldview beyond one's heart was a gigantic category mistake, thus allowing White to categorize any number of bizarre superstitions together and be treated not just as an eclectic and unrelated set of oddities but as single tableau of examples exemplifying the same superstitious theological mentality plaguing Christianity through the ages. Baconianism, to be sure, had many flaws. Yet, it was not so much that God was killed and Baconianism vanquished; it was rather that a particular theology of God's relation to the world—one that, it must be added, was already buckling under some of its own decadence—was simply defined into irrelevance for science alongside the Baconian method that had lionized it. And, to be sure, there were also arguments leveled against Baconianism. But, at large, "anti-religious scientists [and historians] slowly created a new vocabulary for discussing natural events and objects [that] challenged implicit (yet vital) religion-friendly assumptions [and interpretations] about the relationship [between science and religion] that could occur in the natural world."[43] On the one side, those like Huxley were rewriting the history of science to erase its theistic past. On the religious side, similar positivist sensibilities rewrote the history of religion to pigeonhole it as exemplary of the absurd and irrational aspects that our species was finally escaping. Thus, precisely because Baconianism "set up an extremely permeable boundary between [religious

---

41. See: Daston, "The Naturalistic Fallacy is Modern," 579–87. She makes an excellent case through some historical detective work that the so-called "naturalistic fallacy" (namely, that statements of nature or natural fact cannot be translated into statements of value) is deeply contingent and historical rather than a matter of fact itself. The separation of facts from values nonetheless played a large role in the positivist colonization of Baconian discourse, which as we have shown relied on a deeper interconnection of facts and values revealed through the common person's intuition

42. Josephson-Storm, *The Myth of Disenchantment*, 14.

43. Garroutte, "The Positivist Attack," 202.

and scientific] discourses"[44] it presented an especially vulnerable target to positivists exactly because it could be rewritten from the sides of the new categories of "science" *and* "religion" simultaneously. Overnight, design arguments became irrational, mere dreams of God lost among a forest of disenchanted fact; religious hypotheses were just a faint wind, by which no bare rock of science would be turned.

When he finally could not take the sniping of the Protestant university establishment any longer, White already had his strategies at the ready. By hijacking the terms and concepts of Baconian science and religion, White's "[positivism] shifted the grounds of argumentation, by attacking the deep structures of discourse." Positivism filled Baconian concepts with new meaning and so did not so much win by arguing against Baconianism, as much as by rendering its discourse inherently empty and meaningless. Indeed, within the new air of positivism "traditional [Baconian] arguments may not even be sensibly stated." This furthered many of his own liberal Christian sensibilities. By dissolving Baconian arguments, positivism allowed White to dispense with the "superstitious dogma" that undergirded Baconianism and, once dissolved, White could show that the true Christianity hidden beneath the Baconian veneer was a moral disposition of the heart toward God. Science was the vehicle whereby this dissolution and purification occurred, and so ultimately both science and faith were in complete harmony with one another. History was a convenient meeting place for White to make his case and display it in both breadth and depth.[45]

Ultimately White's strategy of attacking the Baconianism of his Protestant university opponents by rewriting them and, really, the entire history of Christianity for his war was his curious and roundabout way to enact a new Protestant vision for a synthesis of science and religion against the old Protestant syntheses. His goal, as he himself openly stated, was "to aid in letting the light of historical truth into that decaying mass of outworn thought which attaches the modern world to medieval conceptions of Christianity, and which still lingers among us—a most serious barrier to [true] religion and morals, and a menace to the whole normal evolution of a society."[46] The accusation no doubt stung even more because he had just claimed his opponents were acting like the medieval Catholics they so abhorred. One of the key factors for the amnesia regarding the Protestant—or broadly Christian—elements of the warfare thesis was ironically the very positivism White wielded, which at large was defining religious more broadly as an irrational object ready made for ridicule. Much as Huxley defined theology into irrelevance by rewriting scientific history, religious history was being rewritten from the other side to be little more than a menagerie of failed explanations of the world, taboos, superstitions, misdescribed physiological responses, all of which were irrational. As we have seen, White was on board with this to a great extent. In his quest to purify Christianity, though, White was using an extremely

44. Garroutte, "The Positivist Attack," 200; Bozeman, *Protestants in an Age of Science*, 166.

45. For the quotations in this paragraph, see Garroutte, "The Positivist Attack," 200.

46. White, *Warfare* v.1, v.

double-edged sword, a weapon that would end up deeply cutting both him and his allies as well. Ultimately the pressures of the broader (often X-Club) related movements would undo his synthesis that looked so much like their own anti-Christian arguments, and White was swept along with the flood.

The New Inventions of Religion

"I believe, because it is absurd"—the most famous statement that the great theologian Tertullian of Carthage, never said.[47] Modified by seventeenth-century Englishman Thomas Browne, and then transmuted again into its familiar form by the polemicist Voltaire, the *credo* is an invention of the Enlightenment.[48] Sadly, it also embodies what is today a fairly standard opinion regarding the history of faith and science—namely that Christianity as a faith, or religion, is characterized by the embrace of the ridiculous precisely because it has no evidence. It is hard to overstate how much of this is a product of the rewriting strategies we have been investigating. Since its inception, of course, Christianity has always had its cultured detractors who find in it not the faintest whiff of rationality or good taste. Celsus, a conservative intellectual in Rome in the second century, for example, wrote a withering tract against what he saw as Christianity's irrationalism and disruption of conservative Roman politics and religion. "Some [Christians]," he wrote, "do not even want to give or to receive a reason for what they believe, and use such expressions as 'Do not ask questions; just believe' and 'Your faith will save you.'"[49] While Celsus was conjuring his criticism because of "some Christians," his alarm is nonetheless palpable that such people existed at all. Criticisms like this have therefore always been around, and with varying degrees of truth to them. The Tertullian quote was unique, however, insofar as the context of its birth signals a historical transformation where the *essential* unreasonableness of Christianity as a religion was named as the chief shadow darkening the corridors of history. As the historian of science and religion Peter Harrison writes, "these modifications [to Tertullian's saying] were not the result of careless mistranslations, but signal a new way of understanding religious faith and the beginnings of its characterization as an epistemic vice."[50]

One must not overplay the causal role of the quote from Tertullian specifically. Nonetheless, even as a passing anecdote its use and subtle influence in a huge variety of works helps sharpen our attention upon sweeping shifts in the concept of religion that have occurred over the past few centuries. "It has become clear," writes Brent Nongbri, "that the isolation of something called 'religion' as a sphere of life ideally

47. Osborn, *Tertullian,* 28.

48. The hilarious and fascinating tale of this apothegm is traced by Harrison, "'I Believe Because It Is Absurd.'"

49. Quoted in Wilken, *Christians as the Romans Saw Them,* 97.

50. Harrison, "I Believe," 340.

separated from politics, economics, and science [e.g.] is not a universal feature of human history. . . .[T]he act of distinguishing between 'religious' and 'secular' [for example] is a recent development. Ancient people simply did not carve up the world that way."[51] Just so, once thought to encompass and affect the whole of human life as it oriented everything to worship of God, when it came to be identified in the newly limiting category of religion, Christianity shrunk into a sphere specified not just by what its newly inscribed circle contained. Just as importantly, it was deeply defined by its new contrasts to things once included and even created within it, but now removed to a different sphere named the secular (politics, economics, science, and so on). In this sense the content of Christianity under the transformed category of "religion" began slowly to be evacuated of rationality by definition, which was now seen as solely the province of secularity.

In turn, the more people paid attention to Christianity as a "religion," the more it now appeared to be a phenomenon in the process of shrinking from the world, perhaps even to vanish. This was of course because the new definitions of religion were quite literally shrinking it. As one scholar puts it—herself no friend to Christianity: "When religion came to be identified as such—that is, in more or less in the same sense that we think of it today—it came to be recognized above all as something that, in the opinions of many self-consciously modern Europeans, was in the process of disappearing from their midst, or if not altogether disappearing, becoming circumscribed in such a way that it was finally discernible as a distinct, limited phenomenon."[52] In other words, the apparent disappearance of religion was in many venues an optical illusion of the new use of the category itself, and its connotations. It was, regardless, a useful illusion that made it that much easier to tell stories of the inevitable passing away of religion in the face of science. The same scholar continues, "the modern discourse on religion and religions was from the very beginning—that is to say, inherently, if also ironically—a discourse of secularization."[53]

The illusion was reinforced by how many newly emerged disciplines internalized the new use of religion and also reinforced it by their own categories. Around the time of the Oxford debate between Wilberforce and Huxley, for example, psychology arose as a discipline and began imagining a portion of human activity under the newly minted category of "the emotions." Psychology also arose because it went through a "positivist apotheosis,"[54] which was directly responsible for differentiating it as a "scientific" category from the broad streams of discussion that led up to it. What was previously spoken of in terms of intuitions, or passions, or affections, or wills—all of which had various Christian pedigrees—this category switch to the use of "emotions"

---

51. Nongbri, *Before Religion*, 1–2.

52. Masuzawa, *The Invention of World Religions*, 18.

53. Masuzawa, *The Invention of World Religions*, 19; W. C. Smith, *The Meaning and End of Religion*, 19, 124.

54. Reed, *From Soul to Mind*, 144–67.

was meant to separate the discipline of psychology from the depths of a Christian past that had in fact deeply nurtured it. Baconian facts, as we saw above, often crossed the divide of the subjective and objective in such a way that because of the strict subjective/objective split in positivism certain qualities of how a fact appeared (its aesthetic form, apparent connections with other phenomena, its usefulness to humanity, and so on) became strictly torn in two with the subjective sequestered to one side, the objective to another. Things like usefulness, or aesthetic quality, or even the ends toward which a thing tended, were considered by Baconians to be an intrinsic piece of what constituted a fact. Just so, many things that they took for granted as a factual part of reality were, by definition, moved by positivism into a purely subjective realm of fancy and imagination.[55] Moreover, the new category of emotions had largely negative connotations at the time—feminine, bodily, irrational, involuntary—that were likewise seen as synonymous with the newly irrationalized borders of religion.[56] Science could be equated not just with objectivity, but an objectivity laced with the "masculine" ethos of unwavering courage in the face of unsettling facts, and contrasted to a "subjective" religion that was the effeminate clutching of pearls at a big, mean world that had hurt its feelings. As one historian trying to complexify the history of the emergence of psychology has written, "By treating their own metaphysics as science, positivists gained a license to ignore (by shrugging them off as metaphysical and therefore unworthy of being answered) the difficult questions and problems raised by others about positivist assumptions."[57] While anecdotal, it is nonetheless interesting to note that three of the most powerful criticisms of the positivist paradigm in psychology and sociology that arose in the mid-twentieth century came by way of three philosophers—Alasdair

55. This movement toward the privatization and subjectivization of religion of course had many causes, and was hardly restricted to the story told here. Cf. for example Cavanaugh, *The Myth of Religious Violence*, where, particular to our story, he describes the rise of the category of religion and its relegation to a private sphere occurred in order to magnify the prime agency of the newly created nation states. Hahn and Morrow, *Modern Biblical Criticism as a Tool of Statecraft*, also tell a fascinating and largely parallel narrative to Cavanaugh, and the one we are relating about the rise of the sciences, and describes how biblical criticism fashioned itself as mirroring the natural sciences, being an "objective" discipline over and against purely "subjective" theological or religious readings of scripture. Biblical criticism, as Hahn and Morrow demonstrate in detail, was often a vehicle of the same university movements that we are describing in relation to White and others. And, paradoxically, the rise of "historicism" so often seen as an enemy of religion, since it reduces it to a purely historical phenomenon, was reliant upon covertly theological pedigrees which continued often into its most "secular" manifestations, as Thomas Albert Howard, *Religion and the Rise of Historicism*, has argued. This is an enormous element of the broader story of the warfare thesis that, while exemplifying many of our claims in this book while also being a vital component of any story of secularization in the West, sadly must be passed by for space considerations. Cf. as well Sheehan, *The Enlightenment Bible*; Legaspi, *The Death of Scripture and the Rise of Biblical Studies*.

56. Dixon, *From Passions to Emotions*; Meador, "'My Own Salvation'"; Taylor, *Sources of the Self*; Siedentop, *Inventing the Individual*.

57. Reed, *From Soul to Mind*, 159.

MacIntyre, Charles Taylor, and René Girard—who gained their critical resources in part because they all converted to Christianity mid-career.[58]

What occurred in psychology in many instances thus followed the double rewriting of history we have been seeing: a deletion of theology from history, and a rewriting of the theology that remained as irrelevant or irrational, in conflict with science by definition. This brings us right back to Tertullian. Sigmund Freud for example used Tertullian's revised saying as evidence bolstering his argument that religion is merely an infantile wish fulfillment that embraces any means necessary to avoid reasoned scrutiny.[59] In a 1990 psychology text (actually, an *abnormal* psychology text!) this avoidance of rational accountability that Freud mentioned is turned into a motto— with reference to Tertullian—which is used to characterize the psychology of the whole historical era of the early church.[60] The philosopher and historian Ernst Cassirer in a similar—albeit more nuanced—vein suggests that Tertullian and his saying represent a religious "type"[61] (as would psychologist Carl Jung),[62] while the great sociologist Max Weber used it to characterize what he thought was the inevitable and universal conflict of science and religion.[63] Even more recently, the curious phrase still occurs in textbooks and reference volumes. Simon Blackburn's 1996 *The Oxford Dictionary of Philosophy* is a good example of this, and he describes it as "Tertullian's Paradox," indicating "the very impossibility of a proposition becomes (mostly in theology) a kind of motivation for belief in it."[64]

When Thomas Huxley—who, it should be mentioned in passing, quoted the Tertullian paradox, along with the more genial Darwin[65]—let slip his own dogs of war as we saw in the last chapter, his powerful strategy of rewriting the history of scientific naturalism thus intersected with similar movements reinterpreting the domain of religion and its role in the university and in public life generally. Not just psychology, but the emerging disciplines of sociology, anthropology, philosophy, and history of religion all quite often saw themselves as actively anti-Christian. To be sure, it is impossible to generalize about this: the new disciplines were often taken up or promoted by Christians or in support of Christianity, while those opposed to the new disciplines often did so for reasons other than religious. Yet, for the story of how the warfare thesis came to sink into our imaginations at large, the anti-Christian elements were prevalent and influential enough to take center stage. They not only

---

58. Blakely, *The Demise of Naturalism*; Palaver, *René Girard's Mimetic Theory*.

59. Sigmund Freud, *The Future of an Illusion*, 49. I am deeply indebted to Harrison, "'I Believe,'" for the initial tracking down of the psychology texts noted in this paragraph, which I subsequently verified.

60. Weckovicz and Weckovickz, *Abnormal Psychology*, 38.

61. Cassirer, *The Philosophy of the Enlightenment*, 180.

62. Jung, *Psychological Types*, 12–16.

63. Weber, "Science as a Vocation," 29.

64. Blackburn, *Dictionary of Philosophy*, 88.

65. Huxley, "Agnosticism," 5:224; Darwin, *The Autobiography of Charles Darwin*, 57.

wanted to discredit Christianity or religion as a whole, but at the same time supplant and replace the functions Christianity and religion had previously discharged.[66] One of their prime strategies were similar forms of historical renarration. This is in no way to continue the unreasonable taboo many Christians have against counseling and psychology, for example. What it does do is call our attention to very real conditions of the emergent disciplines that primed them as well as "religion" for the warfare thesis. "These were not men who accidentally slighted religion," writes sociologist Christian Smith, "these were skeptical Enlightenment atheologians, personally devoted apostles of secularization."[67] Nor is this the embittered judgment of religionists. It is rather words from the mouths of the early practitioners themselves. In a survey of the thirty-three most influential sociology texts between 1883 and 1920 it was discovered they all had nearly identical polemical intentions and argument structure because, in part, "these authors relied so heavily on Comte and [X-Club member Herbert] Spencer . . . [Thomas] Huxley, [Friedrich] Nietzsche, . . . and writers in *Popular Science Monthly* [which we will turn to in the next chapter]."[68]

"Religion" for the story of the warfare thesis was, in other words, not just under fire from these newly christened sciences in the period of time leading to Huxley, White, and Draper. Like the newly professionalized and demarcated category science, religion was being re-described, its borders redrawn. If it seemed these sciences were scoring bullseyes against their targets, often it was because the target boards were being drawn after shots were fired, so to speak. Religion as a category was being primed—consciously or not—as an object ready-made for conflict, and ultimately defeat. The detective work of many historians in the latter half of the twentieth century has revealed the extent that religion was quite literally becoming defined as the fabricated Tertullian Paradox made it out to be. This can be referenced quickly by a number of summary statements used to represent the notions of religion in the same thirty-three most influential sociology texts we mentioned above:

1. *Science and religion are different ways of knowing concerned with different orders of reality* but . . . *they are actually absolutely incompatible and antagonistic sources of knowledge.*

2. *Sociology is an immature science,* but . . . *it will surely deliver the knowledge necessary for social salvation.*

3. *Religion is concerned with the spiritual realm, which is beyond sociology's ability to examine,* but . . . *all religions are finally reducible to naturalistic, material, and social causes, and are clearly false in their claims.*

---

66. Smith, "Secularizing American Higher Education"; J. Roberts, "Psychoanalysis and American Christianity."

67. Smith, "Secularizing American Higher Education," 111.

68. Smith, "Secularizing American Higher Education," 115–17.

4. *Modern religion has advanced well beyond primitive religion, but . . . all religions are essentially identical in being based on the fear and ignorance of savages.*

5. *Religion remains intrinsically important to the mass of humanity, but . . . religion's only real potential value is in instrumentally promoting social harmony.*

6. *Religion is in the business of promoting morality, but . . . in actuality religion has been history's primary source of oppression, immorality, conflict, and error.*

7. *Religion has always been an important force in social life, but . . . its influence and credibility in the modern world are for good reasons rapidly declining.*

8. *Religion has historically been engrossed in politics and public culture, but . . . true religion in the modern world should confine its social role to the private life of individuals.*

9. *Sociology is indifferent to religions concerns per se, but . . . the modern church must renounce the making of truth claims and instead emphasize positive, subjective individual feeling and human idealism*

10. *Religion is a well-meaning agent of social reform, but . . . it is dangerous and irresponsible unless it submits itself to the knowledge and authority of the social sciences.*[69]

Anthropology got in on the action as well, and with some gusto. X-Club acquaintance Sir Edward Burnett Tylor (1832–1917) for example—widely acknowledged as the father of anthropology, certainly in its British variations—by describing all religions no matter how sophisticated as the work of the "savage" mind contemplating nature, made religion as a category appear to be an unsophisticated attempt at a proto- or quasi-scientific explanation of the world in mythological garb. In this manner Tylor was following Comte, and his theory provided a background framework for the claims of Huxley and others that religion—or at least theology—was fighting over the same territory as science and hence inevitably locked into a zero-sum battle that it would lose. Indeed, though "the explanation of religion as an erroneous natural philosophy was hardly an original thesis, Tylor developed it with such skill and persuasiveness that it became more or less a commonplace for scientific publicists like Tyndall, Huxley, and Spencer" to use in their broader strategy as X-Club members.[70] His seminal works like *Primitive Culture,* which set the tone for anthropology as a discipline, also "fueled the warfare model of the relationship between religion and science," by shaping the very categories of anthropology to reflect the terms that the warfare model demanded. Tylor would write, for example, that "whenever scientific thinking was accepted it dispensed with religion."[71] Warfare was not "merely ornamental," to Tylor's

69. These theses are taken verbatim from Smith, "Secularizing American Higher Education," 117–49.

70. Wheeler-Barclay, *The Science of Religion in Britain,* 103.

71. Larsen, *The Slain God,* 25.

theories, either, but rather constituted a major portion of the core of his anthropology. "The most advanced civilizations," like Britain and America, "had already entered into the initial phases of a climactic struggle between ancient beliefs and modern positive science."[72] The theologians might struggle, but they would soon find that they had already died, even if their bodies had not yet realized it. Tylor, under the guise of a scientifically objective description of religion, was in fact narrating a coroner's report.

To do this, Tylor advanced the notion of what he called the "psychic unity of humanity," which meant little respect need be paid the historical contexts of any given belief, for they represented an assemblage of attitudes perennial to the primitive religious mind-set.[73] What counted as the primitive mindset in many senses appeared to be dictated almost completely by Tylor's own elite Victorian sensibilities. Here the psychological and anthropological disciplines cross-fertilized as both saw themselves as active contributors to religion's defeat. Indeed, later anthropologists would complain that Tylor, much like the emerging discipline of psychology, would slight emotions in favor of his own peculiar demarcations of the rational.[74] Tylor's goal was in fact to understand the primitive psychology, or better the psychology of primitivism.[75] As Tylor himself put it "from the philosophy of the savage thinker to that of the modern professor of theology" was a single continuous history, any part of which could beneficially be held up in comparison to any other part as species of the same primitive, emotion-driven genus.[76] Particularly important in this primitive-to-cultured model was the reversal that had taken hold since at least David Hume, in which Tylor argued that monotheism was not original to humanity. Rather it was a late cultural achievement (which Comte's progressive scheme embodied in its own manner). Regardless of his differences to Darwin, and his concerns to maintain methodological distinctions between biological and cultural evolution, this reoriented framework was used to great polemical effect in completely reorganizing and rewriting the data of the then more orthodox "original monotheism" schemas of history.[77] While direct arguments against religion and theology were used, to be sure, more often than not it was the entire orientation of the theories themselves that put pressure on religion by redescribing it in terms that were essentially irrational, primitive, uncultured.[78]

The irony here, however, is just how much Tylor's anti-religious anthropology bore the signatures of his own upbringing as a Quaker—the Christian religious movement founded by George Fox and devoted to the peaceful work of Christ. Tylor

72. Wheeler-Barclay, *The Science of Religion in Britain*, 94.

73. Wheeler-Barclay, *The Science of Religion in Britain*, 90.

74. Logan, *Victorian Fetishism*, 106.

75. Logan, *Victorian Fetishism*, 100.

76. Quoted in Wheeler-Barclay, *The Science of Religion in Britain*, 100; cf. Logan, *Victorian Fetishism*, 100–3.

77. See the work of Corduan, *In the Beginning God*, 35–62.

78. Larsen, *The Slain God*, 28; Logan, *Victorian Fetishism*, 105.

created the category of what he called "religious survivals" to explain the persistence of antiquated savagery in even the most enlightened of cultures. In turn, quite ironically, his own anthropology constituted in effect one large, complex Quaker survival. As one scholar put it, his anthropology was merely "Quakerism minus Christianity."[79] In this sense we can contrast him to Comte, who built *Catholic* "survivals" into his thought in order to capture the power and glamor of the Catholic Church. In his quest to equate current theology with primitive religion, for example, Tylor in good Protestant fashion often went out of his way to describe Catholic practices in terms he reserved for old heathen practices. This was, though cloaked in the language of secularism, a thoroughly Quaker critique of Catholic priestcraft. Tylor's Quaker heritage also led to the particular shape of his humanitarian efforts to reform society, as well as Tylor's vocal abolitionism against the African slave trade. He even advanced his narrative under the auspices of what Huxley called the "New Reformation." In other words, what had begun as intra-Christian polemics was here reworked, the story retold, history rewritten, as a purely secular critique by a foe fashioning itself as exterior and superior to religion. It seems, then, that often what constitutes a "primitive" religious survival as such depended upon the mood and disposition of Tylor as a commentator. Though Tylor was at pains to decry any number of religious survivals as antiquated nonsense harassing modern civilized society, Tylor's outspokenness on such things went curiously silent, for example, when it came to the knighthood he received from the queen in 1912.[80]

Tylor's analysis of religion was reinforced by the emerging work of Victorian folklorists working in anthropology around the same time, like the famous Sir James Frazer (1854–1941), author of the seminal masterwork *The Golden Bough: A Study in Comparative Religion*. Unlike most other works in this chapter, walking into a bookstore today one can easily find copies of *The Golden Bough*—sometimes even in new print runs. Trying to take into its orbit the whole of comparative magic, mythology, and religion, *The Golden Bough* remains to this day a classic study despite its innumerable fictions and fancies. Though Tylor's thought came under increasing scrutiny after the 1890s, in a sense it—along with many borrowings from positivism—hitched a ride on Frazer's work and survived well into the twentieth century beyond any direct criticism.[81] One encounters in Frazer's writing fetishism, animism, religious survivals, secularized progress, and many other bits and pieces of the ensemble of Tylor's theories and Comte's framework. Tylor's notion of "psychic unity" in practice if not in name is also maintained, where Frazer in almost any given passage can arrange a "kaleidoscope of incidents from Zululand to Iceland that strike him as sharing common characteristics reveal an underlying, shared mentality" to constitute the umbrella

---

79. Larsen, *The Slain God*, 33.

80. Larsen, *The Slain God*, 35.

81. Logan, *Victorian Fetishism*, 104.

category, religion.[82] And that taster of the world consisted merely of a single footnote, to say nothing of the smorgasbord on offer in the full work. Beginning work on *The Golden Bough* in the 1880s, it constituted a project that would consume the rest of his lifetime in publication and revision.

Given its central theme, it is not surprising to realize that nowhere are these rhetorical and literary tactics of Frazer on display more forcefully than they are when he stalks around the subject of the crucifixion of Christ. For Frazer in *The Golden Bough*, the "Judeo-Christian traditions are everywhere, and nowhere." Arrayed in an immense patchwork survey of religions and magic he continuously uses Christian terminology to describe these practices. Like Tylor, Frazer is describing something of a continuous mind-set from animism through the most sophisticated seminary graduate of his day. In the obituary to his mentor, the Christian anthropologist William Robertson Smith, for example, Frazer apparently thought it was an appropriate moment to note his opinion that anthropology should have in principle undermined Smith's faith, as it "calls for a reconsideration of the speculative basis of all ethics as well as of theology." As part of this broader view, in *The Golden Bough*, "while Frazer was ostensibly engaging in these transpositions [of Christian with pagan terminology] in order to make savage practices more familiar and understandable [to the Victorian reader], his covert intention was in all likelihood the reverse: to make familiar [Christian] religious practices that his readers had always accepted as understandable come to appear as strange and savage."[83]

Before his explicit equation between *Saturnalia* and cross, "the author had made his intentions clear . . . as he circles about his target, pauses as if to strike, and then retreats again." Indeed, "no brief synopsis can adequately convey the impact of Frazer's technique," where, in his use of irony, he juxtaposes "without comment" pagan and Christian practices he wants his readers to find similar in their own minds. By "employing the language of the Bible or the Book of Common Prayer to describe 'savage' ceremonies," thereby implicitly encouraging his readers to laugh at such antiquated customs only in turn to realize such laughter, by Frazer's art, now turns "against 'Christian Europe' itself—these are some of his favorite devices."[84] The similarities conjured through this method of juxtaposition, it is now generally agreed, are more illusions wrought by Frazer's masterful art than anything. But this strategy of rewriting history was phenomenally successful, in part because it pulsed with the energies of its own mythological rhythms summarizing innumerable nineteenth-century thought strands and sensibilities. In any given sentence or paragraph—to say nothing of the work as a whole—Frazer can lampoon mythology or religion while nonetheless going on to use its imagery, driving home his own points and lacing them with its power.[85]

82. Larsen, *The Slain God*, 39.

83. Larsen, *The Slain God*, 48; Josephson-Storm, *The Myth of Disenchantment*, 133.

84. Wheeler-Barclay, *The Science of Religion*, 205.

85. Vickery, *The Literary Impact of the Golden Bough*, 5.

Unified not so much by the force of its argument, Frazer's work hits home through its "aesthetic cogency" and in how its own words are brimful with the command of myth showing the impact of the subject upon the anthropologist as many like Clifford Geertz have understood is an inevitable side-effect of the discipline.[86] *The Golden Bough*'s success was also in part achieved because the scattershot juxtapositions forced the reader themselves to make the explicit connections. In turn, this often invited a blurring of the lines between the reader's own ideas, and that of Frazer as author.

Though Frazer never entered into the fray of the warfare thesis head on, in the second edition of *The Golden Bough* he did grow bolder, and everywhere his theories on religion shone with the embers of that very Victorian war.[87] Between the first and second editions of *The Golden Bough*, a particularly famous work of warfare—White's, as it happens—had emerged and emboldened many voices. The new preface to Frazer's second edition evoked warfare metaphors while also adding a very strong sense of the imminent disenchantment of the world, and very clearly welded the two together. All the more then did the warfare metaphor entrench itself in Frazer's work as he promoted it as a description of religion's beleaguered battlefront continually falling back through history. This inevitable movement of retreat would continue until the last faerie and the last god would find themselves totally surrounded, outgunned by test tubes and radio waves and electric lights. Frazer began all the more to consider his comparative work on religion as a movement preparing the front lines of this war for their final push: "At present, we are only dragging guns into position," he says in the introduction to the second edition of *The Golden Bough*. "They have hardly yet begun to speak."[88]

Despite Frazer's lasting influence, scholars have recently concluded just how deep the irony was when Frazer conceded that much depended upon acceptance of his definitions of magic, religion, myth, and theology.[89] When they looked past his attempts to fashion his own terminology as scientific and objective, they discovered with a bit of shock,

> That [Frazer's] disenchantment appears not primarily in the theory of the master folklorists [Frazer was drawing from] but within the folktales themselves that often located fairies, magic, and miracles in a bygone age. In fact, the departure of fairy enchantments was one of the first motifs that the nascent discipline [of folklorists] discovered. Although these tales were not his only source, Frazer seems to have been allowing his theory to be imprinted by folkloric conceptions.[90]

86. Geertz, *Works and Lives*.

87. Larsen, *The Slain God*, 56; Josephson-Storm, *The Myth of Disenchantment*, 134.

88. Frazer, *The Golden Bough*, 2d ed., I, xxi–xxii.

89. Wheeler-Barclay, *The Science of Religion*, 197–98.

90. Josephson-Storm, *The Myth of Disenchantment*, 127.

Frazer's secularization theory was the product, at least in part, of an incredibly unlikely source: the myths of the flight of faerie now transposed into the language of Victorian science. But so, as with Tylor's theories being driven by Protestantism, here the expulsion of the fairie as the inspiration for secularization finds its Protestant impulse: as many Protestant records were often keen to relate, such local magical flora and fauna fled at their arrival. Much as Protestants expelled the superstitions of Catholics, so too did the errant magics of indigenous lands wither before the renewed Lamb of God, now marching arm in arm with the new sciences. In fact, in many cases the fairies were seen not as pagan natives, but as preternatural little Catholics afflicted by Protestant clarity. English Bishop Richard Corbet, for example, penned a poetic ode to the flight of the elves:

> By which were note the faries
> Were of the old profession
> Theyre songs were Ave Maryes [Hail Maries!],
> Theyre daunces were procession;
> But now, alas! They all are dead,
> Or gone beyond the seas . . .[91]

This is not, as is sometimes done, to place secularization and disenchantment squarely upon the shoulders of the Protestant Reformation.[92] To be sure, many Protestants were as keen as Catholics to keep many of these "superstitious" traditions, and a cornucopia of Protestant mysticisms that lived alongside their more mainstream brethren. Rather, the point for now is to emphasize that there were often "important Christian motives for going the route of disenchantment."[93] Once again both secularization and warfare find their bizarre roots in Christian polemics between Protestants and Catholics. It was, revealingly, also the same period in which each faction began arguing over the cessation or continuation of miracles after Christ and the age of the apostles. Science was called upon by both sides to help determine authentic miracle from paganism or charlatanism. It would seem secularization and disenchantment "culminates not in the end of religion, but rather *within* Protestantism" and indeed even Catholicism.[94] Yet, with positivism's increasingly extreme rewriting over time, their works made "religious [concepts] reappear in a way" that positivism often borrowed their concepts from religion while nonetheless fashioning themselves as "establishi[ing] scientific

---

91. Quoted in Josephson-Storm, *The Myth of Disenchantment*, 139; Lehner, *The Catholic Enlightenment*, 125–53.

92. Walsham, "The Reformation and the 'Disenchantment of the World' Reassessed."

93. Taylor, *A Secular Age*, 26.

94. Josephson-Storm, *The Myth of Disenchantment*, 271; McCarraher, *The Enchantments of Mammon* instead of using "disenchantment," instead prefers "misenchantments" as a description noting enchantments have hardly disappeared, but rather have moved locations because Protestants "renegotiated the terms of enchantment" and its implications (36).

mastery," over religious utterances.[95] Most of the "scientific" study of religion, instead of providing evidence against religion, by definition turned religion upon its head by inverting it into a purely anthropological category. "When God is assessed primarily as one more unit within a congeries of cultural units and criteria, the issue of atheism has already been engaged and settled. The god that is one more thing [within the horizon of scientific expectation] does not exist." The denial of God through the scientific study of religion was less an argument but the eventual "working out of what had been implicitly denied," by method and definition. As Jason Josephson-Storm puts it,

> The changing language [of religious studies and its relation to sociology, anthropology, and psychology] marks a critical difference. This shift in meaning seems to represent the vanishing of God, and words like *transcendent*, *infinite*, or *sacred* are attempts to cover for an absence, to describe a shadow. Yet the very category of "religion" was formulated around a Christian concept of God. In talking and writing about religion, it is often mistakenly assumed that religions have a common hidden essence that marks them as "religious." In excluding God from its explanatory apparatus, "religion" remains as a category structured around a hole or a fissure. In other words, we find ourselves in a discipline organized around a core that no longer exists and [that by design we cannot] reconstruct.[96]

For this and other reasons, as one historian recently put it:

> Someone wishing to develop an historical understanding of secularization cannot do so by taking up a position within [the current] field of debate. This is in part because there is no historical evidence that a process of secularization of any of the envisaged kinds actually took place, while there is significant evidence to the contrary. But it is also because these various accounts of a process of secularization are not themselves histories in the empirical sense. Rather, they constitute an array of competing theological and philosophical programs, each advancing what purports to be a history of secularization, but only as a means of prosecuting various factional cultural-political agendas, some dedicated to secularism, others to sacralism.[97]

Aftermaths—At the Origins of the History of Science

Given enough time, of course, almost any cross-section of humanity will do a great deal of enormously stupid things—not least of which is committing some of these things to print. Nonetheless, it was precisely the broader arguments of psychology, sociology, and anthropology reshaping the nature of religion and rewriting the history

95. Garroutte, *Language and Cultural Authority*, 128.
96. Josephson-Storm, *The Myth of Disenchantment*, 121.
97. I. Hunter, "Secularization," 8.

of Christianity that in many ways made White's warfare meant to purify Christianity tip dangerously into being viewed as a wholesale and unbridled attack upon religion writ large. What were once Protestant and Catholic (or even intra-Protestant and intra-Catholic) polemics were turned into absolute terms of religion vs. science, or even more specifically into a schema of nature vs. supernature. But even more than just this, the borders between nature and supernature were themselves often renegotiated and altered either by adherents or enemies to Christianity, thereby making the latter—that is, supernature—all the easier to exorcize.

Defined in a sense as opposites, supposed instances of the supernatural could therefore by definition only be moments of conflict or rupture in the natural order. It would require a book in itself, but needless to say this is certainly not how miracles, or even God's direct action, were envisioned by theologians before the modern period. For the fathers and mothers of the church, a major argument was that there really can be no concept of a miracle in juxtaposition to nature. "The positive definition of grace can only be given through grace itself. God must himself reveal what He is within Himself. The creature cannot delimit itself in relation to this Unknown reality. Nor can the creature, as a theologically understood 'pure' nature, ever know wherein it is specifically different from God."[98] To define "nature" would always also be to tacitly define God, even if it is only as the "not this." But since God always is in excess to our definitions, the concept of "nature" itself—tacitly reliant upon definitions of God—must always likewise be open and not a closed concept. In fact, "nature"—paradoxically—requires elements from outside of itself to be defined. As David Bentley Hart eloquently puts it, many theologians in Christian history therefore argued that "we have no direct access to nature as such; we can approach nature only across the interval of the supernatural."[99] In other words, when speaking of God, or, eventually, "the supernatural" what was envisioned was not an either-or juxtaposition to the natural world, but the conditions of possibility that allowed the world to appear in the first place. But things changed. David Hume was perhaps one of the first to define a miracle or any action of God as a rupture in the order of nature in order to lampoon any report of such as laughably improbable, but for positivism it was even worse. Much as a child might point to the profile on a coin and declare "look, a man!" so too the supernatural existed only as a babbled moment of someone's confusion about nature, the "absurd" of Tertullian's fabulated quote. But such definitions make for incredibly poor insights about the history of theology and its complex interactions with the natural knowledge of any given era. It will come as a deep surprise that "the supernatural" was not frequently used in Christian theology until the thirteenth century, and often had meaning only insofar as the theologies in question were integrating Aristotelianism and its own proprietary use of "nature."[100] Not only this, because they are defining both

---

98. Balthasar, *The Theology of Karl Barth*, 279.

99. Hart, *The Experience of God*, 97.

100. Bartlett, *The Natural and the Supernatural*, 1–34; McGrath, *Nature*, 84–134.

nature and supernature, it is a full-blooded act of theology since it must—however tacitly—include definitions of God and the supernature as the "not this" when also talking about nature. Awareness of such historiographical tricks are precisely what has aided much of the deconstruction of White's narrative and those like it. For example:

> [While Andrew Dickson White and John William Draper's] preconception that, as science has advanced, phenomena once considered supernatural have yielded to naturalistic explanation, is not without support. . . . it assumes a dichotomy between nature and supernature that oversimplifies the theologies of the past. If a supernatural power was envisaged as working *through*, as distinct from *interfering with*, nature, the antithesis [between science and theology] would partially collapse. . . . The significance given to explanations in terms of natural causes depends on higher-level assumptions embedded in a broader cultural framework. In the history of Western culture, it has not simply been a case of nature swallowing supernature. Something had to happen to change the higher-level assumptions if the conflict between science and religion was to achieve the self-evident status proclaimed by Draper, White, and their successors.[101]

Even more than this, the *meaning* of nature conquering supernature (if and where it happened) was assumed as a matter of course to be something intrinsically damaging to Christianity, when in fact it was a favorite tool of many reforming currents in both Protestantism and Catholicism. Though not solely responsible, these two "higher-level assumptions" were in large part positivism's lingering presence in the new scientific disciplines and their perceptions of history. The new genre of positivist discourse placed "both sacred and mundane events into a context [that was by definition] devoid of religious implications, thus showing the sufficiency of scientific explanation."[102] As the historian of science Peter Harrison has recently observed, in other words, "once the constructed nature of the categories [of science and religion] is taken into consideration, putative relationships between science and religion [like their historical conflict] may turn out to be artifacts of the categories themselves . . . determined by exactly how one draws the boundaries within the broad limits given by the constructs."[103] White only mentions Tertullian in passing[104]—unlike Draper, the other main proponent of the warfare narrative, who casually remarked in his own positivist moments that Tertullian's negative attitude toward *science* (no longer just rationality or philosophy!) catastrophically "affected the intellectual development of all of Europe."[105] Nonetheless, the ethos supposedly embodied in Tertullian's quote no

101. Brooke, *Science and Religion,* 47–48.
102. Garroutte, *Language and Cultural Authority,* 129.
103. Harrison, "'Science' and 'Religion,'" 39.
104. White, *Warfare,* 2:230.
105. Draper, *Conflict,* 45.

doubt played a part in conditioning White to believe his own broadly positivist misappropriations of what Christian theology entailed through all of history:

> interference with science in the supposed interest of religion, no matter how conscientious such interference may have been, has resulted in the direst evils both to religion and to science—and invariably. . . . I say "invariably" I mean exactly that. It is a rule to which history shows not one exception.[106]

History showed "not one exception," of course, precisely because it was being rewritten to fit the rule. White's work fit neatly into the broader pattern of the positivist sciences which, as a genre, we have seen rewrote "entire religious histories in specifically [positivist] scientific terms, often without any reference to the original [concepts] of religious speakers."[107] Indeed, in the wake of the new anthropology, White's fear was that "the new generation, finding Myth and Legend insisted upon as essential may throw the whole thing [Christianity] overboard altogether."[108] It is true that there were legitimate worries about any number of cumbersome accretions that had attached themselves to Christianity. However, "myth and legend" as we have seen was a category that took on something of a life of its own as the positivists massaged it into a newly tenderized set of targets, and so defined their bullseye shots into existence. White's biographer Glenn C. Altschuler puts it neatly:

> White's aim, then, was clear. He hoped to affirm a rational, nonmythical religion and at the same time to preserve those religious truths (primarily ethical maxims such as love of God and neighbor) which he regarded as absolutes. Yet he also accepted unquestioningly the results of recent scientific investigation which threatened to destroy religion as a moral bulwark. [The question surrounding the reception of White's work was then] could the weapons of the enemies of religion be used to preserve it?[109]

While "not really" would be the ultimate answer to that last question, early reception of White often admitted that this book was essential—even spiritual—reading for the Christian in the modern world. In the weekly publication the *Outlook*, while White overused the warfare metaphor the publication also praised that White's work "should stand beside the 'History of Doctrine' in every theological seminary."[110] Another periodical, the *Independent*, likewise noticed this distinction, its reviewer declaring that White's central aim was not against religion, but to "convict theology as a vicious tendency to interfere with scientific freedom to the injury of both science and religion."[111] Indeed no less prestigious individuals than the founding president

---

106. White, *The Warfare of Science*, v. 1, 7–8.

107. Garroutte, *Language and Cultural Authority*, 126–27.

108. White, *A History of Warfare*, 1:239.

109. Altschuler, *Andrew D. White*, 204.

110. Quoted in Ungureanu, "Science and Religion," 137.

111. Quoted in Ungureanu, "Science and Religion," 137.

of Stanford University David Starr Jordan, and steel industry magnate, millionaire, and philanthropist Andrew Carnegie likewise echoed this distinction in their hearty praise and recommendation of White's work.[112]

And yet, as a matter of course White's attempted distinction between theology and religion set out in his introduction becomes an afterthought as the body of the work gets underway. Given the tools he was using this is not surprising. Though White was scattering his thousandfold bread crumb footnotes for others to also find his trail, it was actually the felled trees, the trampled underbrush, the clearly carved narrative path of destruction through history that signaled the road one should follow to find White's way. Imposing in its command of the relevant literature, "which was duly cited in an elaborate set of footnotes, the work was hardly a dispassionate search for truth." Rather, "from the first page to the last it deployed metaphors of battle, warfare, attack, and retreat that left no doubt about White's passionate desire to see science smite its foes."[113] Even the liberal *Manchester Guardian* noted White's distinction between dogmatic theology and religion, however solid at first, melted into air and appeared as a wholesale condemnation of religion itself. White failed, they said, to appreciate the "greatness of the Middle Ages," and that rather than actually dealing with the complexities of history White's work was merely "a storehouse of curiosities and superstition," like undecayed peacocks and cheese donkeys. The *American Historical Review* likewise recognized White's attempt to immunize true religion from his escapades and "prevent unnecessary damage to Christianity." In the end, though, while White conceded that the love of science contained in the scientist was the same as that which burns in the theologian, "the verdict of his account always seemed to show otherwise."[114]

White's balancing act via positivism was thus quickly nudged onto the less pious half of its razor's edge both by broader currents that coopted it, and by White's own carelessness. Perhaps in the end no figure was more influential in spreading a subliminal (and not so subliminal) message of warfare than that of George Sarton. Considered the father of the history of science as a discipline, he has been shown to be particularly relevant for this story of the afterlife of White's war. We have already encountered Sarton in his relation to Duhem and his heroic representation of figures like Leonardo da Vinci. Writing many foundational texts in the field, and taking on countless students, Sarton also founded *The History of Science Society* and the journals *Isis* and *Osiris* to relieve some of the burden of longer, sustained study. "Sarton's dependence on positivist philosophy is unmistakable" writes historian James Ungureanu.[115] Sarton's "New Humanism," directly recalls Comte's "Religion of Humanity," for

---

112. Quoted in Ungureanu, "Science and Religion," 138.
113. Schaefer, "Andrew Dickson White and the History of a Religious Future," quote at 7.
114. Quoted in Ungureanu, "Science and Religion."
115. Ungureanu, "Science and Religion," 1116.

example, and Sarton was greatly influenced by Mach's history of science. But one does not have to rely just on subtext here.

When speaking of the beginnings of the history of science, Sarton points to Comte, who "must be considered as the founder of the history of science, or at least as the first who had a clear and precise, if not a complete apprehension of it."[116] This is no surprise given the extent Comte influenced visions of the Scientific Revolution, as we examined in chapter 3. As such Comte's ghost—if not his whole system—continued to haunt the pages of Sarton. The "noble dream" of pure objectivity that had affected the American historical discipline, now doubled down as it rooted itself into the first generations of historians of science. So, while promoting a type of purified objectivity, Sarton was, in the words of John F. M. Clark "[building] on eighteenth- and nineteenth-century traditions of positivism and universal history."[117] Sarton could write in this spirit that:

> The history of science is the story of an endless struggle against superstition and error; it is not a vivacious and spectacular struggle, but rather an obscure one—obscure, tenacious and slow. The resistance of science against every form of unreason or irrationalism is so firm and yet so quiet, that it is almost as gentle as non-resistance would be, yet unshakeable.[118]

This passage has the character of a patient but slightly exasperated "objective" historian, conjuring a wizened and patrician tone for his readers. And yet, as Charles Homer Haskins joked, here Sarton's work sometimes reads like it is trying to "distribute medals for modernity."[119] In fact, Sarton's positivism reveals in full display its subterranean use of Christian theology and its themes even while attempting to prune history of the slightest hint of productive uses of such religious ideas: "the progress of science is ultimate dependent upon its emancipation from non-scientific issues," he wrote. But what is more, the history of science seen as a journey away from religion writ large—Sarton here has dropped all pretenses to maintaining the distinction between religion and dogmatic theology—is also "the history of mankind's unity, of its sublime purpose, of its gradual redemption."[120] Sarton could not be more ironically religious here if he had simply cribbed a few lines from a hymnal somewhere. The inversion of humanism away from its Christian roots as examined by the Renaissance scholar Charles Trinkaus and others is here in full force, in turn aiding in the revision of the history of Christianity and its relation to science.

Something of a paradox emerges in Sarton. On the one hand, for example, he could call Comte "crazy," he nonetheless lauded the idea of the positivist calendar of

---

116. Sarton, "The History of Science," 30.

117. Quoted in Ungureanu, "Science and Religion," 1116.

118. Sarton, *The History of Science and the New Humanism*, 10; 43–48; 179.

119. Quoted in Shapin, *Never Pure*, 4.

120. Quoted in Ungureanu, "Science and Religion," 1120.

secular saints, and Comte's general approach to the history of science—even traveling to Comte's "sacred domicile" in Paris to "commune" with his hero's spirit.[121] But this paradox runs deeper. While he wanted to display the gradual overcoming of religion as a "gentle non-resistance" that is hardly a "vivacious or spectacular struggle," nonetheless like the Christian biographers of the saints, Sarton insisted on the other hand that "it was *right* for a historian [of science] to be a hagiographer."[122] Hagiography, for those unfamiliar, is a genre of writing especially associated with early biographers writing on the lives of Christian saints. These writers often represented their subjects in overly glamorous, idealized terms. The lives here are less that of men and women, and more akin to the ethereal, luminous stained glass that would eventually also commemorate their existence. Sarton's secular saints were no different.

Even if he saw science as a general movement away from religion and toward the redemption of humankind, this redemption was won for us only by a handful of these stained-glass messiahs. For the history of Science, as we might recall with his portrait of Leonardo, "is largely the history of a few individuals."[123] The one-sided portraits of genius men and women of a pure science we have already frequently encountered were for Sarton a feature, not a lamentable deficiency, of the historian's quest: "The heroic scientist adds to the grandeur and beauty of every man's existence," he wrote.[124] We might recall Sarton's own attempts to inoculate Leonardo from being implicated by his obviously religious predecessors upon which he was so deeply reliant. Again, the moral and even religious features here are shining through. Science, Sarton wrote, "is the very anchor of our philosophy, of our morality, of our faith" indeed "truth itself is a goal comparable with sanctity."[125] Sarton had no sense of the irony of the religious qualities of his position. However much religion in the form of his New Humanism pervaded his texts, whenever he caught whiffs of religion in other writers his attacks were swift, and often brutal. As Ungureanu writes:

> As the gatekeeper of the incipient history of science discipline, Sarton also decided who and what was important to the field. Censorship was a common complaint by the authors trying to publish in his journals. For instance, when Sarton offered to publish Robert Merton's work [which would become a watershed in the twentieth century on how to picture the historical relation between science and religion] he asked [Merton]—rather curiously—to reduce his discussion of religion. Merton, who described his relationship with Sarton as the "exigent and angry master and I the brooding and unruly apprentice" was astonished when Sarton asked him to condense his section on religion, a

---

121. Sarton, "Auguste Comte, Historian of Science."

122. Shapin, *Never Pure,* 4.

123. Sarton, *The Life of Science,* 61–63.

124. Sarton, *The Study of the History of Science,* 45.

125. Sarton, "Knowledge and Charity," quote on 10.

striking proposal indeed since Merton's presentation of Puritanism in relation to the rise of science became the most celebrated part of his monograph.[126]

Though Sarton wanted to condense Merton, he would write elsewhere that Draper's *History of the Conflict* and White's *History of the Warfare* were "important guides to the whole subject [of the history of science]." And this coming from something with the dry and thoroughly academic sounding label of *Guide to the History of Science*. Instead of books that openly announce their polemics, the warfare thesis had officially migrated and taken root in the unassuming soil of textbooks for future generations. Sarton would later make White an official patron of his journal *Isis*, and Sarton's work would go on to inform the work of the notorious Logical Positivists and the so-called "Vienna Circle" that would, in turn, be one of the greatest concerns for religionists over the first half of the twentieth century.[127] While we have much to thank Sarton for—his Herculean efforts in the history of science must be gratefully acknowledged—it remains true as the famous historian and philosopher of science Thomas Kuhn put it a generation later that "the image of [historian of science's] specialty which [Sarton] propagated continues to do much damage even though it has long since been rejected."[128] Sarton, under the veil of objectivity, created a "painfully naïve" hall of good guys and bad guys to sort out a clear and concise path for science to have traveled through the byways of time.[129] "A hard critic," wrote A. C. Crombie, a mid-century historian of science, would say that "Sarton's approach could have easily killed the study of the history of science" as much as give it birth.[130]

Nonetheless, it was through the pen and personality of Sarton that the open polemics of the warfare thesis drifted into the sawdust of textbooks as simply the way things were. Even as late as 1956 Sarton could still write that "the history of science is the story of the endless struggle against superstition and error." Indeed, religious people were right to hate science "for the scientific spirit is the very spirit of innovation and adventure—the most reckless kind of adventure into the unknown."[131] This was the final solidification of his earlier opinion written in 1916, directly dependent upon White, that "the interaction between science and religion has often had an aggressive character," and that one is quite right to say that "most of the time a real warfare" existed between them.[132] White's now increasingly "secularized" war had migrated into the very birth of the new discipline of the history of science through Sarton. It had now only to be popularized and distributed in its new, "objective" form. To that we now turn by looking at the figures of John William Draper and Edward L.

126. Ungureanu, "Science and Religion," 1120.

127. See for example Knight, *Liberalism versus Postliberalism*.

128. Kuhn, *The Essential Tension*, 146.

129. L. Pierce Williams, quoted in Ungureanu, "Science and Religion," 1110.

130. Crombie, "The Appreciation of Ancient and Medieval Science," 164–65.

131. Sarton, *The History of Science and the New Humanism*, 10.

132. Sarton, "The History of Science," quote at 333.

Youmans, who popularized the warfare narrative by mastering the new waves of print culture that came to the fore after the calamity of Civil War disrupted the reigning print orthodoxies of the nineteenth century.

—Chapter Six—

Warsongs

John William Draper and the Spreading of the Myth

[W]e are here reminded that founding myths also require a kind of negation—an amnesia about what came before, and a forgetting of historical realities that might challenge the integrity of our new conceptions. Indeed, Karl Deutsch's similarly unflattering definition of a nation—"a group of people united by a mistaken view about the past . . ."—is not an altogether unfitting description for those who in recent times have sought to foment hostility between science and religion.

—Peter Harrison, *The Territory of Science and Religion*

THE GREATEST TRICK THAT the warfare narrative ever pulled was to convince the world that, as *narrative*, it does not exist.[1] It is the clear waters of truth bubbling up from beneath the hard stones of historical facts; it is the dust of history itself crying out for justice; it is a meticulously collected and arranged museum display curated behind the transparent glass of newly won historical objectivity. It is a clear picture, a snapshot of what came before. Anything. But a story, arising out of specific historical conditions that lent it plausibility?[2] No. As we have begun to see, however, "one of

1. With my apologies to the movie *The Usual Suspects*.

2. An important part of the rise of the warfare narrative that we have to leave out due to space considerations is the role that fiction, particularly the novel and science fiction, played. The newly invented genre of the novel, as George Lukacs remarked, "is the epic of a world that has been abandoned by God" (quoted in Pecora, *Secularization without End*, 3). The novel's unity of structure stood in place for the unity previously thought bestowed by providence, and so humankind became the inventor of its own continuity. In her *Genres of Doubt*, Elizabeth Sanders details the fascinating way in which "fantasy and science fiction allowed authors [in the Victorian period] to ponder the future of Christianity without naming Christ, and to consider the morality of God's authority without explicitly discussing God" (147). Important, too, as Alan Gregory explores in his book *Science Fiction Theology*, again in the Victorian period science fiction began to displace and continue the theological concept of "the Sublime" by making it completely immanent and this-worldly, a function of science or some

the most vigorous enterprises of the nineteenth and twentieth centuries has been the creation of the mythology of science." By this we do not mean lies per se, but rather "a compression of experience and dogma into symbols," of those on the march to war.[3] In more ways than one, a counter-religion. Notions of rigorous investigation and objectivity, laudable as they might be, veil the reality beneath where an entire chapter "of nineteenth century history [is] pressed together in the phrase 'the warfare of science and religion.'"[4] However well intentioned (or, perhaps, not) "objectivity" here became a rhetorical device, a way to frame and present this picture—"a picture [that] holds us captive" in an oft-quoted phrase from the turn of the century philosopher Ludwig Wittgenstein.

As we turn to John William Draper, we can take Wittgenstein's quote in both its intended, more metaphorical sense, but also woodenly literally. There is something of a historical irony that Draper claimed to be the first to have created a photographic portrait of a person (one of his three sisters, Catherine Draper) using the chemical exposure process that had recently been pioneered by Louis Daguerre. For a certain analogy exists between how Daguerre's process of photography worked, and how Draper's own mentality toward doing history operated. Because Daguerre's process needed incredibly long exposure times to gain a clear image of its subject, he himself had publicly doubted whether a proper portrait of notoriously fidgety human beings was even possible. Any motion would cause irreparable blurring to the image, and so the subject had to sit still, as if their essences were frozen even before the final enduring form of the picture itself came to be. When Daguerre produced his first image through this process he declared "I have seized the light! I have arrested its flight!"[5] Similarly, Draper saw history through his own preconceptions about the necessary essences of things called "Science" and things called "Religion" (note the capitalization). These objects in his picture of history, we might say, already posed in the curated postures Draper insisted upon in his mind. Draper's picture in this sense was already composed far before his subjects ever even sat down for the portrait. His "synthesis [of history was] wagging the data behind it."[6] In this manner Draper, perhaps, should be seen more as a director, or an artist, than a historian proper: "The reader who enters sympathetically into Draper's books on history and politics will find . . . [that] his writings belong to the secret life of the poem—and the sermon." His biographer then

---

other aspect of creation such as an alien civilization now given the prerogatives previously reserved for God. Thus, religious awe became transferred to the sciences (11–40). Cf. also Gross, *The Scientific Sublime*, who notes that not just science fiction, but popular science communication as well took over the sublime as a key thematic that drove interest in science as a quasi-religion (or perhaps a religion full stop).

3. Fleming, *John William Draper*, vii.

4. Fleming, *John William Draper*, 1.

5. O'Hagan, "Capturing the Light."

6. Fleming, *John William Draper*, 76.

adds "But they are meant to belong to the history of science, and this is the intellectual tragedy of Draper's career."[7]

Tragedy or not, it is in this way that Draper stands alongside the likes of Huxley, the X-Club, George Sarton, and Andrew Dickson White as the primary vessels by which the warfare narrative was unleashed upon the world. The association of these figures is more than mere analogy, as we have been seeing. The warfare narrative did not arise independently among them (despite many key and even extensive differences in tones and targets) but through an emerging network of like-minded scholars who were swapping trade secrets and shared similar goals. In fact, though we do not know for sure they met at the Oxford debate between Wilberforce and Huxley (where, we might recall, Draper was the keynote speaker), Draper and John Tyndall became fast friends who shared a deep correspondence through letters and hardships. As Tyndall was dealing with the fallout of his Belfast lecture that we described at the end of chapter 3, Draper wrote to him saying "I am at this moment finishing a book entitled 'History of the Conflict Between Religion and Science.'"[8] Draper had been asked to write this volume by another main character we shall meet in this chapter, Edward L. Youmans for his *International Scientific Series*. Over one hundred volumes found themselves rubbing elbows in Youmans' line-up, but not one sold even half as well as Draper's. In the United States, Draper's *Conflict* went through fifty printings in the same amount of years, while in the UK a mere fifteen years after its initial publication it found itself in its twenty-first edition. It would subsequently be translated into French, German, Italian, Japanese, Spanish, Polish, Russian, Portuguese, and Serbian.[9] "I say to you," he wrote to a discouraged Tyndall, "Stand fast. Your address is doing great good. If you need help, let me know." And Draper then closes: "the friends of science will stand by you in England as they are standing by me in America. Let us all fight shoulder to shoulder in our fighting—not for ourselves, but for posterity."[10]

"But for posterity." The irony as we have already begun to see even with Huxley and the X-Club, but especially with Andrew Dickson White and now with Draper, is that the posterity for which Draper encourages Tyndall to fight soon lost interest in all but Draper's picture of conflict. "Clever metaphors die hard,"[11] as we have already quoted from the historian James Moore, and like a "B-Movie" villain the metaphor with Draper kept coming back for more. Like White—even like Huxley and Tyndall—Draper was no critic of religion per se. In White's own words he deplored those who succeeded in fostering antagonism by "thrusting still deeper into the minds of thousands . . . that most mistaken of all mistaken ideas: the conviction that religion

7. Fleming, *John William Draper*, 64.

8. The correspondence between Draper and Tyndall is quoted from Lightman, "The Victorians" 76–77.

9. Chadwick, *The Secularization of the European Mind*, 161.

10. Lightman, "The Victorians," 77.

11. Moore, *The Post-Darwinian Controversies*, 19.

and science are enemies.["12] The son of a Methodist minister, Draper similarly sought to preserve a purified religion by "always transposing the formulas of Christianity into the key of materialism."[13] Despite being a Deist (that is, believing in a remote God who started the show and then left it well alone), like his friend Tyndall there was a peculiar strain of pantheism in Draper as he "emptied God out upon 'nature'"[14] and gave to nature the attributes and prerogatives that the theologians would previously attribute to the Almighty.

We encounter again a now familiar paradox: methodological naturalism, historically speaking, was making its emergence upon the back of theology and God's providence, and not in opposition to it. In some sense it represents a movement of thought where the Christian doctrine of providence became conscripted to do the heavy lifting as "human progress,"[15] or "the invisible hand of the market," or evolution, and the like. While Draper's particular flavor of this strategy will not find much sympathy amongst conservative Christians, nonetheless historically it is indispensable to once again remind ourselves that neither naturalism nor evolution were initially or even essentially about denying theology or exulting in godlessness, though to be sure they could be turned to those purposes. Nature herself was providential for Draper, urging humankind through a necessary sequence of evolutionary stages as its invisible hand guided it. But this was not meant impiously as a replacement for God, but rather as "the only view of the world [to Draper's mind] compatible with the 'wisdom' and 'nobility' of God."[16] Where White's work found its genesis as he was trying to free himself from his overbearing Protestant contemporaries, Draper's anger was specifically with the Roman Catholicism of his day. This burned especially hot because of the then-current and quite totalitarian Pope Pius IX. It is only a slight exaggeration to say that Draper's picture of the history of science and religion came about as he "project[ed] his condemnation backwards" from his own time, seeing "the whole [of] religion in the image of Pius IX."[17]

The specific limitations of both the origins and the targets of Draper's polemic faded in the memory of the next generation's reading—or hearing about—Draper's ideas. In part as we will see toward the end of this chapter, as with White, this came from a more radically secular wing of historians who picked up the story and trimmed it of Draper's rationalized religious ornamentation. Positivism in its various forms had become a key tenet of many avenues of popular science communication—newspapers, magazines, journals—and so Draper and White's stories were cheerfully spread with various degrees of trimming regarding their broader goals of harmonizing science

---

12. White, *A History of the Warfare*, 1:410.

13. Fleming, *John William Draper*, 1.

14. Fleming, *John William Draper*, 80.

15. Löwith, *Meaning in History*.

16. Fleming, *John William Draper*, 80.

17. J. B. Russell, *Inventing the Flat Earth,* 40.

and religion. Much like White, however, Draper did himself no favors if his intention was truly to foster a certain kind of harmony between religion and science. Because he "came to terms with all of history by an exaggerated response to his own day," this led in turn to his works—despite often referring to Catholicism specifically—to bury this theme underneath the impression that all of religion was under the secular inquisitor's eye. "It had" in other words, "the double effect of making his temper profoundly unhistorical and his book a quarry for the historian of 19th century ideas."[18]

His book on the conflict of science and Christianity in fact opens with such conundrums. On the one hand he states quite clearly that "In speaking of Christianity, [as such] reference is made to the *Roman Church*, partly because its adherents compose the majority of Christendom, partly because its demands are the most pretentious, and partly because it has commonly sought to enforce those demands by the civil power."[19] Draper, unlike White, had quite a few nice things to say about Protestants. On the other hand, one is quickly dazzled by much more forceful condemnations, where "the history of science is not a mere record of isolated discoveries," but rather it is "the conflict of two contending powers, the expansive force of the human intellect on one side, and the compression arising from traditionary faith and human interest on the other."[20] Indeed, Draper goes on a few pages later to say "faith is *in its nature* stationary; Science is in its nature progressive; and eventually a divergence between them, impossible to conceal, must take place."[21] One recalls the Tertullian quote from the last chapter (which we can also remind ourselves Draper accused of damaging European engagement with the sciences for nearly two millenia). Taking with the left hand what he gave with the right, Draper's narrative quickly spiraled out into what appears—even by a charitable reading—to be an attack on the nature of religion itself, despite protestations otherwise. And this is what branded itself into the collective memory of generations after.

Further complicating matters, Draper consistently—and not just in his *Conflict* but also in his work at large—appeared allergic to the use of footnotes to refer to his sources. Whereas White obscured by the shock and awe of (often misused) overabundance, one is often left quite flummoxed at who, exactly, Draper was reading. This further lent itself to readers glossing over the fact that Draper was projecting his particularly time-bound beefs with Catholics backwards as a grid to read the historical record at large. His story appeared, for lack of a better word, as the "natural" course of things. Draper "essentialized" the nature of contemporary Catholicism, seeing their actions as springing up from the very nature of what they were as an institution. By arguing that the "endangering of the [Catholic] position had been mainly brought about

18. Fleming, *John William Draper*, 78.

19. Draper, *History of the Conflict Between Religion and Science*, x–xi. Italics on "Roman Church" have been added for emphasis, and are not present in Draper's original text.

20. Draper, *Conflict*, vi.

21. Draper, *Conflict*, vii.

by the progress of science," Draper not only made hash of the historical evidence, he ignored the state of political affairs that led the Church to take the positions it did in Draper's day.[22] Indeed, before we turn back to Draper's story, and its spread through Edward Youmans and other positivist publishers, we must turn back the clock to see why the Catholic Church was on edge in Draper's time. Much as White's narratives can only be understood against the background of the general Enlightenment narratives of progress and the organization of knowledge and reason, along with the university conflicts of White's own day—so too, Draper's animus against Catholicism can only be understood by taking a running start with—surprisingly enough—Napoleon Bonaparte. For with Bonaparte we have a figure who set off a series of historical cataclysms in politics and religion—particularly among Catholics—that would set the tone of the Roman Church even into the twentieth century. We will see again that as soon as one peers beneath the veil of the genesis of the warfare narrative, an entire labyrinth of political, economic, institutional, and personal intrigue appears.

Napoleon Bonaparte and the Pope Who Became Infallible

At the beginning of the nineteenth century, northbound caravans, shrouded by storms, moved with purpose as they dared to pick the teeth of the Alps through the Mont-Cenis pass between Italy and France. Menacing the air, these great Alpine jaws of the world ground rime from the stars above against rock from the great mouth of the earth below. And between, shadows of men passed between this chew of air and earth and ice with ancient manuscripts in tow, precious works of art, and artifacts from times long past. Above, a knowing gaze spilled down over this procession. Upon his stallion covered in heavy wool and armor, leering out from underneath one of the shadows cast by these great toothy palls of earthwork and stone, the frost-shorn silhouette of a young Napoleon Bonaparte considered his spoils of war. They moved steadily along the road he himself had ordered built through the passes. Far behind him, Rome trembled, growling prayers heavenward after this spoliation, led by "the ogre of the Tiber" Pope Pius VI (in office 1775–1799). All of Christendom was now felt to be moving.

Forty-hour prayer vigils were the order of the day, and in the Vatican they rattled and buckled beneath the winds of change. The only answer received was further mortification, however, as the pope was forced to sue for a peace that culminated in the Treaty of Tolentino (February 19, 1797). Adding insult to injury, the treaty forced Rome to renounce its claims to Avignon along with all of its lucrative holdings in northern Italy. Further, as the historian Thomas Albert Howard writes, Rome had to pay "30 million livres . . . and [among many other conditions] to hand over numerous works

---

22. Moore, *The Post-Darwinian Controversies*, 27. Cf. also Hasler, *How the Pope Became Infallible*, 61f.

of art and pay for their transport to Paris."[23] Ironically, the livre was a unit of currency implemented originally by Charlemagne to help unite the territories of Christendom under his command in the Carolingian empire. Now, Napoleon's victorious procession was being financed by livres to dismember Christendom (both literally and figuratively). Apart from the obvious value of the currency involved, the symbolism was perhaps not lost on Bonaparte. Nor was this limited to the Vatican. Across central and northern Italy the Church's belongings were systematically stripped, artwork was taken, shrines robbed, and statues of Mary moved under Napoleon's droll observation that "we are [now the] masters of our Lady of Loreto."[24]

Then in 1810 the order came: the whole archive of the pope—the largest in all of Europe—was to be sent to France. A giant train of wagons moved again through the Alps from Rome to Paris. Costing 600,000 francs, the operation in the end brought 3,239 chests full of 102,435 "registers, volumes . . . bundles"[25] moving their secrets both banal and supernal to where a new world was to be founded. While this operation was overall a success, Owen Chadwick notes several chests full of documents were lost, including two carts worth when flood waters at Borgo San Donnino took them.[26] Among these documents were some of Leonardo da Vinci's lost notebooks, from which Duhem would attempt to rewrite the anti-theological origins of science.

This had not been the first time the papal library in the Vatican had been raided. Originally commissioned by Pope Sixtus IV in 1475 because the printing press had exponentially increased the number of papers associated with the business of managing the great machine that was the Church, it instantly became the best stocked library in Italy. Enough of an object of fascination even to the uneducated, when the Emperor's army sacked Rome in 1527 soldiers looted the archives mainly in an attempt to turn the parchment into bedding for their horses, and transform the many seals into material for making bullets.[27] In this earlier raid "the soldiery had no idea of the value of what they tore or burnt."[28] Recollections still exist of great masses of manuscripts simply littering the streets of the Vatican, having been abandoned as the soldiers moved on. This latest raid, however, was different and the value of things hidden in the depths of the Vatican archives were precisely the reason for the incursion by Napoleon. Humiliated, the ramshackle papal forces were routed, and now the memories of the See of Rome moved from their homestead in boxes battling snows under the glacial roof of God's world. Many of these artifacts were later to be placed at the *Biblioteque Nationale* at Paris as symbols not just of Napoleon's victory, but as one French general put it these items represented the coming of the modern age:

23. Howard, *The Pope and the Professor,* 16.

24. Howard, *The Pope and the Professor,* 16.

25. Chadwick, *Catholicism and History,* 15.

26. Chadwick, *Catholicism and History,* 15.

27. Chadwick, *Catholicism and History,* 5.

28. Chadwick, *Catholicism and History,* 5.

"[art] which the French have taken from the degenerate Roman Catholic [is] to adorn the museum of Paris, and to distinguish by the most notable trophies, the triumph of liberty over tyranny, and of philosophy over superstition."[29] On the horizon, then, rumbled the coming rise of the modern state, a great mess of Catholic anxiety due to trauma, and indeed a forgotten but major catalyst for what would eventually be known as the warfare of science and religion.

The traumatized Roman Catholics saw things differently than the French revolutionists, of course. Soon these "doctrinaire [French] cannibals were running around, catalogues at the ready, in museums and galleries and libraries," recalls one in the papal court. But this was only the latest in a recent series of humiliations. With France on the brink of bankruptcy, in the fall of 1789 the newly gathered National Assembly made the decision to confiscate church lands by force for resale. As Howard notes, the "sale of lands started the next month and continued years thereafter, saving the treasury of France and financing an increasingly anticlerical Revolution."[30] Then, almost overnight in February of 1790 monasteries and convents were closed—with the exception of a few who were doing work others cared not to, such as attending to the poor and maintaining certain hospitals. Monks who forsook their vows were promised a monetary reward in the form of a pension. On July 12, 1790 the Civil Constitution of the Clergy was invoked, abolishing fifty ecclesiastical sees, and pressing the rest into service of eighty-three newly created political offices. Four thousand priests were instantly dismissed, while other bishops and priests were forced to become civil servants (or, as they might prefer to call it under their breath, puppets of the state) if they were to continue their religious duties. An oath of loyalty to the state above all and an acceptance of the new Constitution was likewise mandatory. Depending upon one's perspective, in either a staggering act of stubbornness in the face of progress or an instance of great courage against tyranny, only seven bishops and roughly half the clergy agreed to the oath. The others declined and faced the consequences of their decision, often with imprisonment, often with death.

It did not end there. The Gregorian calendar was replaced by a French republican one, with the feasts of saints expunged and swapped for figures important to Republican ideals; street names changed; religious references were systematically eradicated; anti-clerical parades were held, and many of the historic churches and monasteries— including the famous Abbey of Cluny, and the Church of St. Genevíeve where Descartes was buried—were destroyed, their ruins repurposed for other building projects. As one might suspect this left a series of increasingly traumatic memories among Roman Catholics, "bitter and enduring memories that shaped Rome's attitude toward what we today generally call 'modernity.'"[31] Many in support of the French revolution saw these "refractory" priests and bishops who declined the oath as seditious elements

29. Howard, *The Pope and the Professor*, 18.

30. Howard, *The Pope and the Professor*, 20.

31. Howard, *The Pope and the Professor*, 18.

in need of purging, especially in the light of increasing counterrevolutionary forces in France. In addition, the pope had released a vicious denunciation of the oath to all Catholics in his encyclical *Quod aliquantum*. Calling the oaths an explicit attempt "to annihilate the Catholic religion, and, with her, the obedience owed to kings" this spurred a bitter anti-state sentiment among Catholics that would result in the nineteenth century's rabid ultramontanism. Ultramontanism was a position that emphasized a centralized power for the pope that cuts across all secular politics and borders—a view that one will not be surprised to learn did little to ingratiate Catholics to their already irate opponents. Ironically, while these measures of the French Revolution were meant to either stamp out or cow the Church, the trauma caused by these very measures would fuel Catholic revivalism the next century. This legacy of Napoleon and the French Revolution would also simultaneously fuel the idea of the warfare of science and religion though on the surface it initially appears to have nothing to do with it. Driving forward *philosophe* and Enlightenment narratives of Catholics hellbent to retard the growth of progress that we glimpsed at in the last chapter, while simultaneously retrenching Catholic reactions and reactions against Catholic reactions, thunderheads of a quite different war were gathering. And upon this storm rode a young Pio Nono—later to become the infamous Pope Pius IX.

Amidst the calamity arising in France, things grew particularly bad when a quarrel erupted between papal loyalists and the republican revolutionaries. After one of the revolutionaries was killed by an errant bullet—a general, no less—retaliation was demanded and Napoleon happily obliged. After Rome itself was seized, Pope Pius VI was given three days to vacate. After the aged pope asked to be left to die in peace in Rome, the response was that "one can die anywhere."[32] Howard writes:

> The image of the frail Pope being escorted from Rome on an uncertain journey to his likely death proved jarring and enduring. In the short term, it helped inspire the counter-revolutionary Sanfedist (Holy Faith) uprising—an armed, peasant crusade . . . In the long term, the Pope's plight helped awaken European-wide sympathy for the papacy—as would the kidnapping of [Pius VI's] successor, Pius VII in 1809—and thereby assisted in shaping the ultramontane imagination.[33]

As the procession of the aging pope passed towns along the way, many both curious and pious came out to see the spectacle, and word continued to spread about the injustice the pope was facing at the hands of his enemies. At a dilapidated *Hôtel du Gouvernment*, too sick to continue his journey to the north, the pope died, reportedly praying for his enemies and sealing a wave of pro-Catholicism throughout European hearts.[34] Despite Napoleon realizing that his past anti-clerical policies were destabiliz-

32. Howard, *The Pope and the Professor*, 22.
33. Howard, *The Pope and the Professor*, 22.
34. Howard, *The Pope and the Professor*, 23.

ing politically, his attempts at a what was for a time a peaceful coexistence with the next pope, Pius VII, rapidly degenerated when Napoleon reached the height of his power in 1805. As was already hinted, when the pope refused to break neutrality in terms of Napoleon's new quest for conquering, Napoleon dealt with his feelings on the matter by again kidnapping a pope. In 1808 Napoleon sieged Rome and annexed the Papal States and began what one French historian has called their long "war against God"[35] as the French began systematically dismantling and reorganizing the Church. On June 9, 1812, under the cover of darkness and a particularly underwhelming cover story of welcoming an honored guest, Napoleon had the pope forcibly removed from Rome to Fontainbleau. As Napoleon's last days approached, he restored the pope to power. Too little, too late. The damage had been done, and a Catholic fury at yet another manhandling by secular states burned into their core. When the pope returned to Rome, he found a statue had been erected depicting him sitting imperious on the papal throne, with Napoleon naked and beggarly at his feet. If one were to zoom in to the crowd in its ecstasy at the return of Pius VII, one could find the figure of a young boy enthralled by the ceremony and no doubt saturated by its meaning for Catholics against the secular encroachment they had undergone. This boy was named Giovanni Feretti, and he would later become Draper's nemesis, Pope Pius IX.

Having had two successive popes kidnapped, along with the forcible closing of innumerable monasteries and schools, and the wholesale of vast acreages of Church lands, the notion that the papacy must also maintain some form of temporal power to protect its spiritual mission reasserted itself. It helped that the European monarchies in general considered the pope to be one of them, and a fellow victim of Napoleon. As such the temporal and the spiritual found themselves once again in alignment—however temporary. As historian Eamon Duffy put it, "the restoration of the Papal States is the single most important fact of the nineteenth-century papacy . . . If the Pope did not remain a temporal king, then it seemed he could no longer remain the Church's chief bishop. That perception colored the response of all the nineteenth-century popes to the modern world."[36] Indeed, this intellectual legacy is extremely important to our immediate purposes of recounting the genesis of the warfare narrative:

> These lessons helped fashion a thoroughly anti-modern synoptic through which the nineteenth-century papacy viewed the emerging "modern age." In light of the despoliation and persecution that the Church had experienced, it should be emphasized, these were far from unreasonable conclusions to draw. Accommodation with modernity simply did not appear like a workable option; it was the task of the Church . . . to double down; to pit papal supremacy against the revolutionary era's challenge to authority, to defend the temporal powers of the papacy as a safeguard for its spiritual function, and to promote constant vigilance against ideas rooted in the eighteenth-century

35. Broers, *The Politics of Religion*.
36. Duffy, *Saints and Sinners*, 272–73.

Enlightenment and its offspring, "the godless [French] Revolution." Such trains of thought helped turn the trauma of the Revolution into the combative ultramontanism of the nineteenth century.[37]

And yet, even with all of this conservative backlash, Pius IX was initially selected as a moderate, and his first acts as pope made "onlookers wonder if the world was witnessing the birth of a liberal pontificate."[38] He gave amnesty to former revolutionaries, set up a commission to allow railroads built in the papal states, an agricultural institution for improving crop production, asked for pontifical subjects to attend scientific congresses, and relaxed censorship laws. All of this was short lived, however. In 1848, Italian emancipationist demands for the expulsion of the Austrian monarchy from Northern Italy reached a near manic state. Crowds of people were calling for Pius to use his power to summon an army and unburden Italy. This would, however, turn the pope against what was in essence a Catholic monarchy, in effect destabilizing all of the rebuilding efforts since 1815. Thus, Pius gave a speech in which he stated that under no conditions would he go to war, thereby taking the same political line as his many predecessors. All hell broke loose.

Overnight, Pius IX went from a newly beloved pope to the most hated man in Italy. Mobs formed, with protestors stabbing Pius's Prime Minister Pellegrino Rossi. One of Pius's secretaries—while standing next to the pope at a window in the papal palace no less—was shot and killed. Pius fled Rome disguised as a normal priest, and his ultramontane supporters likened this to what was in effect a third papal kidnapping. All the trauma of past events again awoke. As the pope departed, Rome drafted a constitution for a new "Roman Republic" declaring the pope's temporal powers had ended while electing a ruling Triumvirate in his place. The pope could return, they said, but only as a spiritual figurehead. Never again as a political power in Rome. Given how extremely reduced political power went for the pope's spiritual mission over the last century—what with all the kidnappings and pillaging—this was unacceptable to Pius IX. In his self-imposed exile, the pope had now to choose between liberalizing his stances further, admitting defeat by "bowing to Babylon" and assuming the limited role the Triumvirate dictated for him. Or, he could condemn the new republic, finding—somehow, some way—avenues to rally Catholic Europe to his cause. For Pius, the choice may not have been easy, but it was decisive. In Howard's terse summary: "He chose condemnation."[39]

The intricacies of what happened next need not detain us. But needless to say, Pius IX had now been radicalized through these events into a conservativism that would echo through all of Catholicism—indeed all of Europe—well into even the next century. "His rule," writes the historian August Hasler, "became reactionary and

---

37. Howard, *The Pope and the Professor*, 31.

38. Howard, *The Pope and the Professor*, 48.

39. Howard, *The Pope and the Professor*, 51.

dictatorial."[40] In 1864 Pius began distribution of the Catholic encyclical *Quanta cura* that had appended to it the now infamous "Syllabus of Errors." These errors included such things as condemning those who demanded that public institutions be free from the oversight of the Church (Prop. 47), or that the Roman pontiff "can and ought to reconcile and harmonize himself with progress, with liberalism, and with modern civilization" (Prop. 80). *Quanta cura* shocked European and American intellectuals of all religious persuasions, but it absolutely drove John Draper mad. Perhaps most damning of all in Draper's eyes was the fact that all of these pronouncements were accompanied also by the denunciation of any who would say of these "Decrees of the Apostolic See and of Roman Congregations," that they "interfere with the free progress of science" (Prop. 13). In other words, not only did Draper see in the encyclical some of his most cherished views being assaulted, according to *Quanta cura* Draper would be holding an "evil opinion" for even thinking that these pronouncements were counterproductive to science. This, as one might imagine, did not sit well.

The encyclical equally unnerved many Catholics, though they took solace in the fact that a General Council of the church was called right after it was issued.[41] The hope was that the Council would create space for interpreting the propositions in a softer light by producing "moderating statements." Fanning this slender flame of hope was the fact that regardless of the bellicose tone of the appended Syllabus of Errors, one could interpret Pius IX as saying that, ultimately, he wanted the Catholic Church to foster the "free progress of science," precisely by these declarations. After all, far from being antagonistic to science, Christian thought and practice incubated and helped establish historically what we now consider to be scientific inquiry. From this angle, the attempt to uphold Catholic authority would also be seen as an attempt to perpetuate the conditions under which science first arose. This way of reading the Syllabus, however, is a bit of a stretch even for an interpretation attempting to maintain some charity, and amelioration of the Syllabus' harshness by the Council "was not to be."[42]

The shift to a militant conservativism also encouraged the return of many "medieval" institutions. Already in the wake of Napoleon the Inquisition and the Index of Forbidden Books (among whose lists the Spanish version of Draper's *Conflict* would later be placed) found themselves reborn, to name but two. But, perhaps most fateful of all, the notion of papal infallibility was put back on the table for discussion by the work of theologian Joseph du Maistre. What is involved in the notion of infallibility is more nuanced than its critics often give credit (though it also tends to die the death of a thousand qualifications). But in essence infallibility states that when the pope is speaking in a mode known as *ex cathedra* (from the throne of Christ), his declarations are both true and irreversible—even by later popes. Though the nuances of the position therefore shy away from the position that everything the pope says is

40. Hasler, *How the Pope Became Infallible*, 109.

41. Moore, *The Post-Darwinian Controversies*, 25.

42. Moore, *The Post-Darwinian Controversies*, 25.

infallible, nonetheless the doctrine even in its more limited form has disturbing and far-reaching consequences. The infallibility of the papacy was the key, du Maistre had argued during the time of Napoleon, to lock down its invincibility in an environment when the wounded Church was circling its wagons. Du Maistre in a single breath drew together the themes that marked the age: "[There can be] no public morals nor national character without religion, no European religion without Christianity, no true Christianity without Catholicism, no Catholicism without the Pope, no Pope without the supremacy which belongs to him."[43]

When du Maistre originally proposed his theology of infallibility, Rome even in the midst of Napoleon "took a rather dim view" of it, not least of which because—perhaps surprising to many—it had almost no historical precedent.[44] Indeed, unlike many notions such as the Trinity which were in Scripture in latent ways that needed to be brought forth and clarified through the first centuries of the church, infallibility as an idea sprang to life all at once and as a whole in the thirteenth century by the Franciscan theologian Peter Olivi.[45] Pope John XXII (1316–1344) however, had no patience about talk of his own infallibility, and called the doctrine "the work of the Devil himself."[46] Ironically enough this was not born out of a place of humility, but because Pope John felt infallibility—which implied the irreversibility of certain decisions once made—impinged upon his rights of judgement as a sovereign. As infallible, John would be bound not just by his own declarations, but by those of past popes now retroactively christened with this strange glory.

And yet, in the newly hardened mind of Pius IX, infallibility rang sweet to his ears like a chorus of angels. In some sense we can take this chorus quite literally. Struck with epilepsy from youth, Pius had always had a large mystical streak that was in this time of crisis ignited amongst himself and his circle. They were all enraptured by visions of Mary coaxing them along their course. He and his closest advisers sought out allies to pass the doctrine at a general council, and so put into play a dogma that would bend history to its will. In fact, it is only recently, when Swiss Catholic historian August Bernard Hasler served for five years in the Vatican Secretariat for Christian Unity and was given access to the documents stored in the Vatican archives, that the intrigues of Pius IX as an individual maneuvering in the papal court were fully understood. Hasler writes in the foreword to the 1978 edition of his *How the Pope Became Infallible*: "Until very recently there had been no historical study of the way the solemn definition of papal infallibility came about, and why this happened precisely in 1870. . . . The opening up of numerous archives and the publication of several historical studies have altered our idea of events which led to the dogma of infallibility."[47] A

---

43. Quoted in McClory, *Power and the Papacy,* 52.

44. Hasler, *How the Pope Became Infallible,* 42.

45. Hasler, *How the Pope Became Infallible,* 36; McClory, *Power and the Papacy.*

46. Hasler, *How the Pope Became Infallible,* 61.

47. Hasler, *How the Pope Became Infallible.*

century of trauma was now playing out in the Catholic Church, pumping through the veins of Pius IX and driving forward an odious decree. At the First Vatican Council starting in 1869, papal infallibility was declared Catholic dogma.

Some of the more grotesque maneuvering at the council must be passed over.[48] To illustrate how far Pius IX had changed from the beginning of his papacy, however, one incident will suffice. The debates against infallibility leading up to the council were severe, with many Catholic theologians and historians—most notably one of the most prestigious historians in Europe at the time, Ignaz von Döllinger—desperately pleaded that neither history nor sound theology could accommodate infallibility. A vocal minority at the Vatican Council itself continued to insist upon this. The Dominican Cardinal Fillipo Maria Guidi, for example, stepped forward in front of those assembled and passionately argued that the pope could not be infallible in himself, but only insofar as the pope's pronouncement's embodied "the view of the Bishops and the tradition of the church." Later that evening Pius IX had the Cardinal called in, and personally chastised him with a line that one might expect reserved for a movie villain: "*I* am tradition. *I* am the church."[49] It was thus in July 1870 with the Vatican Council's First Dogmatic Constitution on the Church of Christ, *Pastor Aeternus* declared "the Roman Pontiff, when he speaks *ex cathedra*, . . . is possessed of that infallibility with which the divine Redeemer willed that His Church should be endowed in defining doctrine or morals."

The Historian and the Pope

With the so-called "Syllabus of Errors," and then the declaration of papal infallibility, one can see how Draper—and others—felt they came by their negative opinions of Catholicism honestly. Yet, Draper cannot be let off the hook for his blatant generalization of Pius IX's temperament as the historically Catholic or Christian mind-set. Draper was, for example, an acquaintance of Catholic historian Ignaz von Döllinger, and records having a lengthy conversation with him on Christmas Day, 1870, just five months after the declaration of infallibility. This is significant because Döllinger's own "vociferous denial of papal infallibility led to his excommunication three months later."[50] Draper's equation of Pius IX's draconian attitude with the totality of Christianity, past and present, as such represents what can only be judged as the total failure of Draper as historian. By arguing that the "endangering of the [Catholic] position had been mainly brought about by the progress of science," Draper not only made hash of the historical evidence, he ignored the state of Italian politics at the time, which as

48. See Hasler, *How the Pope Became Infallible*, 129–46.

49. Hasler, *How the Pope Became Infallible*, 89–91. One thinks of Palpatine in *Star Wars* proclaiming "I am the senate!"

50. Moore, *The Post-Darwinian Controversies*, 27; Hasler, *How the Pope Became Infallible*, 61f; and especially the narrative of Howard, *The Pope and the Professor*.

we have seen surrounded Pius and compressed him like an unyielding mold into his later, unhinged state.

Draper's biographer, Fleming, observes that even before papal infallibility Draper's earlier work *A History of the Intellectual Development of Europe* anticipates nearly all the themes of his later *Conflict*: "The emotional climax of [Draper's] book is the vain effort of Roman Catholicism to hold back the universal dominion of the scientific spirit. The popes appear as the heads of an enormous bureaucracy tyrannizing over the minds of men, and sacrificing the advance of reason and science to the cause of continued faith in the supernatural." The papal pronouncements, as one might imagine, did nothing to quiet Draper's anxiety regarding authoritarian figures. Draper doubled down on this approach in his later work *History of the Conflict Between Religion and Science* and makes no apologies about its extremity:

> It has not been necessary to pay much regard to more moderate or intermediate opinions for, though they may be intrinsically of great value, in conflicts of this kind it is not with the moderates but with the extremists that the impartial reader is mainly concerned. Their movements determine the issue . . . In speaking of Christianity, [as such] reference is made to the Roman Church, partly because its adherents compose the majority of Christendom, partly because its demands are the most pretentious, and partly because it has commonly sought to enforce those demands by the civil power.[51]

Reasoned voices in history are here drowned out by the crazed shouts of zealots—and this is a feature, not a bug, of Draper's method. The import of this, however, is not Draper's problems with Catholicism—as anti-Catholicism was rampant.[52] Rather, what we should pay attention to is that he rewrote history to achieve his ends. "The chapters dealing with earlier history are best approached as a kind of backward extension of this conflict [with the Church under Pius IX]."[53] Thus, where the papacy reinvented history through dogma to ratify infallibility, Draper countered with his own reimagined history. Or, put somewhat more ironically: Pius IX wanted all of Christian history that had come before him to be stamped with his image. Draper merely ensured that the pope's wishes came true. And so, on both sides, history was purified of its intricacies and presented as the clash of homogenous, titanic powers, as if they were the gears of time itself. Ultimately, says Draper, the Church and Science are "absolutely incompatible; they cannot exist together; one must yield to the other; mankind must make its choice—it cannot have both."[54]

51. Draper, *Conflict*, x–xi.

52. On anti-catholicism and its effects, see: Jenkins, *The New Anti-Catholicism*, in particular 113–132 for how the media of this era played its part.

53. Fleming, *John William Draper*, 76.

54. Draper, *Conflict*, 67.

Ignored by Draper was the fact that the Syllabus of Errors that so vexed him was considered "a non-infallible classified index to pronouncements made in the encyclicals,"[55] and the fact that (easily discovered if Draper had cared to look) despite all of his flaws Pius IX was "not averse to employing the latest technological innovations."[56] Quite the opposite. Pio Nono (as he was called before his ascendance to papacy) seemed "almost anxious" to have technology like railways, electricity, aluminum, advancements in wine curating, photography, the water-glass, cement, medicine, and other innovations. In fact, more than the mere want of such advancement, each of these had come by way of a Catholic innovator.[57] Despite this, Draper anticipating the sociologist Max Weber and revealing his own indebtedness to positivism, assumed technology, and human advance was by its very nature toxic for theological dogma. His work in turn became as such

> A paean to science, a hymn, its mighty achievements, a catalogue not exhaustive, telescopes, balloons, diving bells, thermometer, barometer, schools, newspapers, hospitals, canals, sanitation, census reports, cotton-gin, medicines, manures, tractors, railways, telegraph, calculus, air pump, batteries, magnets, photographs, maps, sewing-machines, rifles, and warships . . . all beneficent discoveries of science, but not just science *simpliciter*, of science in contrast, and the contrast is with Catholicism. Draper never stopped to ask himself why anyone who invented a camera or possessed a barometer might be led to think his faith in the God of the Christians shaky.[58]

We might recall an earlier quote from Draper where he writes "the history of science is not a mere record of isolated discoveries," but rather it is "the conflict of two contending powers, the expansive force of the human intellect on one side, and the compression arising from traditionary faith and human interest on the other."[59] Draper's hymns to invention therefore appear as little more than list-keeping for those not beholden to the strange idea that an electric lamp also signals that somewhere there is a defeated religionist moping about in defeat. Yet notions such as this—we might even call it a dogma—were why Draper could at times merely list inventories of invention as evidence for his thesis. These lists were not just a pedantry of discoveries and technological marvels, but were reimagined by Draper as stones plotting the pathway out of a morass of human religious ignorance. These innovations of modernity embodied the clear-eyed essence of modernity itself. Draper no doubt picked up much of this from his positivist and utilitarian education at the newly opened University of London.

55. Moore, *The Post-Darwinian Controversies*, 27.

56. Moore, *The Post-Darwinian Controversies*, 28.

57. Moore, *The Post-Darwinian Controversies*, 28.

58. Chadwick, *The Secularization of the European Mind*, 162.

59. Draper, *Conflict*, vi.

One begins to see another layer of why Draper's animus against Catholicism and Pius IX was so intense. In the stark face of positivism, religious revival at such a level would be seen as a convulsion of history against itself. In the figure of the "infallible" pope was a man with the gall to stand up and throw his dogma down upon the gears of history in an attempt to jam them or perhaps even send them spinning backwards. For history is the inexorable journey of progress into the objective and scientific. Dogmatic religion, while useful for a time, is inherently a product of a past clinging—desperate in its wistfulness—to the shreds of a vanished youth. The myth of warfare as such was less about historical episodes of conflict than it was about a different philosophy of history that organized them, gave them sense, coaxed them to appear: for, given the assumptions of positivism *of course* the Church opposed Galileo. *Of course* humanity in its Christian youth believed the earth to be flat and in the center of the universe. *Of course* Tertullian would believe things because they are absurd. *Of course* Darwinism was now opposed. Close investigation of these events vanished before the zeal of what, given positivism, must have been so. In Draper's *Conflict*, the positivist circle again completed itself. War was allowed to arise not from historical investigation per se, but as a symbol of what history itself was thought to be.

The Fire Spreads: Edward L. Youmans

"What we want are ideas—large, organizing ideas."[60] So wrote Edward L. Youmans, science editor of D. Appleton & Co., to Draper, conscripting him to write a volume for his *International Scientific Series*—the book that would become *The History of the Conflict Between Religion and Science*. Draper delivered. However much actual history occurs in Draper's work, large, organizing ideas are certainly there—and in spades. Youmans, the founder of the magazine *Popular Science Monthly* and a disciple of X-Clubber and famed evolutionist Herbert Spencer,[61] was also a card-carrying positivist intent on delivering its message and promoting the new era of scientific objectivity mankind had achieved. In no uncertain tones, in several of his own publications Youmans speaks of the history of science as moving through inevitable stages—from the metaphysical into the physical. Youmans was personally responsible for publishing spreading abroad the work of Thomas Huxley, John Tyndall, Herbert Spencer, John William Draper, and Andrew Dickson White. Just as important as who he was publishing was when: in the wake of the Civil War print culture and publishing in America were undergoing a revolution, and Youmans was one of its leaders.[62] Establishing new international copyright agreements and capitalizing on the surging waves of interest in literature popularizing the sciences, both his *International Scientific Series* and *Popular*

---

60. Quoted in Moore, *The Post-Darwinian Controversies*, 21.

61. Lightman, "Spencer's American Disciples."

62. For some examples, see Cantor and Shuttleworth, eds,. *Science Serialized*.

Science Monthly were on the front lines priming the American scientific imagination at large starting in the 1870s.

If the X-Club and its friends—however influential—were working behind the scenes to plant their comrades in arms in university positions in a sort of concerted effort to outflank the Protestant and Catholic holds on universities, Youmans provided the full-frontal assault for this new positivist naturalism by way of publishing. We briefly spoke above that it was positivist readings of Darwin—rather than Darwin himself—that provided much of the fodder for the warfare thesis. This was due in no small part to those like Youmans who were constantly making the connection through print. Indeed, tasked with writing the obituary for Darwin, Youmans managed to sneak in a few promotional references to Draper. Darwin, the most scientific man in the world, and Draper "the most eminent man of science in America" were both "the most distinguished representatives of the same school of progressive scientific thought." Both will be "forever associated with that great revelation of ideas for which all modern science has prepared."[63] As with Huxley and the X-Club, Darwin and the warfare thesis were consciously linked together by its promoters. Yet, this is again more of a rhetorical maneuver of the record keepers than it is a strict account of "what really happened":

> [For those having actually read Draper's book] will have noticed that Darwin and the *Origin of Species* were not relevant to Draper's sense of conflict. Though he wrote in the full spate of argument over Darwin amid the decades when everyone . . . debated the conflict between Science and Religion, though he published one of his books four years after Darwin's *Origin*, and the other three years after Darwin's *Descent of Man* . . . [Draper] could have written both of his books equally well or ill . . . if Darwin had drowned in the cam [before publishing a word]. Men afterwards looked back at Darwin and found in him a symbol of the conflict, its center, perhaps even its source. We know of a Harrow schoolboy of the early 1880's who heard that "Darwin had disproved the Bible" and rearranged his faith accordingly. . . . This is bringing us near the heart of the problem over secularization. When we come down to the axioms which intelligent boys of fourteen years learn from less intelligent schoolboys of fifteen, we come near to the point where the cloudy apprehensions of what is known as intellectual history . . . can be shown to affect the attitudes of a whole society.[64]

From his youth, Youmans was stricken with two conditions whose friction against one another set him on his course: the first was a pathological, insatiable desire to know; the second was an actual pathology, opthalmia, which rendered him nearly blind. His sister was his constant companion in his quest for knowledge. Fatefully, after his eyesight had improved through therapy, this quest led her to take the young

---

63. Quoted in Ungureanu, "Youmans and the 'Peacemakers.'"
64. Chadwick, *The Secularization of the European Mind*, 164.

Edward to D. Appleton & Co.'s bookstore on Broadway in New York City. As they were perusing the books one day, William H. Appleton, the successor to his father Daniel Appletone's bookstore, introduced himself. This turned into a fateful encounter, and Appleton befriended the ambitious Youmans, their friendship eventually leading to Youmans spearheading the *International Scientific Series*, which was one of the most ambitious printing projects in America to date.

To further his publication efforts with the admirable goal of communicating science to the broader public, Appleton also supported Youmans when he founded the magazine *Popular Science Monthly*. Both the magazine and the *International* series were designed to bridge the divide between scientists and an interested public. In this manner, however, they also spread to the masses the notion of warfare, of which Youmans—through his association with Huxley, Spencer, Tyndall, Draper, and White—was a full-throated supporter. Like all of the people in our previous chapters, however much a history of warfare was supported, for Youmans this was meant as a vehicle to be used against *establishment* Christianity and doctrinal orthodoxy. Similar to his associates, Youmans believed that in pushing the warfare narrative through his various substantial channels he was actually a sort of John the Baptist crying in the wilderness for these scientific heavy-hitters he was promoting. The large-scale distribution he was enacting was viewed as preparing the way for an ultimate harmony between a (liberal Protestant definition of) religion, and science. "Youmans argued that Draper had demonstrated that the history of science was a long conflict with 'theological authority' between the agencies of 'intolerance and liberalization.' He claimed that Draper's [*Conflict*] was an answer to increasing tensions between Catholics and Protestants in the late nineteenth century, particularly with the ultramontane party of English Catholics."[65] As such, in the environment which radicalized Pius IX and his ultramontane supporters, Youmans's proselytizing seemed all the more plausible.

Yet such historically specific contexts, and the attempt at precision when critiquing not religion but dogmatic theology were all nuances quite easily lost or discarded by the posterity that later picked them up. Youmans's pretense at nuance was swiftly dissolved by his own double-faced treatment of the subject matter, as when he insisted that the conflict between dogmatic religion and science was "natural and inevitable." As one might suspect, coming from someone who was not only viewed at large as an interpreter of science for the public, but also held the reins curating the list of publications for mainstream distribution that included some of the most ardent proponents of war, it is not too surprising that harmonies quickly shifted into shrill warsong. Despite Youmans's ultimate intentions—he published a good deal of theologians who generally agreed with his liberal Protestant sensibilities[66]—when he writes one is left only with the battle cries and hardly with his soothing asides.

---

65. Ungureanu, "Youmans and the 'Peacemakers.'"
66. Ungureanu, "Youmans and the 'Peacemakers.'"

The "fact" is, he wrote for example, that "the history of Science [note again the capitalization!] has been throughout a struggle with the theologians, and that the Bible has been used by devout believers in its infallible inspiration to crush out the results of scientific history." Indeed "bigotry and superstition still offer too vigorous a resistance to the advance of rational inquiry to make it desirable that we should forget the painful lessons of the past." Total war seems back on the table for the eager reading public to place in their minds and in their hearts. It doesn't help that the title of this screed given by Youmans was "The Conflict of Ages."[67] Neither did it help that the so-called "Third Great Religious Awakening" had entered full force, sprinkling religious zealotry across Britain and the United States. However anecdotal it might be, few things give more plausibility to a stereotype than your crazy neighbor appearing to do exactly what the stereotype predicted. Though, as Robert Lightman correctly points out, after 1880 the *International Scientific Series* began to take a less naturalistic direction, in terms of priming the reception of the warfare thesis the damage had already been done, and as Lightman also states the first decade of the massively distributed series while being run by Youmans, Tyndall, and Spencer really did embody and expand the same ethos in its communication as Draper's *Conflict*, even if everyone wanted to avoid the charge of atheism.[68] In the end, concludes historian Leslie Howsam in her fascinating study on the publication history of Youmans's *International Scientific Series*: "[Youmans's] series" evangelized the public at large with a "vision of modern, secular science."[69] Every household in America lining its shelves with the bright future of science would now also be filled with the warsong of a deep dark past. Our point, to reiterate, is not that this science was therefore poor—it was for the most part excellent. It is rather to point out that contained in the background of this "neutral, secular" definition, communication, and popular familiarization with science is the thoroughly incorrect presupposition of historical warfare with religion—specifically Christianity—which gave the host of new developments a sort of unity and excitement it otherwise would not have had. With "the deletion of religious references," and the rise of practices like the "depersonalization of scientific rhetoric," science established itself in the emerging popular literature more and more as a universal discipline that had escaped its roots. Science, in other words, mimicked the pathway of the American Dream, a dream of reason born in the Enlightenment, and paved with a hall of saints and heroes.[70] However much the science was well represented, the publications were often filled only with the movements of shadowy figures that resembled little of their actual historical counterparts. As such, every scientific finding gave one the impression explicitly or otherwise that it was yet another nail in the coffin of traditional Christianity.

67. Youmans, "Editor's Table."
68. Lightman, "The International Scientific Series."
69. Howsam, "An Experiment With Science."
70. Numbers, *Science and Christianity*, 132–33.

Part Three

Legendarium

When we come down to the axioms which intelligent boys of fourteen years learn from less intelligent schoolboys of fifteen, we come near to the point where the cloudy apprehensions of what is known as intellectual history . . . can be shown to affect the attitudes of a whole society.

—Owen Chadwick, *The Secularization of the European Mind*

The problem [facing us] *is that the revision away from the reason-versus-religion thesis has arguably included insufficient reassessment of the historical record from which historians (at least in part) make their generalizations.*

—S. J. Barnett, *The Enlightenment and Religion*

—Chapter 7—

Flat Earths and Fake Footnotes

In fourteen hundred and ninety-two,
Columbus sailed the ocean blue.
He took three ships with him, too,
And called aboard his faithful crew.
Mighty, strong, and brave was he
As he sailed across the open sea.
Some people still thought the world was flat!
Can you even imagine that?

– A traditional children's poem

MOST OF US WILL remember quite vividly the first representation of Christopher Columbus that we were taught; like the glimmering icons of the Orthodox Church, his portrait was given to us less a man, and more an ideal. He was a thin veil through which a portentous light of Enlightenment found quite early access to key into the world. Against the stubbornness of a backwards society, Columbus was a man, so it was said, singularly convinced of the world's rotundity, and to settle this geometric bet he set sail. Behind his lambent form was the pitch abyss of a thousand years of medievalism, ignorant or deceived by the priests' wooden reading of "the four corners of the world" in Scripture. It was from this morass that Columbus strode forward resolute, now tinged with the sort of heroic loneliness American mythology devours. "If no one ever challenged the status quo, the earth would still be flat" as a 2013 Infiniti car commercial put it—making it sound like Columbus literally remolded the earth into a sphere, despite popular taste preferring to take their planets flat. It was a breathtaking tale; both because of his resolution in the face of such apparently astounding religious ignorance, and because no doubt at least a few of our young minds fascinated upon the uncanny image of a horizon line of waters churning over that last, vital edge.

Nor was this some idiosyncrasy of our educations. The flat earth has been a convenient staple invoked in order to emphasize how humanity has advanced out of an age of superstition and religious ignorance for a while now. The historian of science Lawrence Principe records that over the course of a decade nearly 70 percent of his students—mainly American—were taught in grade school that Columbus set sail to prove the world was round.[1] Another historian, Michael Newton Keas, notes that for the last quarter century he has taught around 1,200 American students, a majority of whom by show of hands believed that medievals were ignorant regarding the roundness of the earth until Columbus.[2] While he admits his method is anecdotal, broader evidence would suggest that his experience is, sadly, statistically typical. A best-selling history of science text mentions the flat earth in order to demonstrate how curiosity about the natural world was replaced by fear and bizarre superstition in the Middle Ages. A main culprit in this debacle of the human spirit was the Church, which "redirected the worries of 'educated' people toward abstract theological questions," so that knowledge of nature "was considered superfluous and dangerous." The condition of astronomical knowledge regressed from that established by the Greeks for seven hundred years, and "the Earth was once again considered to be flat!"[3] This observation regarding the flat earth as symbolic of the general decline of knowledge caused by Christians is emblematic of its broader use. The Church set science back centuries, writes Timothy Ferris, and "the proud earth was hammered flat; likewise shimmering in the sun" while the heavens were wheels, pushed by angels in the courses of their perihelion.[4]

Unfortunately Christians have imbibed these claims, and as David Kinnamin and Gabe Lyons reported in their book *Unchristian: What a New Generation Really Thinks About Christianity* the activities of a church listing five things Christians needed to apologize for—"We're Sorry For Saying the Earth Is Flat," being among them.[5] Of all the things Christians should actually apologize for, this one is both entirely wrong and—even if true—seems fairly superfluous given other atrocities supposedly done in Christ's name. It does go to show in an exemplary way, however, how the power of myth can affect even the groups who should be the least inclined to believe them.

Nonetheless, in the name of fairness, it is hard to expect many to know much different when this sort of thing has been part of education for a long while. A widely distributed textbook for middle schoolers, for example, notes that: "Many Europeans still believed the world was flat. Columbus, they thought, would fall off the earth."[6] And a popular fifth-grade text from around the same time repeats this nearly verbatim: "The

1. Principe, "Transmuting History," figure at 786.
2. Keas, *Unbelievable*, 42–43.
3. Gleiser, *The Dancing Universe*, 59.
4. Ferris, *Coming of Age in the Milky Way*, 45.
5. Kinnaman and Lyons, *Unchristian*, 55–56.
6. Bidna, *We The People*, 28–29.

European sailor . . . believed . . . that a ship could sail out to sea just so far before it fell off the edge . . ."[7] It is very hard to fault people, therefore, for retaining their education! As Christine Garwood puts it, "as children across America chewed their pencils and stared out of . . . windows, two conjoined 'facts' were absorbed: medieval people believed the earth to be flat, [and second, that] Columbus was the first to prove it was a globe."[8] This was certainly how it has been presented in popular media as well:

> *Columbus*: The earth is not flat, Father, it is round!
> *Roman Catholic Priest*: Don't say that!
> *Columbus*: It's the truth; it's not a mill-pond strewn with islands, it's a sphere.
> *Roman Catholic Priest*: Don't, don't say that; it's blasphemy![9]

It is no coincidence that the way this scenario is depicted mirrors precisely how depictions of the Galileo affair has been passed down in popular lore. Yet, the actual history is quite different. As Jeffrey Burton Russell puts it in his book-length study on the myth of the flat earth: "in reality, there were *no* skeptics [of a round earth]. All educated people throughout Europe knew the earth's spherical shape and its approximate circumference."[10] C. S. Lewis (who, we sometimes forget because of his popular fiction and apologetic works, had an actual day-job holding the Chair of Medieval and Renaissance Literature at Cambridge) wrote in his survey of the Medieval period: "Physically considered, the earth is a globe; all the authors of the High Middle Ages are agreed on this. . . . The implications of a spherical earth were fully grasped."[11] The list of who knew this is so vast as to constitute essentially every educated person for the last two and a half millennia: Aristotle, Plato, the Venerable Bede, St. Augustine, St. Thomas Aquinas, and on and on.[12] "We can state categorically," says British historian of science James Hannam, "that the flat earth was at no time ever an element of Christian doctrine, and that no one was ever persecuted or pressured into believing it."[13]

To be sure, many of humanity's earliest commentators reasoned back and forth regarding the ultimate shape of the ground we awoke upon as a species. The Sumerians and Babylonians claimed that beneath the sky, but above the underworld, the earth stretched like a flat plane or disk, interrupted only by mountains and seas.[14]

---

7. Schreiber, *America Past and Present*, 98.

8. Garwood, *Flat Earth*, 2.

9. Quoted in Grant, *God & Reason in the Middle Ages*, 345. Though it will become quite evident later, it should be noted Grant is here debunking the myth, not supporting it.

10. J. B. Russell, *Inventing the Flat Earth*, 2.

11. Lewis, *The Discarded Image*, 140–41.

12. Regarding the globular view of earth in Plato's *Phaedo* and why, despite a few ambiguities, Plato most certainly did believe in a round earth, see Calder III, "The Spherical Earth in Plato's *Phaedo*," 121–25; Augustine, *On Genesis*, 1:9–10; 1:19; 1:21; 2:9; Bede, *On the Nature of Things*, ch. 3, 5, 6–10; Aquinas, *Summa Theologia*, Ia.q.68.a2.

13. Hannam, *The Genesis of Science*, 28.

14. Walton, *Ancient Near Eastern Thought and The Old Testament*, 172.

Many have argued that Thales of Melitus (c. 625–c. 547 BC), the Ionian geometer and natural philosopher, whom Aristotle records as being something of an original thinker,[15] and who is often credited with being the first philosopher of the West, believed the earth was a circular disk, like a piece of wood afloat a vast and shoreless sea. "This opinion," writes Aristotle, "is the most ancient which has come down to us, and is attributed to Thales."[16] This may have come from descriptions made by many like Herodotus, who in his *Histories,* describes a "floating island," Chemmis, northeast of Naucratis, an Egyptian trading post.[17] On the other hand, Cicero attributed to Thales the earliest idea of a celestial globe.[18] There is in fact an ambiguity in Aristotle, who may also be describing Thales as a "globalist" when he notes some have thought the earth is spherical.[19] Regardless, Aristotle does ascribe a flat-earth view to Anaximenes, Anaxagoras, and Democritus.[20]

By and large, however, the round earth was the position taken by the Greeks. Accordingly, to maintain the illusion that Christians propagated the myth, it is sometimes argued on the basis of Christian maps that the Greek view of the round earth was lost—even suppressed—by the rise of Christianity. A 1988 textbook, for example, reads:

> The maps of Ptolemy . . . were forgotten in the West for a thousand years, and replaced by imaginary constructions based on supposed teachings of Holy Writ. The sphericity of the earth was, in fact, formally denied by the church, and the mind of the Western man, so far as it moved in this matter at all, moved back to the odd confused notion of a modulated "flatland" with the kingdoms of the world surrounded by Jerusalem, the divinely chosen centre of the terrestrial disk.[21]

And in 1986, William O'Neil could still write of the church fathers that:

> Without differentiating amongst the details of their several views it may be said that they rejected the Hellenistic notion of the sphericity of the earth and of the universe in favor of a layered, flat, square scheme as suggested by Genesis. Indeed to varying degrees they tended to support the view that the Mosaic Tabernacle represented the shape of the universe.[22]

---

15. Aristotle, *Metaphysics,* 983 b20–28.

16. Aristotle, *De Caelo,* 294 a28–30.

17. Herodotus, *The Histories* II.156.

18. Cicero, *Republic* I.XIII.22.

19. Aristotle, *De Caelo,* 293 b33–294 a1.

20. Aristotle, *De Caelo,* 294 b14–15.

21. Holt-Jenson, *Geography,* 12–13.

22. Quoted in J. B. Russell, *Inventing the Flat Earth,* 47.

Yet this is utterly false. The Venerable Bede (672–735), the famous scholar and monk at the monastery of St. Peter in Northumbria, later takes up a similar argument in his work *On Times* (*Bedae opera de temporibus*):

> The cause of the inequality of the length of days is that the earth is round, and it is not in vain that in both the bible and pagan literature it is called "orb of lands." For truly it is an orb placed in the center of the universe; in its width it is like a circle, and not circular like a shield but rather like a ball, and it extends from its center with a perfect roundness on all sides.[23]

Here we are left with little doubt, for Bede spends the time to speak of the difference between the words "ball" and "plate" with the earth being the former. For other arguments regarding the spherical earth, in the Western tradition, we can take Aristotle (384–322 BC) as typical of thinking on the matter in his work *de Caelo* [*On The Heavens*]:

> [T]he evidence of the senses further corroborates [a spherical earth]. How else would eclipses of the moon show segments shaped as we see them? As it is, the shapes which the moon itself each month shows are of every kind—straight, gibbous, and concave—but in eclipses the outline is always curved; and since it is the interposition of the earth that makes the eclipse, the form of this line will be cased by the form of the earth's surface, which is therefore spherical. Again, our observation of the stars makes it quite evident, not only that the earth is circular, but also that it is a circle of no great size. For quite a small change of position on our part to south or north causes a manifest alteration of the horizon. . . . All of which goes to show not only that the earth is circular in shape, but also that it is a sphere of no great size; for otherwise the effect of so slight a change of place would not be so quickly apparent.[24]

Others like Eratosthenes (c. 276–c. 194 BC), a Libyan astronomer and Librarian of Alexandria, used mathematical reasoning to prod from the earth the secret of its waistline. Utilizing "trigonometry obtained from observations of the sun's declination at different latitudes,"[25] Eratosthenes devised an ingenious method to calculate the world's shape and size. He heard reports from a town called Syene that during the summer solstice, the sun cast no shadow because it was directly overhead. At the same time in Eratosthenes's home city of Alexandria, Egypt, the sun still cast a shadow at an angle equivalent to one-fiftieth of a circle (or seven point two degrees). Assuming the sun's rays are basically parallel, and that Alexandria was approximately (what we would measure as) 530 miles due north of Syene, the calculation based on the radius of a circle could be made to estimate that the circumference of the earth was in the ballpark of 29,000 miles. We do not know the exact figure, since Eratosthenes

---

23. Text taken from J. B. Russell, *Inventing the Flat Earth*, 87n55.

24. Aristotle, *De Caelo*, 2.14, 297b–298a.20.

25. J. B. Russell, *Inventing the Flat Earth*, 25.

was working with "stadia" as a unit of measurement—the earth, he said, being about 250,000 of them—and no one is completely certain what precisely this length indicates. The problem being that "Stadia" literally indicates "the length of one Roman stadium," which runs into the immediate problem that stadiums had no real set length. Regardless, from a shadow in a well, this is a pretty imaginative method to say the least, and one that is not too far off from the length we now know to be a circumference of around 24,860 miles.[26]

This amazing feat of geometry turns out to be an important bit of trivia for how Columbus's arguments with the Spanish court and the scholastics at Salamanca actually went. Several estimates of the circumference of the world—including Eratosthenes's—had been passed down, many of which were much larger. Dazzled by the travel narratives of Marco Polo and Pierre d'Ailly, Columbus was hellbent upon making the journey, and so through either sincere belief or a bit of sleight of hand Columbus not only picked the smallest available number, but ended up reducing that again by something along the lines of 20 percent. Many were not convinced by Columbus's miniaturization. The highest irony is that not only was the flat earth not an issue raised, the opponents of Columbus held to a round earth so implicitly it was one of the major premises in their arguments against the viability of his proposed journey! Aristotle had estimated that the spherical world was around 400,000 stadia, nearly twice the size that Eratosthenes had calculated, for example. As such, because the world was round the physical implications of this were thought by many to indicate that it was simple too enormous for Columbus to successfully execute his journey without starving everyone on board. Though there are no records of that exchange, we do have accounts written by Fernando, Columbus' son, and by Bartolomé de las Casas:

> . . . the replies and reports that the geographers gave their Highnesses were as varied as their grasp of the subject and their opinions. . . . [Some] who based themselves on geography, claimed the world was so large that to reach the end of Asia, whither the Admiral wished to sail, would take more than three years . . . To this they added that of this inferior sphere of land and water only a small belt or cap was inhabited, all the rest being sea that could be navigated only near coasts and shores. And even if learned men admitted that one could reach the end of Asia, they did not say that one could go from the end of Spain to the extreme West. Others argued . . . that if one were to set out and travel due west, as the Admiral proposed, one would not be able to return to Spain *because the world was round* [emphasis added]. These men were absolutely certain that one who left the hemisphere known to Ptolemy would be going downhill and so could not return, for that would be like sailing a ship to the top of a mountain, a thing that ships could not do even with the aid of the strongest wind.[27]

26. See the account in Garwood, *Flat Earth*, 20–21.
27. Colon, *The Life of the Admiral Christopher Columbus by His Son Ferdinand*, 39.

Now admittedly, that last bit about getting stuck at the bottom of the world will no doubt bring a smile to the reader's face for being so quaint. But we will do well to remember that they were following well entrenched principles of Ptolemaic science, and it is hard to blame this subset of men arguing such things for not being up to date on gravity or the relativity of motion nearly two centuries in advance of Isaac Newton and four before Einstein. For the Ptolemaic layout, up was up, and down was down. Hence the notion that moving along the circulature of the earth's sphere would bring with it as one traveled a shift in the frame of reference did not occur to them. Columbus was not immune to this opinion. In his travel journals transcribed by de las Casas, Columbus remarks that as he entered the mouth of the Orinoco river the vast pressures against the boat led him to believe he may have already run into the hill of the earth's curvature, and was beginning to sail upward. Far from invoking fear, however, a thrill came over Columbus in part because of the idea put forward by the theologian Peter Lombard, and later by the poet Dante, speculating that the earth was in fact more shaped like a pear hanging from a tree—with the lost paradise of Eden sitting atop the bulging crest of the world. The thought of trying to storm Eden, inaccessible because of the currents stemming from the cataract of waters glissading downward off the roof of the world, was no small motivator for Columbus to continue forward, whatever the cost.

Maps and Monsters

Imagine a map of our present . . . what's the shape of the existential terrain in which we find ourselves in late modernity? Where are the valleys of despair and mountains of bliss, the pitfalls and dead ends? What are the sites of malaise and regions of doubt? Where are the spaces of meaning? Are they hidden in secluded places, or waiting to be discovered in the mundane that is always with us? Where should we look for the "thin places" that still seem haunted by transcendence? Or have they disappeared, torn up to make way for progress and development? . . . Could we imagine an existential map of our secular age that would actually help us to locate ourselves and give us a feel for where we are?

—James K. A. Smith, *How Not To Be Secular*

[Eustace] *read only the wrong books. They had a lot to say about exports and imports and governments and drains, but they were weak on dragons.*

—C. S. Lewis, *The Voyage of the Dawn Treader*

A question thus emerges: how and why did the myth *of the myth* of the flat earth, start? Part of this has come down to us through the widespread and near systematic misinterpretations of a certain genre of medieval maps as being useless for navigation, the so-called *mappae mundi*.[28] Drawn on scraps of parchment, or occasionally on vellum (calf skin, sometimes goat or other animals) treated with lime and scraped clean by a crescent-shaped blade called a *lunarium* (or *lunellum*), medieval maps drew the world with the Holy City of Jerusalem at its center. Typically oriented in a famous "T and O" pattern, the circular portrayals (hence the "O") orient themselves with East at the top and Asia sitting above the lid of a "T" trisecting the map. Europe and Africa are then placed on either side of the plunging T-stem, Europe to the North (and so, left) and Africa, South (and so, to the right). These *mappae mundi* ("charts of the world") such as the Hereford map, tantalizingly place the lost Edenic paradise bound in circular flame at the top-most eastern edge, while the Pillars of Hercules line the Strait of Gibraltar at the map's bottom, and mark the western extreme of the habitable world with the words "nothing further beyond." And everywhere else along the circular courses of the *mappae*, monsters roam.

At times, the strange burlesque of these creatures almost appear to portend an invasion, straddling as some do the far borderlines which circumvallate the world of the map, refusing to be either solely inside or wholly outside; neither wholly natural nor wholly supernatural.[29] The lines of sacred and secular also appear blurred. One finds amongst the maps like the Hereford scriptural events like the Tower of Babel, or the Jewish exodus out of Egypt. But as the viewer's eyes journey through the colorful landscape, they will also run upon the golden fleece of Jason and his Argonauts, or the Cretan Labyrinth that Greek mythology tells us was built by Daedalus and populated by a particularly pugnacious minotaur. The Ebstorf map in particular gives us a clue to such juxtapositions. Named for a nunnery in Ebstorf in lower Saxony, it is an enormous drawing, sewn from the hide of thirty goats, which most scholars now estimate to have been crafted in the thirteenth century. Sadly, fires destroyed the original in 1943 during the Allied bombings of Hanover, Germany. Yet, there remain several recreated colored facsimiles, along with a black and white photograph taken in 1893.

In these photographs we see at the map's eastern top Christ's disembodied (yet still rather jovial looking) head, while the northern and southern edges portray his equally disembodied hands, and the southern lip (as one might now guess), his feet. Upon further examination, these parts are not as lacking of a body as appears at first glance. Rather, as you stand back the realization hits that the whole map itself is Christ's form. His navel is Jerusalem, again at the center of the world;[30] his head is Rome and the See of Peter. In other words, the map portrays the whole world along

---

28. On this see the definitive work of Woodward, "Medieval *Mappaemundi*." Woodward, it seems to me, makes an irrefutable case both that the *mappae* do not represent a flat earth, and do not represent geographical ignorance. Like icons, the *mappae* were more about communicating important Christian stories and events to those who couldn't read.

29. Cohen, "Monster Culture (Seven Theses)." See 7: "The monster's very existence is a rebuke to boundary and closure."

30. See Ezekiel 5:5: "I have set the city of Jerusalem in the midst of the nations and their peoples."

the symbolic lines of the eucharist wafer in which the world participates in Christ,[31] and the map allows one a sort of "proxy pilgrimage" through the earth seen via the economy of salvation and lit by the light of divine life.[32] From their incorporation into, and impending (re)birth from Christ's body, both the material and spiritual worlds are brought together, overcoming the lines of sacred and secular, temporal and eternal, past and present and future, nature and the supernatural. And yes, this applies even to the monsters. Thus for example in a twelfth-century tympanum of the Abbey church Vézalay in Burgundy, next to Christ's command "to preach the gospel to all creation" stands not just men and women, but "pygmies using ladders to mount their horses, the Panotti, with ears so huge they can use them as blankets, and, of course, the dog-heads [*cynocephali*]."[33]

Such medieval maps often offend modern cartographic sensibilities in the manner that they don't appear to be useful for referencing anything *real*. And what is real appears, well, flat. Embroidered as these maps are with their theological, philosophical, spiritual, and allegorical ornamentation; mixing together as they do time with space, present with past and future; distorting—as is their wont—topography to suit ethical and cosmological sensibilities, one could only imagine the reaction when handing this map to a boat captain or a team of explorers, and asking them to make their way in the world with it. They would want to discard all the elements that they deem fanciful, give a weak nod of approval to things that vaguely approximate the general layout of the world as we know it, and really just damn the whole thing with faint praise, perhaps by calling it "an important example of historical map-making." Or, perhaps, something much worse.

Daniel Boorstin—who, as a former Librarian of Congress, is another prestigious and well-read individual who should have known better—in his bestselling *The Discoverers*[34] paints the pitiable portrait of a world almost too hapless to know how to move to and fro as he gestures toward the *mappae*: in their efforts to navigate the world travelers "did not find much help in Cosmas Indicopleustes' neat box of the universe . . . The outlines of the seacoast . . . could not be modified or ignored by what was written in Isidore of Seville or even in Saint Augustine . . . The schematic Christian T–O map was little use to Europeans seeking an eastward sea passage to the Indies."[35] Such oafish ineptitude was born, he says, by "a Europe-wide phenomenon of scholarly amnesia . . . afflict[ing] the continent from A.D. 300 to at least 1300. During those centuries the Christian faith and dogma suppressed the useful image of the world that had been so painfully, and so scrupulously drawn by the ancient geographers."[36] This he called the

31. Kupfer, "Reflections on the Ebstorf Map."
32. Kupfer, "Ebstorf Map," 119.
33. Bartlett, *The Natural and the Supernatural,* 100.
34. Boorstin, *The Discoverers.*
35. Boorstin, *The Discoverers,* 146–49.
36. Boorstin, *The Discoverers,* 100.

"great interruption" of ancient learning, caused by the glut of Christian geographical fanatics rushing to touch the hem of their master's garment, a sixth-century eccentric known as Cosmas (mentioned in the quote above). "After Cosmas," he says, "came a legion of Christian geographers, each offering his own variant on the scriptural plan [of the world as flat]."[37] Being no less than the twelfth officially appointed Librarian of Congress, it seems reasonable that Boorstin should have known better. Yet, he did not; and so neither do we. An initial trouble here, we should point out so we may bracket it until a bit later, is that Cosmas Indicopleustas was completely unknown until he was rediscovered in 1706. And even then it would be nearly 200 years before it was argued in hindsight that Cosmas was a representative of the Middle Ages on this view: "No medieval author knew Cosmas, and his text was considered an authority of the 'Dark Ages' only after its English publication in 1897!" as Umberto Eco remarks.[38] A nonexistent mob of flatland geographers are thus argued by Boorstin to be led by a ghostly figure they would not have recognized.

Regardless, interpreting these medieval maps as failed early attempts at topography (or repression of Greek learning) is extremely wrongheaded from the start.[39] Charles Raymond Beazley dismisses the *mappae* for example, by charging that one need only look at the "monstrosities of the *Hereford* or the *Ebstorf*" to see that they are "of . . . complete futility."[40] One is reminded of J. R. R. Tolkien's famous essay on the poem *Beowulf*.[41] Tolkien chides many critics who take *Beowulf* merely as an item of philological and historical significance while being practically embarrassed at the presence of Grendel, or the dragon, as a sort of aesthetic mistake that we moderns must condescend to overlook. Tolkien states, rather "the monsters are not an inexplicable blunder of taste; they are essential, fundamentally allied to the underlying ideas of the poem, which gives it its lofty tone and high seriousness."[42] Contemporary scholars suffer by "placing the unimportant things at the centre, and the important on the outer edges,"[43] he says. Just so, when the poet of *Beowulf* writes of "the mighty men upon the earth," this is not, in the taste of future science, somehow at odds and obscured by a monstrous presence best pruned by we who, in later ages, are wise and see the hard earth and the bare stone for what they really are. The sense of the phrase "mighty men upon the earth" depends on the sinister presence of monsters out in the world; it is meant to summon the image of "*eormengrund*, the great earth, ringed by *garsecg*, the shoreless sea," and beneath "the sky's inaccessible roof." Thus these men, tormented by Grendel, gather "as in a little circle of light about their halls, great men

---

37. Boorstin, *The Discoverers,* 107–9.
38. Eco, "The Force of Falsity," 5.
39. Tattersall, "Sphere or Disc?," esp. 41–44 on the *mappae.*
40. Quoted in Woodward, "Medieval *Mappaemundi,*" 288.
41. Tolkien, "*Beowulf,*" 5–48.
42. Tolkien, "*Beowulf,*" 19.
43. Tolkien, "*Beowulf,*" 5.

with courage as their stay went forward to that battle with the hostile world and the offspring of the dark which ends for all, even kings, in defeat."[44] Tolkien thus concludes, "That even this 'geography,' once held as material fact could now be classed as a mere folk-tale matters very little . . . [For] astronomy has done [nothing] to make the island feel more secure, or the outer seas less formidable."[45]

However odd such symbolic references might seem to us, then, it is not as out of the ordinary as we think. As Mark Johnson and George Lakoff write, even our routine conceptual systems (to say nothing of monsters on maps) "in terms of which we both think and act, [are] fundamentally metaphorical in nature. The concepts that govern our thought are not just matters of the intellect. They also govern our everyday functioning, down to the most mundane details. Our concepts structure what we perceive, how we get around in the world, and how we relate to other people [and other communities]."[46] Which is to say, even our modern conceptual maps have monsters, regardless of how often we acknowledge that fact. Today, the "monsters" of theology, religion, the historical period of the Middle Ages (and so on) are put on the edges of our contemporary "scientific" conceptual maps for the same reason that the older monsters of medieval maps roamed the edges: they signaled "a category crisis," as Jerome Cohen puts it. We merely happen to be much less forthcoming on such matters than our ancestors. When theology and religion are understood to constitute a vital part of the total apparatus by which the adventure of human exploration and understanding peer into the world, they thereby threaten the modern "purity" and separation of our disciplinary categories. This is, indeed, exactly what has happened in many cases. Kathleen Biddick minces no words here about the general trend: "One sign of trouble in recent histories of medieval studies is their tendency to be paradoxically yoked to the scientific victors of the nineteenth century."[47]

While it isn't Biddick's point per se, for our story the fascinating twist turns out to be that not just the flat earth, but indeed the warfare metaphor blanketed over history are themselves one more map full of monsters. Both the flat earth and the warfare metaphor have common origins as artifacts of the emerging professional battles to seize social power and define the role of the "scientist" in Victorian England, and across the pond in America. Perhaps to no one's surprise the nineteenth century debates over evolution are in one sense a key factor responsible for the concept of warfare between science and Christianity, but not in a way one would initially expect. In the effort to seize the rhetorical high ground as we saw with Huxley, Tyndall, White, Draper, Youmans, and others, a campaign to turn evolution—and science at large—against institutional theology was one of the key wedges in the power shifts

44. Tolkien, "*Beowulf*," 5.

45. Tolkien, "*Beowulf*," 5. Cf. Woodward, "Reality, Symbolism, Time and Space in Medieval World Maps."

46. Lakoff and Johnson, "Conceptual Metaphors in Everyday Language."

47. Biddick, *The Shock of Medievalism*, 2.

of professionalization began, of which absurd pictures like the flat earth were also marshalled in order to perform a flanking action to demonstrate the absurd pedigree of the church's decision making. Thus, while the flat earth is in no sense uniquely causative of the notion of warfare, it provides one unique and fascinatingly weird angle to enter into the construction and deconstruction of the warfare metaphors organizing our historical maps.

The Myth of the Myth

In a remarkable piece of detective work, Jeffery Burton Russell traces the emergence of the myth through history. The real perpetrator comes with a man named Washington Irving (1783–1859), who was the author of both *Sleepy Hollow* and *Rip Van Winkle*. Irving's work was the *Da Vinci Code* of its day: he was writing fiction with historical research thrown in, in order to satirize his distaste for pedantic historians. "Irving knew how to use libraries and archives, and the public was fooled into taking his literary game as history."[48] In this work, a very artful and elaborate account of Columbus standing before the Inquisitors attempting to convince them of his journey appears.

> [The Council] was comprised of professors of astronomy, geography, mathematics, and other branches of science, together with various dignitaries of the church, and learned friars. Before this erudite assembly, Columbus presented himself to propound and defend his conclusions. He had been scoffed at as a visionary by the vulgar and the ignorant; but he was convinced that he only required a body of enlightened men to listen dispassionately to his reasonings. . . . Columbus was assailed with citations from the Bible and the Testament: the book of epistles of the apostles, and the gospels of the Evangelists. To these were added expositions of various saints and revered commentators: St. Chrysostom and St. Augustine, St. Jerome and St. Gregory, St. Basil and St. Ambrose, and Lactantius . . . Mathematical demonstrations was allowed no weight, if it appeared to clash with a text of scripture, or a commentary of one of the fathers.[49]

This is all quite literally made up. Irving wanted to turn Columbus into a mythical figure, "the hero of a romantic novel, or an epic modern Odysseus or a Faust . . . or an American Adam, the First Man of the New World . . ."[50] Irving's work was a sensation, but its intent, it seems, lost on everyone. While Irving spread the myth at a more popular level, Jean-Antoine Letronne (1787–1848) secured it as an academic commonplace. Learned in Latin, Greek, Egyptology, and mathematics, the intelligent and charming Frenchman was adored by his contemporaries, and those who eulogized

---

48. J. B. Russell, *Inventing the Flat Earth*, 50.

49. Irving, *Christopher Columbus*, 61–62, 47–51.

50. J. B. Russell, *Inventing the Flat Earth*, 56.

him declared him a "secular saint."[51] He got on well with all manners of government, supported his widowed mother, engaged in secret acts of charity that were not known to be his work until after he passed on, fathered ten children, became director of the *Ecole des Chartes*, then Inspector General at the University of Paris, and eventually would obtain the chair of history at the College de France. Living the life of what seems to be ten men undoubtedly leaves one with little spare time, but Letronne still managed to squeeze in some blistering polemics against Christianity. In particular our interests turn us toward his essay "On the Cosmographical Opinions of the Church Fathers" (1834).

Its polemical stance is made known immediately. In the very first sentence Letronne recounts acidly that until recently "all science was to be based on the Bible."[52] Astronomers were "forced to believe" that the earth is flat, and though a few theologians like Augustine and Origin knew better, Letronne makes sure to emphasize that they were marginal figures. Our friend Cosmas, whom we met earlier, also makes an appearance. In the course of six detailed pages oozing with all of Cosmas's bizarre opinions including the idea that the earth is a tableland, flat beneath the vaulted heavens, Letronne gives the impression that Cosmas was well known, influential, important. The reader may recall, however, that Cosmas had only just then in Letronne's own time been rediscovered and unceremoniously baptized by nineteenth-century historians as representative of that backwater they declared the "Dark Ages." We have a record of only one individual who had read Cosmas, St. Photius, the Ecumenical Patriarch of Constantinople. Photius, who had garnered a widely reputed honor of being the most well-read scholar of his age and who is referred to in a recent history as "the leading light in the 9th century renaissance," had in turn nothing nice to say about Cosmas.[53] "The style is poor," he says, "he relates much that is incredible from a historical point of view, so that he may fairly be regarded as a fabulist rather than a trustworthy authority." Indeed, Photius then goes on to mention that it is specifically Cosmas's view on the flat earth that he finds so bizarre: "the views on which he [Cosmas] lays special stress are: that neither the sky nor the earth is spherical, but that the former is a kind of vault, and the latter a rectangular plane."[54] In other words the only person to have apparently read Cosmas sees him as something of a misguided fantasist notable for holding such an outlandish view. It is, to say the least, a bold strategy for Letronne to then hold up Cosmas as exemplary of an entire era while simultaneously making the argument that two of Christianity's most seminal theologians—Augustine and Aquinas—were in this case ignored as the black sheep of Christ's ramshackle flock.

Nonetheless, this haphazard backwardness, for Letronne, essentially summarized the whole Christian legacy: "The flat earth theories . . . dominated up to the time

---

51. J. B. Russell, *Inventing the Flat Earth*, 59.
52. J. B. Russell, *Inventing the Flat Earth*, 60.
53. Louth, *Greek East and Latin West*, 159.
54. Quoted in Cormack, "Flat Earth or Round Sphere?," 381n19.

of Columbus and Magellan, and even persisted afterward, but finally the discoveries of Kepler, Huygens, and Newton erased the childish ideas that the theologians had defended inch by inch as orthodox."[55] Yet despite the fact that much of this could have been discredited by a slight glance at many of the sources to which Letronne was supposedly pointing, by the time he wrote this essay his sterling reputation made checking his footnotes apparently unnecessary. His focus on Cosmas as singularly significant was passed on as a bad habit to many like Charles Raymond Beazley, for example, as well as his general impression of the church fathers as a confused band of miscreants who, despite the stupidity of their views, "had three irresistible arguments: persecution, prison, and the stake."[56]

Both Irving and Letronne's works found their way into the pages of two men with whom we are already quite familiar: John William Draper and Andrew Dickson White. Given that the title "Flat Earther" is still used to indicate anyone who holds dogmatically to an outlandish view in the face of science, one can imagine the sort of currency Irving's and Letronne's tale had for men who wanted to portray the length of history as one of the heroic struggles of science in the face of dogmatic repression. "[White and Draper] saw the Flat Error as a powerful weapon."[57] And so, they made it a mascot of their story. White took Irving's portrayal of Columbus and weaponized it to create the popular anti-science image of institutionalized Christian cosmography:

> Many a bold navigator, who was quite ready to brave pirates and tempests, trembled at the thought of tumbling with his ship into one of the openings into hell which a widespread belief placed in the Atlantic at some unknown distance from Europe. This terror among sailors was one of the main obstacles to the voyage of Columbus.[58]

A better scholar than Draper—who portrayed Columbus as assailed by the "Grand Cardinal of Spain," berating Columbus with flat earth arguments supposedly from "St. Chrysostom and St. Augustine, St. Jerome . . . St. Basil and St. Ambrose"[59]—White knew that those like Augustine and Aquinas, quite inconveniently for his thesis, were in full support of a round earth. So, he painted them, much as Letronne did, as unique lights, lost in the smugly self-satisfied murk of the majority of Christian faith. Yet, again, if your argument relies on painting Augustine or Aquinas as a "minority" in the Christian thought of the West, and "those two pipsqueaks"[60] Lactantius and Cosmas as representative of the orthodox, something has gone wrong.

55. J. B. Russell, *Inventing the Flat Earth,* 61.

56. J. B. Russell, *Inventing the Flat Earth,* 60.

57. J. B. Russell, *Inventing the Flat Earth,* 43.

58. A. D. White, *Warfare,* I:97.

59. Draper, *Conflict,* 65.

60. Gould, "Late Birth of the Flat Earth," 43.

The story takes an unexpected and interesting turn, however. White cites Irving in support of the Columbus tale. When Russell traces this citation in White's work back to Irving, he finds a footnote supporting Irving's story for the Flat Earth and Columbus that reads in total: "Mss. Bibliot. Roi. Fr." For those confused by the phrase, this is academic shorthand, though it may as well have been a sorcerer's incantation for all the good it did. When translated into full script, it reads: "manuscripts in the French royal library."[61] Which is to say, at this point, Irving is having a laugh, hardly bothering to cover up his ruse. He is, in essence, saying with this fake footnote: "*somewhere* in the French royal library there are unnamed, unspecified documents which totally support my story." As Russell kept his detective hat on, the mystery unfolded back further, to an intriguing origin: Copernicus, in a rare moment of self-promoting polemic, likened those who did not believe his assertion that the earth traveled around the sun to Lactantius, "though Copernicus was careful not to blanket either ancient or medieval Christianity with Lactantius' error [as he himself was a devout Catholic]."[62] Such caution did not last. As recently as 1998, historian R. Youngson claimed that Giordano Bruno was burned at the stake for, among other things, denying the belief held so dearly by the church that "the earth was flat and supported on pillars."[63] And, as we mentioned at the outset of this chapter no less a commentator than Thomas Jefferson ends up simply conflating the flat earth with the notion of heliocentrism (that is, the earth rotating around the sun) when he asserts that Galileo was put before the Inquisition for claiming the earth is a globe.[64] This idea of history fit well with Jefferson's Enlightenment notion that modernity was the continuing history of an "unsparing sunrise" burning away the fog of ancient ideas and institutions.[65]

Even stranger, the sparse instances of the flat earth, the strange nature of the fake footnotes that account for it, were all overlooked because, as it happens, modern forms of flat earth belief began to arise contemporary to Draper and White, Irving and Letronne, giving their histories, no doubt, the air of plausibility. In fact, as we have mentioned, far from being a storied legacy of Christianity the flat earth is largely an artifact from what came to be understood as the debates over Darwinian evolution and the professionalization of science in the latter half of the nineteenth century. And this origin is no mere piece of trivia, for it is precisely in the same period that the first few waves of writers authoring histories of science—and so giving legitimacy to their new profession—were putting pen to paper. In this way not just the flat earth—but the notion of the warfare it often represented—came to nest themselves firmly into the self-perceptions of scientists, even religious ones.

61. J. B. Russell, *Inventing the Flat Earth*, 96n148.

62. J. B. Russell, *Inventing the Flat Earth*, 70.

63. Youngson, *Scientific Blunders*, 282.

64. Jefferson, *Notes on the State of Virginia*, 165–66.

65. Quoted in Neem, "The Early Republic," 37.

The Recent Inventions of the Flat Earth

It was the decade of the 1920s, and Samuel Shenton had a good idea.[66] Born the son of an army sergeant, he recalled after the First World War he would often stand in awe as zeppelins and other "bird-shaped" aircraft refused gravity and took flight. Already marvelously ingenious vessels, Shenton was curious how he could improve upon this technology for the benefit of humankind. Having learned—again to his amazement—that the earth rotates around its axis somewhere in the vicinity of 1,600 kilometers per hour (or approximately 1,000 miles per hour), the idea struck him: why not build a craft that would combine the power of gas and engine to float into the atmosphere carrying cargo? Resting anchored at high altitude, it would have but to wait as the earth spun its practiced course westward. Aided by this frantic twirl of the earth, after the proper amount of time the cargo craft would then simply descend upon its new destination far faster than anything yet designed. "Think of the possibilities," he later gushed. "It was staggering!"[67]

Of course, Shenton's plucky idea completely disregarded the fact that the atmosphere and everything in it rotates with the earth. "Overlooking this crucial fact," writes Christine Garwood, "Shenton wondered why no other individual had hit upon this simple but ground-breaking idea."[68] As it happened, the apparent uniqueness of his invention so startled him that he grew increasingly suspicious. Surely someone else had also thought of this design? When he was not plied with grant offers for what he took to be his obviously world-changing idea, the inkling that some sort of cover-up or conspiracy was afoot took hold of him, though he couldn't yet say regarding exactly what. Styling himself a heroic lone figure seeking truth, he set out to prove the merit of his invention. While one might assume the rest of the story involves a somewhat melancholic Shenton getting on with his life after readily available information concerning inertial rotation popped his money-making fantasy, the truth is, as ever, so much stranger. In the course of his quixotic search, much to Shenton's delight he ran across an aircraft design quite similar to his own in the literature of the then almost defunct Zetetic Society. In a panic of joy, Shenton recalls frantically setting out to buy every piece of Zetetic material he could get his hands on. He was in awe with what he found, which would set the course of the rest of his life: no one would patent or purchase the rights to his machine because it would reveal a dark secret—the world not only did not rotate on its axis, it was completely flat, "just as the Scriptures described it." When the space race came, Shenton, in turn, was one of the first to vocally suggest it was a hoax, because "the scriptures describe an impassible dome";

---

66. My recounting of Shenton's tale is here indebted to the phenomenal work of Garwood, *Flat Earth*, 219–80.

67. Quoted in Garwood, *Flat Earth*, 221.

68. Garwood, *Flat Earth*, 221.

"circumnavigation of the globe is in reality making a circle above a flat plane"; and claims otherwise "are basically anti-God."[69]

For those unfamiliar (perhaps blissfully so) with Zetetics, as the reader might have guessed by now, they were a society founded in the mid-nineteenth century and were dedicated to promoting and proving the flatness of the earth. Fancying themselves a guild of those seeking truth ("zetetic" means "seeker"), they were established by one Samuel Birley Rowbotham, and later given new life by the generous patroness Lady Elizabeth Blount, who changed the name to the Universal Zetetic Society.[70] A showman to his core, for his public persona Rowbotham named himself "Parallax" (only one of several monikers he was to adopt through his life, another being "Dr. Birley, PhD" when he turned to hawking cures for that minor irritation we have named "mortality"), and marketed himself as a new Francis Bacon.[71] Though today the epithet "Flat Earther" is typically meant to invoke those who cling to dogma in the face of obvious evidence, Rowbotham saw himself in stark contrast as an anti-elitist scientist, a Baconian Robin Hood trying to put knowledge back in the hands of the people, as opposed to it being the sole property of the Royal Society. The highest irony here is that the use of the flat earth as an *insult* was, it appears, tossed up by none other than Rowbotham's hero Francis Bacon who, as a good Protestant, taught not only that the Catholics held to the idea, but that they put many dissenters to trial for not holding it.[72] Regardless, as an exceptional orator and intensely shrewd debater, in the guise of Parallax, Rowbotham swayed, confused, divided, and often generally convinced crowds as he toured around Britain with his entertaining mixture of the ludicrous and the ostensibly scientific.

Despite the eccentricity of Zetetic views, they were not the only ones glorying in their own confessedly "alternative" sciences. In the Victorian scramble to institutionalize the newly minted concept of "the scientist," a variety of conflicts and exceptional trajectories like mesmerism and phrenology sprang outward like free radicals under the centralizing pressures exerted via professionalization.[73] Perhaps the most famous example of this was the anonymously penned *Vestiges of the Natural History of Creation*, which we now know was written by the Scotsman Robert Chambers. Though we cannot tell the remarkable tale of the *Vestiges* here,[74] it (far more than Darwin's *Origin of Species*) was largely responsible for priming the shape and course of later debates over evolution and religion. "Some people read the *Vestiges* as the epitome of scientific expertise; others dismissed it as the product of a dilettante. It all depended on what

---

69. Garwood, *Flat Earth*, 221.
70. Garwood, *Flat Earth*, 154–87.
71. Garwood, *Flat Earth,* 36–79.
72. Bacon, *The New Organon*, 87, section 89.
73. Turner, "The Victorian Conflict Between Science and Religion."
74. Secord, *Victorian Sensation*.

one thought profound knowledge really was."[75] Attempts to pry open spaces outside of the all-seeing eye of institutional science also helps account for the rising fascination with magic and the occult in the same period—which could, much like Rowbotham, quite ironically find precedent in the "Father of Modern Science," Francis Bacon, who unabashedly described his project in terms of "purified magic."[76] Indeed here Andrew Dickson White's project of rewriting Baconianism received a particularly juicy set of examples. For with the democratic elevation of "common sense" in Baconianism, the flat-earthers (while admittedly on the far fringes) could describe themselves as more or less within the pale of Baconianism, which because of its populism often had any number of uncontrolled sensibilities and agendas attached to it.

Rowbotham and his somewhat roguish disciples understood full well how to leverage populism against professionalization, and saw in the flat earth a wedge to drive into the still-tender trunk of elite science. "Working men have brains too," as William Carpenter, one of Parallax's first disciples, was fond of saying.[77] To reinforce his point, with the controversy stirred over evolution, science, and theology by the bombshell that was the *Vestiges of the Natural History of Creation*, Parallax made sure his performances emphasized, by way of the image of the flat earth, that there was a war not just between stodgy professionals who thumb their noses at the common man's attempts to understand the world, but between religion and (professionalized, institutional) science. In Parallax's mind, these two wars were really one and the same. Thus, despite the fact that with the 1865 publication of *Zetetic Astronomy: Earth not a Globe!*, Parallax wanted to emphasize that his extensive compilation of scriptural proofs regarding the earth's flatness obtained only after his mathematical and observational data had been rigorously demonstrated, in truth, "Wherever he went, the same line of argument followed: science and the scriptures were at war, and both could not be right."[78] Thus, in a masterful (if not consistent) stroke, Parallax leveraged both true science and true religion to his cause. He combined them, or turned them against one another, as it suited his rhetorical purposes.

All of this seems, perhaps, like a bit of a sideshow from more serious matters that usually attend science and religion debates. Yet this is revealed as untrue when we look at how the flat earth intersected with the Victorian controversies of the day surrounding the professionalization of science, class conflict between "elites" and "commoners," debates regarding the use of Scripture, and broader concerns over the legacy of Christianity. As it turns out, flat-earthers were some of the first to popularize the "warfare of science and Christianity" narrative. "Why did the battle rage [over whether or not Christianity held to a flat earth]?" asks Lesley B. Cormack. "Because belief in a flat earth was equated with willful ignorance, while an understanding of the

75. Secord, *Victorian Sensation*, 21.
76. Josephson-Storm, *The Myth of Disenchantment*, esp. 41–66.
77. Garwood, *Flat Earth*, 63.
78. Garwood, *Flat Earth*, 71.

spherical earth was seen as a measure of modernity."[79] But from the flat-earther's per-spective, a "globate" earth represented the haughty and unproven claims of scientists who hated God and didn't read their Scriptures.

All of this was just too good for White and Draper to ignore. Few images encap-sulate, represent, and indeed perpetuate the concept of an epochal Christian "Dark Ages" better than the image of the flat earth.[80] In fact, as it turns out, the coining of the very term *scientist* was accompanied by the myth of Medieval and antique belief in a flat earth. William Whewell (himself a Christian, mind you), who wrote one of the first histories of science, minted the term in the 1830s,[81] and indeed there in print ap-pears the accusation of flat-earthism.[82] This image of a flat earth also accompanies the idea of a thousand-year darkness, the dogmatic servility of Christianity, the forlorn, aeonic stillness of the world of the human intellect under the boot of the church. The entire apparatus that is the flat earth stands for the ignorance of a Christian age emerges precisely with the very first usage of "scientist" in Whewell,[83] which swells to characterize the flatness of the entire age:

> We have now to consider more especially a long and barren period, which in-tervened between the scientific activity of ancient Greece, and that of modern Europe; and which we may, therefore, call the Stationary Period of Science. . . . In speaking of the character of the age of commentators [i.e. the Middle Ages], we noticed principally the ingenious servility which it displays . . . the want of all vigor and fertility in acquiring any real and new truths. . . . speculative men became Tyrants without ceasing to be slaves; to their character as Com-mentators, they added that of Dogmatists. . . . We have thus rapidly traced the cause of the almost complete blank, which the history of physical science offers, from the decline of the Roman Empire, for almost a thousand years.[84]

His use of the flat earth gave distinctiveness to the newly minted category of "the scientist" by differentiating it from a past, backward epoch. Simultaneously, it rein-forced the continuing professionalization of science in Whewell's day, thus removing a distinct source of intellectual authority from Victorian clerics. The flat earth could not be allowed not just because it represented ignorance, but it represented the most of-fensive thing of all to Whewell: mixing ecclesiastical authority with scientific inquiry. When someone like Whewell puts a spot like "the Dark Ages" on his conceptual map of history, he is asserting a "do not cross" line. But, at the same time, he is thereby also asserting the nature of what he wants to be the "normal" terrain of his map. As such,

79. Cormack, "Myth 3," 29.

80. On early history of the myth of the flat earth, see Cormack, "Flat Earth or Round Sphere."

81. See Ross, "Scientist," 66–67; H. F. Cohen, *The Scientific Revolution,* 27–39.

82. J. B. Russell, *Inventing the Flat Earth,* 31–32.

83. Whewell, *Philosophy of the Inductive Sciences,* on the flat earth, I:195–97.

84. Whewell, *Philosophy of the Inductive Sciences,* II:181, II:237, II:271

however, paradoxically the "do not cross" is inscribed and made a part of how we represent ourselves to ourselves.[85] Without a dark age to put behind it, without a monster to slay, the bright light of science would not be so pure, or so clear. More often than not the mental maps we use, as moderns, want to appear devoid of such cartoonish things; as it turns out in practice, however, this is not the case at all. "The 'Middle Ages' is a mobile category," remarks Kathleen Davis, "applicable at any time to any society that has not 'yet' achieved modernity, or worse, has become retrograde."[86] We are, as we have mentioned, merely much less honest than the ancients in representing our monsters explicitly, while we nonetheless fill our conceptual maps with them:

> Just as Grendel [in *Beowulf*] frequents the borders of the Danish moors, the Middle Ages as a period continually threatens to disrupt modernity from its position on the edges of history: if the Middle Ages is popularly imagined as a time full of monsters, then it can also be said to operate itself as a *kind of historiographic monster* [emphasis added], challenging ideas of modernity as radically different.[87]

One can dial the time for the emergence in from another angle as well. None of the great eighteenth-century polemicists against Christianity—Edward Gibbon, David Hume, Denis Diderot, or others—ever accused the scholastics of believing in a flat earth. No doubt had they even whispered of this exotic belief, it would have been turned over the spit with morbid relish. We have seen how quick many were to pounce upon Lactantius and Cosmas's opinions on the matter. But it wasn't a point of polemic for the philosophes, because the scholastics didn't hold to it. Early medieval theologians didn't hold to it. The patristic fathers and mothers didn't hold to it. Flat earth belief did not exist in earnest until the nineteenth century, as we suggested above, and was largely an artifact of what would eventually be called in hindsight the evolutionary debates. The flat-earthers, however much they represented a strange and extremely small group of oddballs

> were convenient symbols to be used as weapons against the anti-Darwinists. By the 1870's the relationship between science and theology was beginning to be described in military metaphors. The philosophers (the propagandists of the Enlightenment), particularly Hume, had planted a seed by implying that the scientific and the Christian views were in conflict. August Comte (1798–1857) had argued that humanity was laboriously struggling upward toward the reign of science; his followers advanced the corollary that anything impeding the coming of the kingdom of science was retrograde. Their value

---

85. Morgan, *The Monster in the Garden*, 170–71.
86. Davis, "The Sense of an Epoch," 41.
87. Bildhauer and Mills, eds., *The Monstrous Middle Ages*, 3.

system perceived the movement toward science as "good," so that anything blocking the movement in that direction was "evil."[88]

The point of this admittedly eccentric historical excursion? As Herbert Butterfield noted in his seminal *The Whig Interpretation of History*, when we prioritize the present as the context in which to understand the past, this leads to the "over-dramatization"[89] of certain events due to overemphasis on qualities that supposedly led to where we are today. In our case, this means the many nuanced ways Scripture has interacted with science or natural philosophy in the past get lost in exchange for a few sensationalist examples—and these are themselves typically also misunderstood. The flat earth is one such sensation. As it turns out, Parallax and his followers had exceptionally bad timing, as they burned their sideshow into the minds of scientists beginning to write their own history textbooks, fashioning for themselves their sense of self-identity.

Ten years after "Darwin's Bulldog" Thomas Huxley's own mythologized debate with, and supposed besting of, Bishop William Wilberforce over the concept of evolution,[90] came the infamous "Flat Earth Wager." One of Parallax's more radical followers, John Hambden, challenged one of Huxley's compatriots, Alfred Russell Wallace—the co-discoverer with Charles Darwin of the principle of natural selection—to prove that the earth was round. Hambden put £500 of his own money on the line for anyone who could prove to him the earth was not flat. Trying to muster up some good press, Hambden decided to target Wallace in order to land a big fish in the scientific world and show him a thing or two. Though one might expect someone with the prestige of Wallace to refuse such an outlandish contest, he was in between book projects and hurting financially. Unlike many of Wallace's colleagues, such as Darwin, he did not have a private income to fall back upon. So, he took the bet, much to the delight of Hambden and the absolute perplexity of Wallace's colleagues. Writing to his friend Alfred Newton, a professor of zoology at Cambridge, Wallace confided that he had taken on "a heavy wager" with one "of those strange phenomena" who do not believe in the earth's roundness, and who is "willing to pay to be enlightened." His light tone indicates Wallace had no inkling this wager was something he would regret for the rest of his life.[91] To cut a long story short, after a few false starts to find appropriate judges for the contest, Wallace was—perhaps to no one's surprise—eventually declared the victor. The world was round! However, a furious Hambden spent the rest of his days hounding Wallace with threats upon him and his family. It escalated to such a point that the mild-mannered Wallace had to get a restraining order, and eventually he had Hambden arrested and later even committed to an asylum to protect himself from himself.

---

88. J. B. Russell, *Inventing the Flat Earth*, 35n33.

89. Butterfield, *The Whig Interpretation of History*, 53.

90. See chapter 3 in this book.

91. Quotes taken from Garwood, *Flat Earth*, 88.

This was too little, too late. The damage had been done, and the image of a fanatical Christian touting the flat earth against scientists stuck. Far from being confined to the fictional imagination of Irving or the academic networks of Letronne, after the publication of Darwin's *Origin of Species* the flat earth migrated into the polemical toolbox of evolutionists (some Darwinian, some otherwise) to use as a bludgeon against any who doubted the way the new evolutionary winds were blowing. They cannot be wholly to blame for this. Parallax and his disciples as we saw were in many ways responsible for using the flat earth as rhetorical fuel for the warfare metaphor's fire in the nineteenth century, only from the opposite side. Promoting a biblical literalism that would even make later six-day creationists blush, the flat-earthers were primarily a distraction, a weird and vocal minority that struck at what turned out to be an inopportune moment and took just enough pressure off of historical investigations pertaining to whether Christians actually believed in a flat earth that it stuck at a very formative time of historical writing done in the name of self-fashioning and carving out professional and personal identities. As Russell summarizes the history, "the reason for promoting both the specific lie about the sphericity of the earth and the general lie that religion and science and in natural and eternal conflict in Western society," went hand in hand, with the ultimate end game of this mythological partnership being "to defend Darwinism."[92] The social and political forces that were shaping the newly minted concept of "the scientist" internalized the threat of the flat earth as yet another wicked device of churchmen meddling where they had no business, and so felt the need to produce polemical antibodies for it.

One recent historian, after surveying a sample of 130 textbooks, notes the first college textbooks incorporating the myth began to appear at the turn of the century—though several astronomy textbooks following Letronne appeared slightly earlier. But especially in the 1960s, during the heyday of secularization theories in the human sciences, the appearance of the flat earth accelerated even further. "Adopting the anti-clerical posture of Washington Irving and other nineteenth-century polemicists, textbook authors began to depict ancient and medieval Christians as exceedingly anti-intellectual about the earth's shape, and more."[93] Indeed, to this very day it sticks as a point of argument paired to evolutionary theory:

> [If Christians] insist on teaching your children falsehoods—that the earth is flat, that 'Man' is not a product of evolution by natural selection—then you must expect, at the very least, that those of us who have freedom of speech will feel free to describe your teachings as the spreading of falsehoods, and will attempt to demonstrate this to your children at our earliest opportunity. Our future well-being—the well-being of all of us on the planet—depends on the education of our descendants.[94]

92. Russell, "The Myth of the Flat Earth."
93. Keas, *Unbelievable*, 47.
94. Dennett, *Darwin's Dangerous Idea*, 519.

The pairing of the falsehoods Christians teach—that of the flat earth, and the denial of evolution by natural selection—are no arbitrary pairing but, as we have seen, historically generated. Yet with this pairing that Dennett invokes is a strange lopsidedness. For, as we have seen, natural selection was rejected by many card-carrying evolutionists—even Huxley for a time. The representation of Christians in the pairing is not just mythology, but one that is frozen into a particular form, while what constitutes legitimate Darwinism is allowed the growth and hindsight of time. This is a common rhetorical strategy, and an important one to keep in mind as we move to the next chapter. For there are fewer things about Christianity considered quite so frozen and immovable as the Dark Ages, a thousand yawning years of no learning or advancement that Christianity foisted upon the world.

—Chapter 8—

A Thousand Years Lay Dreaming,
Or Half Awake: Rethinking the Dark Ages
(Part One)

All that is designated by "the medieval" is never overcome and rarely superseded, but rather continuously posited as that necessary anachronism that paradoxically generates "the modern" as we know it. To forget "the medieval" is to conjure a modernity that can never be known.

—Andrew Cole and D. Vance Smith, *The Legitimacy of the Middle Ages*

ON THE CUSP OF the end of the first millennia AD, a figure in monkish robe sat ornamented by firelight as he hunched over a writing desk with quill and parchment. Not two years before, the line of Charlemagne had failed. In just nine months a new millennium would commence, and rumors of the end times "filled almost the entire world." Here every mist threatened to bring with it the dreaded beasts of Revelation, and every moan of wind may well have been the brass section of the heavens warming up. And yet, the notations this monk—an Archbishop, more precisely—was scribbling in the chiaroscuro of the room were not apocalyptic musings, nor esoterica theorizing upon the unutterable things St. Paul refused to speak of in his journey to the third heaven (2 Cor 12:2–4). "On the eve of the Apocalypse," writes Nancy Marie Brown, "the archbishop of Ravenna and his friend [were in a letter] discussing the best method for finding the area of a triangle."[1] This archbishop was Gerbert of Aurillac, soon to be known as Pope Sylvester II, and to write his history, says Brown "is to rewrite the history of the Middle Ages."[2]

---

1. N. M. Brown, *The Abacus and the Cross*. Our intro paragraph here is based on the account given on pages 1–4 of Brown's book. The quote comes at page 3.

2. N. M. Brown, *The Abacus and the Cross*, 3.

In [Sylvester's] day, the earth was not flat. People were not terrified that the world would end at the stroke of midnight on December 31, 999. Christians did not believe Muslims and Jews were the devil's spawn. The Church was not anti-science—just the reverse. Mathematics ranked among the highest forms of worship, for God had created the world as scripture said, according to number, measure, and weight. To study science, was to approach the mind of God.[3]

With his typical wit, the theologian David Bentley Hart remarks that when it comes to this period of history it appears that "the ghastly light of a thousand inane legends burns with an almost inextinguishable incandescence."[4] It is not enough merely to set one's aim to extinguish a few of these "almost inextinguishable" fires. Just as important the question of who lit them, why, and what was sought in such flame must constantly occupy us. Otherwise, like the hydra heads severed by Hercules, for every one snuffed, two more spring into place. As the former Curator of the Louvre Regine Pernoud writes in her delightful book *Those Terrible Middle Ages!*: "Now the medievalist, if he has in mind set on collecting foolish quotations on the subject, finds himself overloaded by everyday life. Not a day goes by that one does not hear some reflection [invoking the Dark Ages, or using "medieval" as an insult]."[5] Recalling an occasion where she accompanied her nephew to one of his classes "so that afterward [parents] can work with their children," she goes on to note her astonishment at the catechism her nephew's class is put through:

> *Teacher*: What are the peasants of the Middle Ages called?
> *Class:* They are called serfs.
> *Teacher*: And what did they do, what did they have?
> *Class:* They were sick.
> *Teacher*: What illnesses did they have, Jerome?
> *Jerome* (very serious): The plague.
> *Teacher*: And what else, Emmanuel?
> *Emmanuel* (enthusiastically): Cholera!
> *Teacher*: You know history very well; let's go on to geography.[6]

Pernoud then recalls another time from her many experiences, when she received a call from a TV research assistant who began by saying "I understand that you have some transparencies. Do you have any that *represent the Middle Ages*?" "Represent?" Pernoud asked. "Yes, that give an idea of the Middle Ages in general: slaughter, massacres, scenes of violence, famines, epidemics . . ." She immediately burst into laughter, and then promptly regretted it.[7] She notes it was unfair of her to expect popular

3. N. M. Brown, *The Abacus and the Cross*, 3–4.
4. Hart, *Atheist Delusions*, 36.
5. Pernoud, *Those Terrible Middle Ages!*, 9.
6. Pernoud, *Those Terrible Middle Ages!*, 10.
7. Pernoud, *Those Terrible Middle Ages!*, 10.

knowledge to somehow surge beyond that of the general level of knowledge of the Middle Ages by non-specialist academics, who for example in 1964 came together for a meeting of the *Cercle catholique des intellectuels Francais* (Catholic Circle of French Intellectuals) on the topic "Were the Middle Ages Civilized?," which was a question posed, she notes "without the least bit of humor." What is more, she writes, hardly able to hide her astonishment, is that "it involved intellectuals who were for the most part from universities, and for the most part, employed."[8]

Indeed, things today are hardly different. When Mel Gibson's *Braveheart* (1995) received an "R" rating it was for "brutal medieval violence." Much more recently, the best-selling fantasy novel series turned HBO sensation *Game of Thrones* likewise revels in a grotesque notion of medieval viciousness.[9] On the other hand, whatever myth-busting scholars might do, when thousands—if not millions—can turn on their TV to the *History Channel* and watch a program still titled *The Dark Ages* directed by Christopher Cassel in 2007, the presumption for the viewer would be that not much has changed among scholars, either. The Middle Ages are all around us as a sort of vague stream of impressions, set pieces, and an endless flood-tide of pop culture drawing upon stereotypically "medieval" imagery. "It is a culturally constructed 'medieval' which shapes our understanding of the Middle Ages," writes historian Winston Black. And this is "even for serious students of history."[10] In fact, as another historian David Matthews expertly records, an entire field of study known as "Medievalism" has arisen over the last three decades especially to "hunt these revenants," these apparitions of memory that are responses to the medieval, real and (mostly) imagined.[11]

Swerving into the Dark Ages

While we will get to how they were constructed in the first place, opinions about the "Dark Ages"—especially in terms of its relationship to science—began to change in the latter half of the twentieth century. In fact, the Middle Ages (perhaps somewhat counterintuitively) have been something of a hot topic in studies that are attempting to reevaluate the significance of religious contributions to modern science, or indeed in simply overturning the mythology that serves as a placeholder for actual engagement with this intellectually rich and diverse period. Margaret Osler notes that such reevaluations of contributions in the previously denounced "Middle Ages" have the simultaneous effect of calling into question certain interpretations of the "Scientific Revolution" and as such are a major part of the sea changes in the historiography of science and Christianity that we have already seen. And since both periods—the Middle or "Dark" ages, and the "Scientific Revolution"—are in general taken as pictures

8. Pernoud, *Those Terrible Middle Ages!*, 12.

9. Marsden, "Game of Thrones."

10. Black, *The Middle Ages*, ix.

11. Matthews, *Medievalism*.

and symbols for religion and science in general (much as Wilberforce and Huxley were taken as representing ideal types), these reevaluations of the "periodization" of history simultaneously lead to affecting "general assumptions of the relationship between science and religion."[12]

The historian of medieval science Edward Grant serves as one of the more interesting cases to report regarding a renewed attention to the broader religious, theological, and institutional contexts that set the stage for later scientific theorizing. Grant himself underwent an apparent "conversion" as self-described in the introduction to his *The Foundations of Modern Science in the Middle Ages*. It would be going too far to call him a convert of Duhem, though the trajectory Duhem started certainly influenced Grant, as he notes. He describes how he came to disagree with his own earlier work, *Physical Science in the Middle Ages*: "*Physical Science in the Middle Ages* was written with the conviction that [non-religious interpretations of the rise of science] were essentially correct, and that the Middle Ages had not contributed significantly to the scientific revolution of the seventeenth century." And yet, he says, as time went on

> It occurred to me that perhaps we—historians of medieval science and the scientific revolution—had interpreted the medieval contribution too narrowly in terms of the specific influences it might have exerted on this or that science, usually physics, and on whether it had played a role in reshaping scientific methodology. . . . My attitude changed dramatically, however, when some years ago, I asked myself whether the Scientific Revolution could have occurred in the seventeenth century . . . if the massive translations of Greco-Arabic science, and natural philosophy into Latin had never taken place? The response seemed obvious: no, it could not.[13]

Grant goes on to say that "it is in the Latin Middle Ages in Western Europe" that we must look for the answers to such questions as: why it was that science as we know it today materialized only in Western society? What made it possible for such prestige and influence to be given to scientific activity? The establishment of science, argues Grant, depends on more than simple expertise or technical achievement in experiments. "After all, science can be found in many early societies."[14] Grant therefore argues: "the exact sciences are unlikely to flourish in isolation from a well-developed natural philosophy."[15] We have, for example, at our far end of history, become numb to the metaphor of "natural law" and the complex sets of presuppositions about the world it entails. Yet when natural philosophers and theologians spoke of "natural law," they "were not glibly choosing the metaphor."[16] Laws were expected to be in place

---

12. Osler, "Religion and the Changing Historiography of the Scientific Revolution," 71.

13. Grant, *The Foundations of Modern Science*, xii.

14. Grant, *The Foundations of Modern Science*, 168.

15. Grant, *The Foundations of Modern Science*, 185.

16. Brooke, *Science and Religion*, 26.

even prior to their discovery because of theological theory: "Laws were the result of a legislation by an intelligent Deity." That the scientific endeavor thought it could proceed at all, that the human mind could fathom nature and express it mathematically, found its basis in the assumption that creation expressed God's mind, and that God's mind found a mirror in human thought.[17] Granted that many of the medieval explanations were found wanting and so radically transformed in many instances in the Scientific Revolution. But the problems and modes of inquiry that those like Galileo and Newton worked with—notions of cause, necessity, contingency, infinite space, counterfactual reasoning, void space, nature—were bequeathed as a medieval inheritance: "The revolution in physics and cosmology was not the result of new questions put to nature in place of medieval questions. It was, at least initially, more a matter of finding new answers to old questions . . ."[18]

Nonetheless the sheer amount of high-profile texts not just perpetuating but breathing new literary life into the old prejudices and methods creating and reinforcing the warfare thesis via "The Dark Ages" are legion. One can point to a very recent example in Harvard Shakespeare scholar Stephen Greenblatt, who, in December 2011, ascended the stage at the Cipriani Club in New York City to receive a National Book Award Prize.[19] A 365-page *New York Times* best-seller, the book, *The Swerve: How the World Became Modern*, was ostensibly about the rediscovery of the Epicurean philosopher Lucretius's (d. 59 BC) poem *De Rerum Natura* (*On The Nature of Things*) by a fifteenth-century Italian employee of the pope—the well-traveled Poggio Bracciolini. The poem, representing a philosophical school known as Epicureanism, argued that the world was created by the chance collision of atoms, that the gods do not interfere with our world, and that mankind is therefore free to choose as it will. A brutal summary, but it will have to suffice. Greenblatt's book also won a Pulitzer Prize, and it would eventually net him $735,000 in additional prize money when he was awarded the 2016 Holberg Prize by the Norwegian Ministry of Education and Research.

The epic poem had lain undisturbed on a dusty library shelf in Germany free from the inquisitorial eyes of the monks—who would surely have burned this document (says Greenblatt) if they knew what it was. To be fair, this tale of discovery is worth the price of the book alone. Greenblatt deftly tells this portion of the narrative with his talent for description. It is a truly gripping story of lost history. Less worthwhile, however, is the rest of the book, which is embarrassingly ill-fitted over this fascinating core. Like David stumbling about in his overlarge, donated armor (1 Sam 17:39), Greenblatt's story awkwardly dresses itself in the tired narratives of the Christian Dark Ages, the death of the philosopher Hypatia at the hands of a Christian mob, the repression of Galileo for his science handed down to him via the Enlightenment. His chronicle is also obviously bejeweled with Greenblatt's own extreme displeasure

17. Brooke, *Science and Religion*, 29.

18. Grant, *The Foundations of Modern Science*, 198.

19. Greenblatt, *The Swerve*.

with religion of all sorts. Greenblatt's monks are thuggish layabouts, while others of a more entrepreneurial nature self-flagellate, if only to break up the monotony of their concerted efforts to hold back civilization. Here, scientific thinking is anathema, a matter of indifference, or perhaps, most of all, just a superfluous luxury in a world that considers its time well spent huddling timidly beneath the shadow of an angry God's middle finger.

Unlike David, however, who tells Saul he cannot fight in this bloated frame, Greenblatt's unwieldy epic of history is promoted as the ultimate point of his investigation. Invoking the thrill of discovery that inevitably accompanies stumbling upon some ancient, forbidden secret (if not quite playing the actual theme music of Indiana Jones), Greenblatt argues Poggio's remarkable find hidden in the dust of Christendom began to change everything held sacred in Medieval Europe. In a bracing crescendo of prose Greenblatt describes his book's theme:

> The transformation was not sudden or once-for-all, but it became increasingly possible to turn away from a preoccupation with angels and demons and immaterial causes and to focus instead on things in this world; to understand that humans are made of the same stuff as everything else and are part of the natural order; to conduct experiments without thinking that one is infringing on God's jealously guarded secrets; to question authorities and challenge received doctrines; to legitimate the pursuit of pleasure and the avoidance of pain; to imagine that there are other worlds beside the one that we inhabit; to entertain the thought that the sun is only one star in an infinite universe; to live an ethical life without reference to postmortem rewards and punishments; to contemplate without trembling the death of the soul. In short, it became possible—never easy, but possible—in the poet Auden's phrase to find the mortal world enough.[20]

If this sounds familiar, it may well be because it imitates to a startling degree the same sequence of events and their meaning that Tyndall used in the infamous Belfast address, that was later picked up by his friend Draper. What Greenblatt's book attempts to narrate, in other words, is escape from a world where "curiosity was to be avoided at all costs."[21] This is fantastically—indeed almost systematically—incorrect as one proceeds point by point through Greenblatt's list. Even Greenblatt's choice of quoting Auden at the end is bizarre—the same Auden who wrote that "the contrast between jolly, good-looking, extrovert Pagans on the one hand, and gloomy, introvert Christians on the other was a romantic myth without any basis." Auden continues: "One may like or dislike Christianity, but no one can deny that it was Christianity and the Bible which raised western literature from the dead."[22] While *The Swerve* was

---

20. Greenblatt, *The Swerve*, 10–11.
21. Greenblatt, *The Swerve*, 16.
22. Auden, "Heresies."

warmly received by award committees and the public alike, scholars in the field of Medieval and Renaissance studies found it off base at best, appalling misinformation at worst—what journalist Michael Dirda calls in his review of the book a "strangely unserious" work that reads like a bad attempt to create "a non-fiction pot boiler."[23]

More often than not, Greenblatt turns his wonderful grasp of the English language to the task of gussying up historical interpretations that were beginning to collect dust even a century ago, let alone today. Renaissance scholar John Monfasani, reviewing the UK version, calls *The Swerve's* central premise—that once Lucretius's poem was made available it became a major piece in undoing the medieval picture of the world—a fundamentally "unwarranted assumption," with "virtually no evidence . . . while a massive amount [of evidence] exists for constructing a different story."[24] Greenblatt, he says, has written an "entertaining" but an ultimately "wrongheaded [but] belletristic tale." Historian of Medieval Literature Laura Saetvit Miles writes similarly in her review of the book in *Vox*: "the oppressive, dark, ignorant Middle Ages that Greenblatt depicts for 262 pages [is] simply fiction." In fact, she continues: "it's fiction worse than Dan Brown [author of *The Da Vinci Code*], because it masquerades as fact."[25] She even found its silky narrative taking her along its pathways of sparkling prose, salted as they were with "so many interesting details" and with no uninvited "footnotes to distract me from the story." Amidst the tides of Greenblatt's pleasurable siren song, she found herself "totally swept up." Until she came back to reality. The book is indeed written in "enthusiastic, accessible style" she admits, but it comes "at a devastating, unethical cost: the misrepresentation of a thousand years of brilliant literature, vibrant culture, and actual people."

Not least among Greenblatt's distortions comes by way of painting Lucretius as modern and, in his own way, scientific. In presenting Lucretius in this manner, Greenblatt plumps him—and his rediscoverer Poggio—as lonely heroes heroically subverting the Dark Ages with what he represents as essentially the scientific method. Among any number of problem's Greenblatt's account runs into a very basic one appears nearly immediately: Lucretius was not, by any measure, doing science. His approach to the world actually consisted of a profoundly philosophical—even religious—Epicurean goal known as *ataraxia* (translated broadly: a serene distance or disinterest in the world) that "argued *against* a purely theoretical use of physics."[26] Indeed, "the object of [physical] knowledge, like death, must be 'nothing for us,'" as Remi Brague puts it summarizing the basic Epicurean stance.[27] This meant that understanding the natural world was not a scientific endeavor that sought truly to understand or control, but to accept and set oneself at ease so as to eat, drink, and be merry. In other words,

23. Dirda, "Review of *The Swerve,* by Steven Greenblatt."
24. Manfasani, "Review of *The Swerve.*"
25. Miles, "Stephen Greenblatt's *The Swerve.*"
26. Brague, *The Wisdom of the World,* 38. Italics added.
27. Brague, *The Wisdom of the World,* 38.

Lucretius's physics was a means toward hedonism. Now, some might be all for this—but it still is not science.[28]

This revelation is no doubt frustrating for Greenblatt's vision of the shape of history. The irony is especially thick: Lucretius's atomism only really eventually aided modern science because Christians (who in fact did read the poem before Poggio's discovery)[29] eventually disembedded it from the Epicurean virtue of *ataraxia,* and in the works of those like Pierre Gassendi and Robert Boyle, disabused it of any supposed atheism. However even in the Middle Ages there were many arguing for the use of Epicureanism as well. The French scholastic philosopher William of Conches (1050–1154), when accused of "falling back on the opinion of the Epicureans," responded: "When the Epicureans said that the world consists of atoms, they were correct. But it must be regarded as a fable when they said that those atoms were without beginning and 'flew to and fro separately through the great void,' then massed themselves into four great bodies. For nothing can be without beginning and place except God."[30] Christians had, in other words, already decontextualized many Epicurean ideas from the broader ornamentations of Epicurean philosophy and religion that were actively stopping their "scientific" application to the world. Nor indeed did the medievals live up to the caricature of frantic supernaturalists falling down wells because they were peering up toward God while they walked about. Rather a profound (and theologically justified) naturalism pervaded their thought both in theory and the manner institutional organization of the universities divided subjects and their boundaries.[31] It was not only in this manner that atomism caught on as a great "scientific" idea in Christian Europe, of course, but the efforts of Christian theologians and philosophers were substantial.

Inventing the Dark Ages—Gibbon and the Fall of Rome

Retelling some of the mistakes in *The Swerve* is not meant to pick on Greenblatt, but to use him as an immediately relevant example precisely because his way of framing the Middle Ages is hardly unique. In his famed booked *Cosmos,* for example, Carl Sagan presents a very (in)famous timeline where nothing scientific happened between 415 and 1543. After that great gap of over a thousand years Copernicus—and then, later, Galileo—revived the whole scientific enterprise singlehandedly.[32] We have already seen how common such claims became in Protestant polemics, and later in Tyler and Frazer, but more will come in the next chapter. And the examples continue to expand indefinitely even if one contains the set to bestsellers. In *A World Lit Only By Fire* by

28. Jardine, "Epistemology of the Sciences," 685.

29. Reeve, "Lucretius in the Middle Ages," 207.

30. Quoted in Johnson and Wilson, "Lucretius and the History of Science," 132.

31. Shank, "Naturalist Tendencies in Medieval Science," 37–57.

32. Sagan, *Cosmos,* 171–207.

William Manchester—a prestigious historian and scholar who won both the National Humanities Medal and the Abraham Lincoln Literary Award and who, like Greenblatt, should have known better—really swings for the fences with his concept of "The Dark Ages":

> In all that time nothing of real consequence had either improved or declined. Except for the introduction of waterwheels in the 800s and windmills in the late 1100s, there had be no inventions of significance. No startling new ideas had appeared, no new territories outside Europe had been explored. Everything was as it had been for as long as the oldest European could remember. Shackled in ignorance, disciplined by fear, and sheathed in superstition, they trudged into the sixteenth century in the clumsy, hunched, pigeon-toed gait of rickets victims, their vacant faces, pocked by smallpox, turned blindly toward the future they thought they knew . . .[33]

"They trudged into the sixteenth century in the clumsy, hunched, pigeon-toed gait of rickets victims . . . turned blindly to the future they thought they knew." My goodness. All of this is starting to sound sadly familiar by now, however. Indeed, so widespread are these ideas that they have made their way into most dictionaries and encyclopedias up to as late as the 1990s:

> The 1934 Webster's asserted that "the term *Dark Ages* is applied to the whole, or more often to the earlier part of the [medieval] period, because of its intellectual stagnation." The 1966 Random House dictionary agreed, defining "Dark Ages" as "1. The period in European history from about A.D. 476 to about 1000; 2. The whole of the Middle Ages, from about A.D. 476 to the Renaissance," a description repeated verbatim in its 1987 edition.[34]

So, what happened to cause what we will argue are vast misunderstandings? There is no simple, single path that led to the notion of the Dark Ages, and everything it entails. Given all we have already spoken of in this book, however, there is no need to bury the major reveal: what began as Protestant polemics against Catholics (or even Catholic humanist rhetoric against Catholic scholastics), in turn migrated so that a majority portion of our knowledge "was a legacy of the nineteenth century" and the scholars of the Victorian era that was easily absorbed into the broader debates over the professionalization of science and over evolutionary theory by the likes of Huxley.[35] As one scholar summarizes: "The hostility toward the Middle Ages, indeed the contempt for it that was felt and often expressed by cultural elites during the Renaissance, began in the fourteenth century, but increasingly during the fifteenth, and above all, in the sixteenth century, was intensified in the eighteenth century by the anti-clerical bias of Enlightenment thinkers" and ultimately became a part of the emerging battle over

33. Manchester, *A World Lit Only By Fire*, 27.

34. Gies and Gies, *Cathedral, Forge, and Waterwheel*, 2.

35. Cantor, *Inventing the Middle Ages*, 28.

the concept of "scientist" in the professionalization debates of which we have already spoken.[36] Indeed, while the sparseness of information that they had available to work with is to blame, as historian Norman Cantor suggests, perhaps there is also "something about the Victorian mind—its love for huge entities [we might interject: like "science" and "religion"], vulgarly simple models, hastily generalized and evolutionary schemes—that made it unsuitable for doing lasting work on the Middle Ages."[37] Cantor asks this as a tentative question, but given the evidence we have already accumulated in prior chapters, it seems fair to put it forward as more of an assertion. And so, for example, John William Draper could write:

> The Christian party [in the early Middle Ages] asserted that all knowledge is to be found in the Scriptures and in the traditions of the Church . . . The Church thus set herself forth as the depository and arbiter of knowledge; she was ever ready to resort to the civil power to compel obedience to her decisions. She thus took a course which determined her whole future career: she became a stumbling-block in the intellectual advancement of Europe for more than a thousand years.[38]

In part this came from the German influence upon many Victorian thinkers. With the historians Jules Michelet and Jacob Burckhardt the largely intramural Christian use of the Renaissance and the Dark Ages began to change its nature even further than it had among the Enlightenment *philosophes* who delighted in the battery of the church militant by pointing and laughing at this period of darkness. Both Michelet and Burckhardt narrated the Renaissance's opposition to the Middle or Dark Ages as a secular event emerging out of the Christian West. "To exaggerate slightly," writes one commentator, "Burckhardt was one of the authors of modernity as a literary device."[39] Burckhardt's *The Civilization in Renaissance Italy*[40] is a gorgeous work, written in spellbinding prose. And it is still quite useful in many ways. Like the impressionist painter Cezanne's refusal to try to capture the whole truth of a landscape in a painting, so too Burckhardt refused totality by letting a single element vividly bubble up from the froth of history: Modern Man.[41] "In the Middle Ages both sides of human consciousness—that which was turned within as that which was turned without—lay dreaming or half–awake beneath a common veil," wrote Burckhardt. "The veil was woven of faith, illusion, and childish prepossession, through which the world and history were seen clad in strange hues. Man was conscious of himself only as a member of a race, people, party, family, or corporation—only through some general category. In Italy

36. Le Goff, *Must We Divide*, 59.

37. Cantor, *Inventing the Middle Ages*, 29.

38. Draper, *History of the Conflict*, 52.

39. Josephson-Storm, *The Myth of Disenchantment*, 91–93; cf. Shuger, *The Renaissance Bible*, 2–3.

40. Burckhardt, *The Civilization of the Renaissance in Italy*.

41. Hinde, *Jacob Burckhardt and the Crisis of Modernity*, 216–20.

this veil first melted into air; an *objective* treatment and consideration of the State and of all the things of this world became possible. . . . Man became a spiritual *individual* and recognized himself as such."[42]

It is an amusing exercise to see how much of Burckhardt's famous quote is relying upon the power of rhetoric when we compare it to an equally beautiful quote that is yet opposite in intent. In its famous opening paragraph, Johann Huizinga's[43] *The Autumn of the Middle Ages* announces that: "When the world was half a thousand years younger all events had much sharper outlines than now. The distance between sadness and joy, between good and bad fortune, seemed to be much greater than for us; every experience had that degree of directness and absoluteness that joy and sadness still have for the mind of a child. Every event, every deed was defined in given and expressive form and was in accord with the solemnity of a tight, invariable life style. The great events of human life: birth, marriage, death . . . by virtue of the sacraments, basked in the radiance of divine mystery. But even the lesser events . . . a journey, labor, a visit, were accompanied by a multitude of blessings, ceremonies, sayings." While coming from different backgrounds, we can see how Huizinga and Burckhardt have shaped their stories using similar data. For Burckhardt the childishness of the middle ages is make-believe, and religion is but the dreamlike illusion of a youth that must be cast off for the adult to appear; for Huizinga the child-like nature of the middle ages is rather a clarity and a vividness whose shock and spark ignite by religion's touch upon them. Without glossing many of the differences one can see certain parallels between this emphasis in Burckhardt between the veiled illusion of faith as opposed to the awakened Modern Man, and Tylor and Frazer's notions of humanity coming of age, discarding mythology while embracing science. By placing an emphasis on the uniqueness of the new consciousness supposedly awakened in the Renaissance, a stark division was created between the primitive before and the enlightened after. Ultimately, the changes invoked by Michelet and Burckhardt accelerated even further as Secular Humanists in the twentieth century appropriated the term "Humanist" and so retroactively interpreted the Renaissance Humanists not only as secular, but as explicitly tending toward—or sometimes even fully embodying—a naturalistic atheism that would only later arise at the end of the nineteenth century. Even the *philosophes* and the Enlightenment at large—whose members were in the majority deeply concerned with God,[44] and even with alchemy and magic—were transformed by this later scholarship into more or less prototypical skeptics, even atheists.[45] Even here,

42. Burckhardt, *The Civilization of the Renaissance*, 187.

43. Huizinga, *The Autumn of the Middle Ages*, 1. With these two quotes, we can see the truth in Charles Taylor's analysis that "it was not so much the science which decided things [for the warfare thesis and the loss of faith] as it was a battle between two understandings of our epistemological predicament, tinged with moral import, and related to images of adulthood and childishness" (*A Secular Age*, 388).

44. Trinkaus, "Italian Humanism and Scholastic Theology," 3:327–44.

45. Barnett, *The Enlightenment and Religion*.

however, there was not an abandonment of Christianity so much as its continuation in a different register. The solidification of "the period term 'Renaissance' by the great nineteenth century historians . . . [along with] nineteenth century secularization of historiography . . . raised Christian patterns [of thought] to a new [secular] level by purging them of any specifically religious content or transcendental reference."[46] And so "Modern Man," originally raised to his height by stretching his limbs exploring the transcendent horizon of God, is now self-referential, self-made, the picture of the "genius."[47]

It is from this retroactive interpretation that we inherited the often repeated notion that the turn to man in the Renaissance (or anywhere such a turn happens) was and continues to be a turn away from God. In truth, the Humanists—whatever one might think of their orthodoxy—were never willing to solve theoretical problems by "denying the efficacy or authority of God."[48] As the massive scholarship of Charles Trinkaus has shown, despite the self-description of many Italian humanists about the influence of classic Greco-Roman antiquity, in truth they received their inspiration far more from the church fathers than they did from the pagan ancients.[49] Or perhaps stated in another way: their appropriation of Greco-Roman antiquity was often through the complex theological interpretations that the church fathers themselves had made. The rise of anatomical investigations in this period, for example, has been shown to be part of an intricate demonstration of God's wisdom and glory.[50] Yet, from here parallel arguments can be made involving a particularly secular vision of the Renaissance, small steps can be made in turn to denote the Enlightenment as a purely secular event, the Scientific Revolution as a purely secular event, and so on. Much of this we have seen.[51]

One of the most abiding keys to unlock the invention of the Dark Ages is the image, both ghastly and spellbinding, of the Fall of Rome. From there, it is said, the West bloomed with a darkness that did not relent for a thousand years. This perhaps sounds like the premise of some fairy tale—and indeed it is. There was a darkness of mind in this era, it is said, to rival even the charcoal black remains of the monuments

46. Lupton, *The Afterlives of the Saints*, 4.

47 On the notion of the self as a legacy of Christianity that nonetheless does undergo (often theologically loaded) transformations in the modern period, see: Brague, *The Kingdom of Man*; Dupré, *Passage to Modernity*, 93–166; Pfau, *Minding the Modern*; Siedentop, *Inventing the Individual*; Taylor, *Sources of the Self*.

48. Gillespie, *The Theological Origins of Modernity*, 86.

49. Trinkaus, *In Our Image and Likeness*.

50. A. Cunningham, *The Anatomical Renaissance*.

51. For examples overturning the usual stories, see Bulman and Ingram, eds., *God in the Enlightenment*; Firestone and Jacobs, *The Persistence of the Sacred in Modern Thought*; Harrison and Roberts, eds., *Science Without God?*; Brooke, *Science and Religion*; Funkenstein, *Theology and the Scientific Imagination*; Gaukroger, *The Emergence of a Scientific Culture*; Josephson-Storm, *The Myth of Disenchantment*; Osler, ed., *Rethinking the Scientific Revolution*; Hooykaas, *Religion and the Rise of Modern Science*; Klaaren, *Religious Origins of Modern Science*.

of Roman civilization, the broken teeth of a silenced culture. In the east, where the sight of ruins was less present, Byzantium was associated not with preservation but with visions of a suffocatingly ornate "autocracy, bureaucracy, deviousness, and a stultifying lack of creativity."[52] And so our vision of darkness is completed by incorporating two poles into itself: in the Christian West, the darkness of chaos. In the Christian East, the darkness of an impenetrable labyrinth. Byzantium has the added drawback of appearing to Westerners not only as primitive but also as exotic, and obscure, "relegated to the sphere of negativity." Indeed, "even the very name we use today—'Byzantium'—was a derogatory coinage of the early modern period."[53] Draper, for example, described Byzantine as "profess[ing] to cultivate morals, but [instead] it crushed the mind."[54]

Nonetheless, the fascination with something known as "The Fall of Rome" that set the course for Western civilization as the pattern of every perceived decline, and which was the fault of Christians at large, is a product of the eighteenth and nineteenth centuries. No work is more famous or more influential for this than Edward Gibbon's sprawling, multivolume *The History of the Decline and Fall of the Roman Empire*, the first volume of which landed like a voice booming down a secular Mt. Sinai on February 17, 1776. "Rather strikingly, Gibbon . . . strongly distinguished himself from the theologians, who might have described 'Religion as she descended from Heaven, arrayed in her native purity,' whereas in his view the historian has the 'melancholy duty' to discover the inevitable mixture of error and corruption' here on earth (1.446)."[55] With these words, Gibbon separated himself from "church history as practiced until his time and indicated that his would be a secular history."[56] In many ways we can see Gibbon as a precursor for Tyler and Frazer—in fact Frazer styled *The Golden Bough* explicitly after *The History of the Decline and Fall of Rome*.[57]

Just two days after Gibbon's first volume was published, it was already hailed as a classic. Soon after, the famous philosopher and skeptic of Christianity David Hume wrote to Gibbon commending him on the volume. By way of a slow burn building into his eventual crescendo against Christianity, Gibbon relates his history with an incredibly vast reading, wit, and insight. It is, whatever one thinks, a work of stunning scholarship. There are, all told, 8,350 of the recently invented apparatus of the academic footnote across his tale. And in them are hidden many of his most daring, fiercest fire-tinged lances of playfulness, withering critique, and observation.[58] In the first fourteen chapters Gibbon relates the slow decline of Rome, until the final victory

52. Cameron, *Byzantine Matters*, 10.

53. Cameron, *Byzantine Matters*, 9.

54. Draper, *History of the Intellectual Development of Europe*, 2:129.

55. Bremmer, *The Rise of Christianity*, 121.

56. Bremmer, *The Rise of Christianity*, 7.

57. Wheeler-Barclay, *The Science of Religion*, 189–90.

58. Bremmer, *The Rise of Christianity*, 8.

of the Emperor Constantine through 324 AD, the year before the pivotal Christian Council of Nicaea. Yet, there is, to this point, not a single mention of Christianity.[59] Then, suddenly, a crescendo; the hammer drops. In the last two chapters of the first volume the reader suddenly realizes what Gibbon means by the "triumph of barbarism and religion." The barbarian hordes that Gibbon records conquered Rome were in his mind of a piece with Christianity in eroding the eternal empire. But, being persuaded by the mass of Gibbon's great reading and erudition that had preceded it, suddenly even the Christian had to take his shocking plot twist seriously. The trap, at last, had sprung.[60] Gibbon—his own work something of a theological treatise in disguise[61]—to be sure, described many of the positive qualities of Christianity that allowed it to persevere and cut through the empire like a tongue of fire. The hope given by the resurrection of the dead in Christ, in particular—as it happens the earliest declaration of the gospel in Acts—was a major glory.[62] So taken with this Christian theme was the ancient world, that from the time of Nero (37–68) the theme of resurrection began migrating into Roman literature, which tried to emulate it and synthesize it with its own myths.[63] But these were merely flowers cast upon the grave Gibbon had dug for Christianity.

Gibbon has since then cast a long shadow over historical scholarship. Pulling back on overt polemics due to initial stinging criticisms, his vendetta did not disappear but ran underground and so became all the more dangerous, baptized now with the veneer of objectivity. Despite the fact that by at least 1995—if not sooner—"no responsible historian of the ancient or the medieval world" would want to any longer include the fall of Rome as either "a fact or a paradigm" because "[the notion of the Fall of Rome] is a construction that has its own place in modern history"[64] (one work of scholarship, in fact, recorded no less than 208 different interpretations of the Roman fall or transition), "The Fall of Rome" at the hands of Christians nonetheless continues to fascinate the imaginations of popular historians and popular media alike. Gibbon's shadow has yet to relent its grasp, and it is a shade that—especially in the strange world of the internet—allowed all sorts of nasty misunderstandings to sprout up again and again. How does it cling on so determinedly in the face of so much contrary evidence? We spoke in our introduction and in the chapters that have so far followed that "clever metaphors die hard." In Gibbon's case as well, this holds true. When one visits Rome or Greece today, immediately striking are the ruins that stand in a quiet but broken obstinance against the ravages of time. Walking among them,

---

59. Bremmer, *The Rise of Christianity*, 6.

60. Bremmer, *The Rise of Christianity*, 5.

61. Brown, *Society and the Holy in Late Antiquity*, 49–62; ("In Gibbon's Shade") at 52.

62. On the notion of resurrection in early Christianity, see Bynum, *The Resurrection of the Body in Western Christianity 200–1336*.

63. See the study of Bowersock, *Fiction into History*, esp. 99–119.

64. Bowersock, "The Vanishing Paradigm," 42.

it is easy to feel both greatness and loss, and from there, perhaps, conjure in one's mind a time-lapse view of buildings and life springing up only to be broken down by roving barbarians and unwashed religionists. The great ages of time and their great measures of both magnificence and sadness wash over us all at once. This is in fact how Gibbon first envisioned the scope and shape of his work: "It was among the ruins of the [Roman] Capitol that I first conceived the idea of a work which has amused and exercised near twenty years of my life," he says.[65] And the impact of Gibbon's ruins are hard to overstate. "Modern readers will hardly realize that there had not been any such narrative history before Gibbon, the sole exception being the slightly earlier *History of England* by the great philosopher, and Gibbon's older acquaintance, [the equally anti-Christian] David Hume."[66] Once again, at its origin, history as a discipline sounds a peal of the songs of war. Early Christianity became in these stories a decadence that weighed down not just civilization, but the historical course of the human spirit.[67] This can be seen especially in Gibbon's use of the stories of the burning of the library of Alexandria, and the death of the philosopher Hypatia, and their reception after Gibbon, which still reign as the two dominant token narratives to decry Christian destruction.

The Burning of the Library at Alexandria

It is hard—perhaps impossible—to overstate how influential Gibbon's story about the burning of the wonders of ancient knowledge has been on the last 300 years of opinions about Christianity. Carl Sagan created a chart about the thousand years of scientific stagnation that the Dark Ages wrought that—in the last TV episode of his famous *Cosmos* (and the book of the same name)—paints the tale of the great fall of the Alexandrian library to the forces of superstition and religious madness. Similarly, Jonathan Kirsch in his *God Against the Gods* writes, "In 390 . . . a mob of Christian zealots attacked the ancient library of Alexandria, a place where works of the greatest rarity and antiquity had been collected . . . some 700,000 volumes and scrolls in all. The whole collection of parchment and papyri was torched, the library itself was pulled down, and the loss to Western civilization is beyond calculation or even imagination."[68]

More recently, Catherine Nixey's manifesto volume *The Darkening Age* is a good example of just how embellished such accounts can become. She drops us into the scene of the final hours of the library, to gaze upon the horror as her imagined pagans might:

> Intellectuals looked on in despair as volumes of supposedly unchristian books—often in reality texts on the liberal arts—went up in flames. Art lovers

65. Quoted in Bowersock, "The Vanishing Paradigm," 37.
66. Bremmer, *The Rise of Christianity*, 2.
67. Frend, "Edward Gibbon."
68. Kirsch, *God Against the Gods*, 276.

watched in horror as some of the greatest sculptures in the ancient world were smashed by people too stupid to appreciate—and certainly too stupid to recreate them. The Christians could not even destroy effectively: many statues on many temples were saved simply by virtue of being too high for them, with their primitive ladders and hammers, to reach.[69]

Such melodramatics are sustained for the course of an entire book, in fact. As with Greenblatt, a tiny army of academics came out against this tide of disdain playing dress up as history as the mythmaking it surely is, but again too late. Like Greenblatt, Nixey managed to find herself awarded Book of the Year by *The Telegraph, The Spectator, The Observer,* and even the *BBC History Magazine.* Yet Dame Avril Cameron—one of the leading Byzantine and Medieval scholars in the world—writes for example that "Hearts will sink among historians of early Christianity and late antiquity, as well as medievalists and, needless to say, Byzantinists when they see [Nixey's] pugnatious and energetically written little book" that is reviving the hoary but completely incorrect "blame the Christians" model of things.[70] Cameron continues, "a quick look at the citations in Nixey's footnotes shows what she has been reading," namely "a small group of like-minded historians equally hostile to Christianity." One is not surprised to find a copious amount of references to Gibbon among them.

Historian Tim O'Neill has written an even more devastating and in-depth review of Nixey, who penned "easily one of the worst books I [O'Neill] have read in years."[71] Not only are Christians represented as cretinous, maddened bulls in the China shop of ancient wisdom—to add insult to injury they are not even very good at this, either. One is left a bit mystified at Nixey's decision to further ornament the story with the wholly invented image of Christians—like cavemen, one suspects—clumsily wielding primitive ladders and hammers. Perhaps she honestly believes that a few Christians, armed with their tiny stepping stools, would truly look in awe upon the exotic technological innovation of a tall ladder with many rungs. As O'Neill points out, one suspects that the Christians building the fifty-five meter (180.45 feet) tall Hagia Sophia—completed in AD 537—would find themselves a bit perplexed about Nixey's truly bizarre opinions. Surely in the time that passed after the Serapeum attack such troglodytes must have tortured that arcane technology of ladder-making out of a few hapless pagans? Readers of Nixey's book, sadly, would be left in a position where this question might even be given a moment's pause of consideration. Regardless, however much Nixey and Kirsch embellish the tale further, their accounts in essence repeat Gibbon's original observation: "The valuable library of Alexandria was pillaged or destroyed; and near twenty years afterwards, the appearance of the empty shelves excited

---

69. Nixey, *A Darkening Age,* xxxiv.
70. Cameron, "Blame the Christians."
71. O'Neill, "Review: Catherine Nixey, *A Darkening Age.*"

the regret and indignation of every spectator whose mind was not totally darkened by religious prejudice."[72]

To cut a somewhat lengthy story short, despite some historical records we do not really know who founded the library, or when, or through what strategies of collection, and the like. Nor, in the same vein, do we have an accurate grasp of what, or how much, was contained in its stacks. As such "unless some long-lost catalogue of the library of Alexandria has recently turned up in some pawn shop in Cairo, the list of works that Kirsch claims the library possessed is sheer fantasy."[73] The numbers passed down to us through Gibbon and then popularized again in works like that of Kirsch's are, in effect, an illusion (Greenblatt, we should note, opts for "at least 500,000" volumes in the library).[74] Of course, the larger the library, the more severe the loss can be placed in the lap of the Christians. No one wants to blame their opponents for a tragedy that is only putting up rookie numbers. More poetically, this involves what Roger S. Bagnall calls the "three dreams" of the library that twist historical perceptions of this storehouse of knowledge at Alexandria—dreams about the size of its collection, dreams about who destroyed it, and dreams of this destruction's consequences.[75] For our purposes, we might rather talk about two tricks of rewriting. On the one hand, the vast swath of ancient learning used to contrast the Christian Dark Ages "crudely put, consists of taking the Alexandrian achievements, beginning with Euclid, and spreading them with a trowel over the entire Roman empire until the murder of Hypatia."[76] On the other hand, the outrage of the burning of the Library relies in part on the opposite assertion, that at Alexandria, and only Alexandria, was this ancient wisdom truly preserved. One has their cake and eats it too, it seems.

"It is reasonably obvious," Bagnall observes, "that the ancient sources thought the libraries were enormous, but had no good figure to work with."[77] This plural that Bagnall uses to describe *libraries* is also key. The ability to strictly account for the destruction of the library complexifies considerably when we consider that there was not just one library of Alexandria, but several. Which of course leads to the obvious question: Which library? Whose destruction? A clear problem is that popular accounts like Gibbon or Sagan or Kirsch's often collapse multiple libraries into one when they tell their tales of Christian destruction.[78] We may, in addition, picture a storehouse of scientific and literary knowledge, much like a searchable database or a cumulative research program for posterity. The Alexandrian Library was not, despite the rhapsodic flights of fancy that surround it, a center of either technological or scientific

72. Gibbon, *Decline and Fall*, V:ch.28.

73. Hart, *Atheist Delusions*, 37.

74. Greenblatt, *The Swerve*, 88.

75. Bagnall, "Alexandria."

76. Shank, "Myth 1," 7.

77. Bagnall, "Alexandria," 352.

78. Hart, *Atheist Delusions*, 37.

innovation. Much as Greenblatt overlooks the religious and philosophical contexts of Lucretius's atomism, so too many overlook that the Library, while revolutionizing storage by being filed alphabetically, was understood to be organized at a broader thematic level around the Nine Muses, goddesses of inspiration and insight.[79] Carl Sagan in his program *Cosmos* gingerly gets around the notion of religion by noting that though the Serapeum was originally a temple, it was eventually "reconsecrated to knowledge." This is, as one can see, trying to place a quite stark contrast between anything religious and knowledge. As O'Neill puts it, however, "[Sagan's assertion] is nonsense. The Serapeum was always a temple and was not 'reconsecrated' to anything. Libraries were often established as adjuncts to temples, but it seems Sagan was attempting to distance . . . the Great Library from the temple in which it sat because this did not quite fit his theme of secular knowledge's superiority to 'mysticism.'"[80]

These mathematical and historical gymnastics are, however, ultimately beside the point of our inquiry. Even if there were 700,000 scrolls, even if they contained the height of secular, scientific wisdom, it was not Christians who destroyed them. Even by Hypatia's time the library was but a memory. The most popular ancient candidate is, surprisingly, none other than Julius Caesar who, in a defensive military action in 48 BC burned the Alexandrian harbor in the region of the royal quarter. In fact, ironically enough, Aulus Gellius who is also our major source for the incredible number of 700,000 volumes often used to magnify the Christian crime, quite literally in the next few sentences mentions that "but these were all burned during the sack of the city in our first war with Alexandria."[81] One is tempted to remark that an entire alternate history has been conjured because a few influential folks quit before finishing their read of the whole paragraph. If this location of the library did survive this, other scholars note it probably perished along with the whole museum due to the wars waged by emperor Aurelian in AD 272. There are accounts of a daughter library being founded—perhaps in a structure called the Serapeum that will play deeply into the story of Hypatia's death. But we have no good evidence of a library at the Serapeum aside from perhaps a few scrolls being housed there. When the Serapeum—a pagan temple dedicated to Serapis, a god designed by Ptolemy I Soter to combine aspects of the personalities and functions of Osiris, Zeus, Bacchus, Helios, and Hades to unite Egyptian and Greek subjects—was assaulted by Roman soldiers and, yes, by Christians in AD 391, not even the vociferous anti-Christian Eunapius of Sardis mentions anything about a library also being destroyed when he would have had every reason to pin it on Christians for his own political gain.

At the end of the day the real culprits are probably much more mundane than emperors and their wars, or roving bands of book-burning zealots. While there may have been a few moments that amounted to a grand ending for the library, it was

79. See Casson, *Libraries in the Ancient World*, 31–47.

80. O'Neill, "The Great Myths 5."

81. Aulus Gellius, *Attic Nights*, VII.17.

already being consumed from within by a thousand smaller sorrows. Libraries in the ancient world (and even today) face a constant battle against the destruction wrought by a small and determined army of mice, insects, mold, and fires. Staffing and monetary issues involved in maintaining such an enormous collection also no doubt played their part. As such, while Gibbon did not wholesale invent the story, his particular way of arranging the sources to find Christians lurking in the shadows with their torches and pitchforks has latched on to the popular imagination ever since. Bagnall notes his own somewhat traumatic scholarly experience of redaction that testifies to the staying power of the myth of Alexandria. He had been asked to write a description of the destruction of the Library for a short-lived magazine called *The Dial*. His editor at the time soured considerably over Bagnall's caution about attributing blame for the destruction, and promptly—without telling Bagnall—rewrote the whole article blaming Christians for the affair.[82] Complexity, restraint, and nuance are hard sells in a market where something only leads if it bleeds. But while Bagnall's editor is a particular extreme anecdote, the general impulse is so widespread that, with only slight exaggeration, it can be said to represent a methodology that pervades the entire cottage industry dedicated to perpetuating the image of the "Dark Ages."

The Murder of Hypatia of Alexandria

Another major historical event that Gibbon used in concert with the burning of the library to show the descent of Rome into religious barbarity was the murder of Hypatia. Hypatia was a renowned female Greek philosopher, daughter to Theon Alexandricus, the mathematician and possible curator of a library in Alexandria (sometimes for dramatic effect portrayed as *the* Library, as we have seen). She is, in fact, popularly thought of as "the last Librarian," and "the last Hellene," thus tying her into the broader narrative of the epic of darkness wrought by Christians. This despite the fact that the main outposts of the Library of Alexandria, we might remind ourselves from the last section, were already but a distant memory during her life.[83] She was murdered, so some versions go, in the Serapeum—identified illicitly, as we saw above, with the Library of Alexandria—as Christians tore it down. "Asked who Hypatia was," writes Maria Dzielska, "you will probably be told 'she was the beautiful young pagan philosopher who was torn to pieces by monks (or, more generally, by Christians) in Alexandria in A.D. 415."[84] As the story goes, the precocious Hypatia in the full beauty of her youth blossomed into one of the leading intellectuals in Alexandria and, in so doing,

82. Bagnall, "Alexandria," 357.

83. Watts, *Hypatia*, 14: "The physical Museum and the Royal Library both disappear from the historical record in the mid-third century. The last reference to members of the old Museum appears in materials dating from the 260's. By the year 300, it appears that both the original Museum and the Royal Library had been destroyed, probably during the emperor Aurelian's campaign against the Palmyrene queen Zenobia in 272 CE."

84. Dzielska, *Hypatia of Alexandria*, 1.

attracted the jealousy and the wrath of a particularly significant figure in the history of Christianity, St. Cyril of Alexandria (378–444).[85] Known for his tireless efforts in the theological and philosophical arguments over the identity of Christ as both God and man in the fifth century, Cyril also had a shrewd sense of political power and grew enraged at the rumors of a rising competitor for influence in Alexandria. He was, so it goes, especially wary of Hypatia's sway both as pagan philosopher and as a female.[86] Above all, however, his lust for worldly goods lay at the root of his scheming.

Cyril began refusing to let Christians attend her lectures, and through his intricate connections sowed discord against her. Following the damage caused by a mob war between Cyril's followers and Orestes—the power-hungry Christian prefect of Egypt—Cyril spread a rumor that the resulting damage and unrest were, in fact, caused by the secret conspiracies of Hypatia who was at pains to keep Orestes and Cyril from reconciling because of her hatred for Christianity. A mob of Christians led by a man known as Peter the Reader, enraged at this, swarmed like a froth of ants out from the dark depths of their colonies and fell upon the Serapeum in a tantrum of zeal. Armed with blade-sharp oyster shells (as Gibbon says) or pottery shards (as records the late-fourth-/early-fifth-century Christian historian Socrates Scholasticus), the swarm drug her through the streets, flayed her alive, and then dismembered the body. "The just progress of inquiry and punishment was stopped by seasonable gifts," wrote Gibbon. "But the murder of Hypatia has imprinted an indelible stain upon the character and religion of Cyril of Alexandria."[87]

As historian Bryan Whitfield summarizes the matter, all of these accounts "may be high drama. But they are poor history."[88] Especially filtered through the work of the Victorian Anglican Charles Kingsley, given inspiration by Gibbon's use of Cyril as a display of the lamentable qualities of the Catholics Gibbon took pleasure in eviscerating, the figure of Cyril becomes a thinly veiled caricature of Kingsley's own clerical opponents, the (what Kingsley takes to be) reactionary, authoritarian, and spiteful Oxford Tractarians, rather than the historical Cyril himself. The Tractarians (named such for their distribution of a series of publications from 1833–1841 called *The Tracts of the Times*) were focused upon emphasizing the compatibility of Roman Catholicism with Anglicanism. They were as such often caricatured by their opponents as being heavy handed, closed to "pagan" or secular ideas, and caught the brunt of the still pervasive anti-Catholicism everywhere ready to hand.[89] Like a ready-made voodoo doll, the image of a dirty old and power-hungry man Cyril leering over the beautiful pagan Hypatia soon lost the slight nuances that its origins in intra-Anglican debates afforded it, and once again became a Victorian symbol for authoritarian Christianity

---

85. For an introduction to Cyril, see McGuckin, *Saint Cyril of Alexandria*.

86. Dzielska, *Hypatia of Alexandria*, 1–26; Watts, *Hypatia*, 1–5.

87. Gibbon, *Decline and Fall*, 1:109–10.

88. Whitfield, "The Beauty of Reasoning."

89. See Jenkins, *The New Anti-Catholicism: The Last Acceptable Prejudice*, 177–206

at the ready for use in any conflict of the time—evolution, the professionalization of science, mocking reactionary Catholicism, and so on. Above all, given his place as one of the foremost ecclesiastical figures in the history of Christian dogma Cyril perfectly represented the the violently closed-off attitudes of theologians to strange, new, "pagan" ideas.

Especially in Alexandria among the greats like Clement and Origen, however, Christians were in fact fascinated with pagan philosophy and learning. A major portion of almost any given theological work including at least tacit reflection upon the appropriation, modification, and rejection of pagan thought.[90] Nor, we might add, was Hypatia, despite her obvious talents, the inventor of such things as the hydrometer or the astrolabe as some claim, hoping perhaps to swell her myth further and create a starker contrast between pagan advancement and Christian decay.[91] These Christians often found themselves fast allies or, at least, interested interlocutors, with pagan philosophers, who likewise (to the surprise of many) drew from Christian thought. Telling the whole story of such complicated relationships would indeed be impossible here, but in summary instead of warfare it is in fact a truism of those who work in the history of theology to note that not only is pure and simple conflict not easy to find, it is "extremely difficult to find real-life evidence for the *existence* of both sides of this confrontation as characterized in these two ideal models [of 'theology' against 'philosophy']."[92] Hypatia in particular taught in a "confessionally neutral atmosphere which was neither particularly hostile to Christianity nor dependent upon sacerdotal paganism."[93] As such Hypatia's flavor of philosophy was found to be quite amenable to Christian sensibilities regarding God and the world.[94] Hypatia was, in spite of her portrayal as an enemy of Christianity, in fact good friends with many Christian intellectuals—some of whom were even her students—including Synesius of Cyrene (d. 414) who was the Bishop of Ptolemais.[95] Synesius considered himself a "philosopher-bishop" (a play on Plato's notion of "philosopher king"), and reflected a broader Christian ethos appreciating many facets of Hellenism after they had been appropriated and were no longer pagan.[96] Even in his new faith, and his new position, he would until her death send Hypatia letters, along with works he himself recently wrote, looking forward to each correspondence and her reflections. We have a letter from the year of his death in 413 where, in the thick of grief after the loss of his three sons as well

---

90. For a sampling of this, see Marenbon, *Pagans and Philosophers*; Pelikan, *Christianity and Classical Culture*; Karlowicz, *Socrates and Other Saints*.

91. Hart, "Hypatia Reassembled."

92. Karlowicz, *Socrates and Other Saints*, 7.

93. Bregman, *Synesius of Cyrene*, 24.

94. On this see the study on Hypatia's student, the Bishop Synesius: Bregman, *Synesius of Cyrene*.

95. Hart, *Atheist Delusions*, 46.

96. On the general relation of Christianity and Hellenism at the time, see Pelikan, *Christianity and Classical Culture*.

as the mounting burdens of his office as bishop, he sought comfort in the advice and invention of his great master, Hypatia. While Synesius converted to Christianity only after his student years in Alexandria, many of his fellow students under Hypatia were Christians the whole time.[97]

Despite the widespread image of Christians destroying temples, and pagans violently defending them, recent archaeological evidence has also caused scholars to conclude—in the face of popular opinion—that temples were rarely if ever either destroyed, or even repurposed. Christian sources that used violent imagery to describe the Christian victory over these temples have been shown to be an attempt at a sort of pious hyperbole—often in the name of conveying the spiritual victory of Christianity—but do not describe the reality enacted on the ground.[98] Broadly speaking, temple preservation and restoration—sometimes even directly ordered by Christian rulers—was more common than either repurposing or destruction, especially in Italy, Greece, and Africa.[99]

As such, the dualisms many want to use to tell the story of Hypatia's tragedy begin to collapse. Times were, in a word, complex. And all of this even before we have bothered to mention the small point regarding the actual teachings of Christ to love one's enemies, turn the other cheek, and like Peter, to put away one's sword in his defense (one could go on). But of course, history is a messy, roiling cloud of people acting irrationally, angrily, and often, as it happens, all of this in large and terrifying groups. Amidst a violent age, Alexandria was one of the ancient world's most violent cities. Socrates Scholasticus records in the chapters before his retelling of the death of Hypatia that mounting swells of mob violence of all parties and persuasions were wracking Alexandria, and such squalls could erupt with startling swiftness at even the smallest hint of controversy. These violent delights and their violent ends reflect well on no one—Christian, pagan, Jew, or otherwise. All were involved, and all constituted various currents surging and seizing into one another within the pressure system of the great city. "What one will find [in Alexandria] is that pagans and Christians alike had their scholars and philosophers, who frequently studied at one another's feet regardless of religious adherence," and that both "also had their cruel, superstitious, violent rabble, and that the priests of both traditions were as likely to occupy one class as the other."[100]

In the mid-nineteenth century, however, during the same period of rapidly rising anti-clericalism that we have seen along with the professionalization debates over science the legend of Hypatia perhaps unsurprisingly also reached its zenith.[101] It transformed into striking and deeply mythoogical renditions, and derives largely from

97. Bregman, *Synesius of Cyrene*, 24.

98. Lavan, "The End of the Temples," xv–lxv.

99. Lavan, "The End of the Temples," xxxvii.

100. Hart, *Atheist Delusions*, 40.

101. Watts, *Hypatia*, 139–46.

Gibbon and his nineteenth-century followers. Following their positivist sensibilities, White and Draper along with many others in the latter half of the nineteenth century, slotted Hypatia in to their warfare narratives as, primarily, a scientist who—quite conveniently—also embodies in a rather suspicious way the same conflicts that these Victorian era men felt they had with some of the clergy of their own day. For Draper, writing in his *History of the Intellectual Development of Europe*, there were, we might recall, two powers contending for the European soul—the spirit of free and diligent inquiry, and the smirking repose of arrogant dogmatism holding it back. Hypatia, of course, embodied the first. Her opponents, the second. She was "a valiant defender of science against religion," Draper says. In fact, her death was one of those moments "in which general principles embody themselves in individuals. It is Greek Philosophy under the appropriate form of Hypatia; ecclesiastical ambition under that of Cyril." Again, a clash within history—whatever its nature—is elevated to the world-historical in its significance by Draper.[102] We might recall that Greenblatt, too, sees in Hypatia the incarnation of Greek wisdom in general—despite the fact that she, a Platonist, and Greenblatt's hero, Lucretius the Epicurean, would have been intellectual opponents.

Of course this had a precedent. An earlier and rather rambunctious specimen of anticlericalism is John Toland's comically titled work of 1720, the *Hypatia, Or, the History of the Most Beautiful, Most Virtuous, Most Learned and in Every Way Accomplished Lady, Who Was Torn to Pieces by the Clergy of Alexandria, to Gratify the Pride, Emulation, and Cruelty of the Archbishop, Commonly but Undeservedly Titled St. Cyril.*[103] If that is its title, one can only imagine the florid and sprawling prose awaiting within the book proper. In this style—if not quite with Toland's flourish—many others followed suit. More contemporary to Huxley, Draper, and White is the poem *Hypatie* by Charles Leconte de Lile, where she becomes something like the incarnation of wisdom herself. There are, in fact, two versions of the poem. The first, written in 1847, paints Hypatia as merely the victim of the inexorable and passionless laws of history, detached from any system, religion, or government; here, there is no hint of a Christian plot. Then in 1874, as the X-Club's anti-clerical sentiment was reaching its peak, de Lile reworked his poem to place a distinctly anti-Christian hue upon the death of his fair lady: "The vile Galilean cursed you," he says in the second edition. And he continues, describing how Hypatia lives now, immortal in the Western imagination and the Hellenic inheritance of Europe:

> She alone survives, immutable, eternal;
> Death can scatter the trembling universes
> But Beauty still dazzles with her fire,
> and all is reborn in her.
> And the worlds are still prostrate beneath her feet.[104]

102. Draper, *Intellectual Development*, 238–44.

103. Dzielska, *Hypatia*, 2.

104. Dzielska, *Hypatia*, 5.

The language echoed from Scripture ("every knee shall bow, every tongue confess . . ." Rom 14:11), or perhaps from the Wisdom of Solomon ("Wisdom, She is more mobile than any motion; because of her pureness she penetrates all things . . . in every generation she passes into holy souls and makes them friends of God," 4:24–28) along with the menacing portrait of "the Galilean" means quite literally that the poor woman has been turned into an anti-Christ. Hypatia as such no longer dies just once, one tragic day in 415. Like the "Lamb, slain from the foundations of the world" (Rev 13:8) she bears the weight of doxologies that begin in earnest in the eighteenth and nineteenth centuries, where "all have used the figure of Hypatia to articulate their attitude toward Christianity."[105] Some for good, most for ill. Her bones and flesh are torn now not just by shells, but she is flayed as grist pressed under the millstone of myth, grinding history down into something more perennial. Like the Greek myth of Dionysius, Hypatia's rendered flesh is reassembled and transcends into a new unity, only to be splintered again into many fragments of useful myth; here she becomes a "symbol of . . . sexual freedom"; there she is a sign for "the decline of paganism—and with it the waning of free thought, natural reason, freedom of inquiry."[106] Reborn in our collective memory, she incarnates "the spirit of Plato and the body of Aphrodite"[107] (pay no mind to the fact that she died, not a precocious youth, but in her sixties).

In the end, the best—though perhaps not most emotionally satisfactory—explanation for what happened is that, in one of the many local epileptic fits of mob justice of the time, one responding to other, prior acts of mob justice in the circle of violence that was Alexandria, the great Hypatia became a casualty. Her execution was but one sad example in a long line of rival public purifications that roiled the bowels of the great city.[108] Whoever was ultimately responsible, her death was nearly immediately lamented by Christians as well as pagans as an abomination. Socrates Scholasticus records quite clearly that Hypatia fell "to the political jealousy which at that time prevailed." And though he sees Cyril as having a hand in it (though as scholars note we have no direct evidence of this), Socrates goes on immediately to note that if this is true "that nothing could be farther from the spirit of Christianity."[109] The pagan Damascius, similarly, lamented it as "the most ungodly murder of all."[110]

Hypatia's death, like many, is a surd; a silence—a thing that cannot be grasped, that cannot be reduced to an essence or tamed by a category like "the warfare of science and Christianity." It is an event. The horrible "once" that cannot be a "once upon a time." We must lament her death, and condemn whatever forces (including Christian ones, like Cyril of Alexandria's implacable ego) that may have contributed to it. But

105. Cited in Dzielska, *Hypatia*, 135–48.
106. Dzielska, *Hypatia*, 102; Molinaro, "Hypatia."
107. Dzielska, *Hypatia*, 9.
108. Haas, *Alexandria*, 87–90.
109. Scholasticus, *Ecclesiastical History*, VI.15.
110. Damascius, *The Philosophical History*, 131.

beyond that we must leave her well alone. The fact is, however, that we do not. She has, as Whitfield concludes, "suffered a fate worse than neglect: she has become a symbol."[111] And so from an unquiet grave her spirit is roused again and again as witness against the very Christians she often counted among her acquaintances and friends and students and patrons. Forced back into existence by the necromancy of popular historians, her shadow is charged with the burden of damning men and women who had nothing to do with her death, who would have more likely than not been disgusted by it, with the accusation of breaking an entire epoch of history in an age of darkness. Though this is often done under the auspices of honoring her science, her brilliance, her feminism, she is in fact turned into little more than a grisly marionette for those who, impatient with the messy and often unresolvable complexities of the past, need an easy hero to bandy about. She becomes "a beautiful empty vessel," writes one historian reviewing her case, "whose virtue and learning existed only to be destroyed in a way that invited an author to assign blame."[112]

111. Whitfield, "A Reexamination of Hypatia of Alexandria," 14.
112. Watts, *Hypatia*, 137.

—Chapter 9—

Against the Dark,
A Tall White Fountain Played[1]

Rethinking the Dark Ages (Part Two)

There is one matter that has amused me greatly: every now and then a critic or a reader writes to say that some character of mine declares things that are too modern, and in every one of these instances, and only in these instances, I was actually quoting four-teenth-century texts. There were other pages in which readers appreciated the exquisite medieval quality, whereas I felt those pages were illegitimately modern. The fact is that everyone has their own idea, usually corrupt, of the Middle Ages.

—Umberto Eco, Postscript to The Name of the Rose

AT 2 PM ON October 28, 1998, an auction starring a charred, stained, and barely legible twelfth-century medieval manuscript on prayer—called the *Euchologion*—began in earnest at Christie's Auction House in New York City. Not much to look at to the undiscerning eye, the manuscript felt even more unsubstantial when one considered that earlier that morning Marie Curie's doctoral manuscript, a copy of Einstein's 1905 paper on general relativity, and a first edition copy of Darwin's *On the Origin of Species* had exchanged hands. Later that afternoon, Wilbur Wright's first published account of the test flights at Kittyhawk would be sent on its way by the hammering of the auctioneer's gavel. It was, as Reviel Netz and William Noel put it with some understatement, a busy day.[2] To make matters worse, a lawsuit had been filed just the day before against the auction house by the Orthodox Patriarchate of Jerusalem, claiming that this

---

1. Taken from Nabokov, *Pale Fire,* 59. This poem also occurs in an absolutely stunning way in the movie *Bladerunner 2049.*

2. Netz and Noel, *The Archimedes Codex,* 3–27. The story of the auction in our opening paragraphs is indebted to their wonderful retelling.

delicate little smudge of a codex had been stolen from their collections. In spite of all of this, Christie's pressed on with the auction, and the unassuming, handwritten book scarred by fire and pocked by mold would turn into a quite unexpected sensation.

The codex was discovered originally in 1905 by a Danish philologist named Johan Ludvig Heiberg in the Church of the Holy Sepulchre in Istanbul. Heiberg quickly became stunned as he worked to translate what he could see of a ghostly bottom layer of text. The text was a palimpsest—a manuscript that contains another, erased text beneath it. What was uncovered were a series of texts that happened to be the earliest known copy of the great Greek mathematician Archimedes's lost manuscript *The Method of Mechanical Theorums*, a work entitled *Floating Bodies*, and the oldest known mathematical puzzle in existence entitled the *Stomachion* (or *Ostomachion*).

No one could read the full texts of the palimpsest (though Heiberg's work turned out to be stunningly accurate given the limitations of the time).[3] Many merely knew the older texts were there. And so, with the hope that modern imaging technology would shed light upon what mere eyes could not, the bidding began at an outrageous $800,000. It met this nearly instantly. The price continued to rise, soaring past $1,000,000, to $1,500,000, to $1,900,000, until a somewhat befuddled audience of reporters witnessed this rag of a text find a numbered paddle held high on its behalf, indicating a buyer wanted to try and have the final word at a whopping $2,000,000. It worked. The gavel fell, and what has come to be known as the Archimedes Palimpsest exchanged hands. By itself, it made just under half of all total sales that day. The next morning the *New York Times* ran a front-page story that an anonymous American buyer "who was not Bill Gates" had made this unusual purchase.

In 1999 the Walters Art Museum in Baltimore, in concert with several other institutions began using new Multi-Spectral Imaging technology to see beneath the prayer text and discover what had been erased and obscured. What should have been the tale of an amazing discovery, however, quickly degenerated. The juicy faith and science conflict angle ready to hand in a medieval book of prayer quite literally erasing the writing of one of the most noted mathematicians of all time was not lost on the media, or the public at large. Predictably, outrage ensued. During the nine-year intensive reconstruction of Archimedes' works a firestorm of anti-Christian sentiment sustained a steady, tire-fire like burn—raging with all the stench of the Dark Ages and that long night of reason "that everyone knows" Christians visited upon the world.[4] The Palimpsest bloated from its particular context-bound existence, into a metaphor of ruin draped over the breadth of Christian history; Carl Sagan's infamous chart showing the great interruption of the Middle Ages was dusted off (often in a hilarious, almost crayon-on-napkin like version reproduced by some nameless soul in Microsoft Paint); Gibbon's name was brought up; the Library of Alexandria, Hypatia, stupid

---

3. Netz and Noel, *The Archimedes Codex*, 283–84.

4. For a fantastic review of this, see O'Neill, "The Archimedes Palimpsest."

monks, the flat earth, and a host of almost uncountable anecdotes were marshalled. It was, in other words, business as usual.

The New Middle Ages

The deep irony of the story of the Archimedes Palimpsest, or Greenblatt's narrative of Lucretius's lost poem, and those like it, is that Christians did indeed *preserve* these texts. For Lucretius, to revisit Greenblatt's centerpiece, "enough traditions surface or resurface in the fifteenth century to cast doubt on the common notion that Christian scruples were to blame for the neglect of a poet who preached the mortality of the soul and the unconcern of the gods."[5] For Archimedes, what many immediately wanted to fold into the grand story of the Christian erasure of knowledge—both uncritical and systematic—was the exact opposite: it was evidence of how Christians, in fact, preserved Archimedes's writings, and others like them. One of the first Christian copiers of Archimedes of which we have record was Eutocius of Ascalon (480–540), who not only preserved several of Archimedes's texts, but also wrote involved commentaries on them. Eutocius's preservation of Archimedes is especially ironic given that he—along with a colleague Isidore of Miletus—were the chief architects chosen by Emperor Justinian I to oversee the construction of the Hagia Sophia. This church was used in the last chapter, as you may remember, as a reprimand to Catherine Nixey's particularly silly characterization of Christians as too primitive for good ladders and hammers. Well, it turns out not only were their ladders and masonry tools just fine, they were readers of Archimedes who used his principles to construct one of the most grandiloquent examples of architecture in the ancient world.

We will not belabor the point by enumerating the journeys of various works of Archimedes through the Middle Ages to their flowering in the thirteenth-century translations of William of Moerbeke.[6] The major point for the moment is that Archimedes' texts were being preserved by Christians. In fact, the Archimedes Palimpsest can be traced to scribes working in Constantinople copying the lost works in question, which is the only reason this copy survived to later become a palimpsest. Even in the particular instance of the Archimedes Palimpsest, the evidence is that these manuscripts were prevalent enough to make their way to a small, extremely poor church in Jerusalem (not a monastery). There, the impoverished priest in charge of his congregation had no use for the extremely technical work, and seeing that the pages were fit for reuse—did so. It should be noted that among the erased texts in the Archimedes Palimpsest, many Christian ones can also be counted. This was not a targeted erasure of pagan learning, but an erasure of convenience that arose at a particular time. O'Neill makes the observation that a colophon (that is, a type of publisher's emblem) marks the date of the recycling as April 13, 1229. This is merely one month

5. Reeve, "Lucretius in the Middle Ages and Early Renaissance," 207.
6. See Claggett, "The Impact of Archimedes on Medieval Science."

after Emperor Frederick II wrested control of Jerusalem from the occupying Ayyubid Sultanate, thereby making it extremely plausible that this military action made parchment nearly impossible to acquire either through scarcity or, what amounts to the same, inflated prices.[7] And yet, this single instance ripped from any context that gives it meaning is often taken as paradigmatic for a deeper Christian antipathy or apathy to pagan sources, when it in fact evidences the exact opposite.

The Middle Ages held books and writing up with an almost totemic reverence. Greenblatt, for example, knows this—and thus invents a cover for his own story by saying that most monks copied texts out of sheer dogmatic devotion to command, with no real understanding of the things they were copying. As we have already seen with Eutocius, there is large evidence to the contrary. Yet Greenblatt levels the common characterization of the Middle Ages as a time where "a whole culture turn[ed] away from reading and writing." The evidence is, again, not in Greenblatt's favor. Medieval historian Lynn White Jr. goes so far as to say that there was "no evidence of a break in the continuity of technological development following the decline of the Western Roman Empire."[8] In fact, though he overstates his case, White continues and provides his sweeping judgment that "in technology . . . the Dark Ages mark a steady and uninterrupted advance over the Roman Empire." Literacy may have ebbed and flowed in pockets and eddies throughout the inhabited world, but there was certainly no "turn away." Just the opposite. As a former curator of the Louvre put it "in letters as in the arts, it seems that the populations, liberated from the Roman yoke, spontaneously found once again the originality they had never really lost," but which had been stifled under the reigning canons of Roman academic taste.[9]

In the place of decline, the much more regal sounding phrase "Late Antiquity," has become standard after the pioneering work of historian Peter Brown, who nearly single-handedly invented it as field of study after its general neglect. Indeed, recently historian Ihor Ševcenko has called Brown's publication of a major essay "The Rise and Function of the Holy Man in Late Antiquity" not just a watershed but "the Big Bang of 1971,"[10] setting the path for new views of the era supposedly after Rome. "There is a danger that we [scholars] may exaggerate the height and stability of the Roman achievement, and, as a result, we may exaggerate the depths to which Europe fell once the empire had been removed." To be sure a "drastic downsizing of life" in many of its aspects "cannot be denied." However this new existence cannot be treated as a saggy aftermath, it does not "represent a regression from the more elevated standards of its own Early Christian and late antique past into 'archaic' modes of thinking."[11] Where many scholars previously saw only ruin and waste leading into an age of

7. O'Neill, "The Archimedes Palimpsest."

8. Quoted in Gies and Gies, *Cathedral, Forge, and Waterwheel*, 40.

9. Pernoud, *Those Terrible Middle Ages*, 51.

10. Quoted in Bowersock, "The Vanishing Paradigm," 41.

11. Brown, *The Rise of Western Christendom*, 20.

darkness, Brown brought to our attention the era's dynamism and creativity.[12] "If we look carefully," he says, "we will not see darkness, but rather a world that is slowly but surely becoming more like our own."[13] So deeply have we been soaked through by this subterranean transformation that occurred in the so-called "Dark Ages," that, ironically, "it is where [Christianity] continued to move imperceptibly to the rhythms of a very ancient world that early medieval Christianity often seems most 'backward' to us."[14]

If there were pockets that did turn away from reading and writing (to continue with Greenblatt's specific theme), these were exceptions to the emerging ethos of Christianity. "What modern scholarship has recovered," says Brown in his *The Rise of Western Christendom*, "is the sheer excitement of men of the pen [that is, those traditionally called 'barbarians' of Frankish Gaul, Celtic Ireland, Saxon Britain, and later Germany and Scandinavia] of the sixth, seventh, and eighth centuries" as they received the tools to record an orderly and useable notion of their own past and identity. "The tools with which they did this," Brown continues, "were provided by the new religion [Christianity]."

> Not only did Christian intellectuals bring the skills of writing to previously non-literate societies. They brought ways of writing historical narratives, which derived from the Old Testament and from the historical traditions of the Roman world which Christians had already adapted to their own needs in earlier centuries. We owe almost all we know of the history and literature of pre-Christian northern Europe to learned clergymen who set to work with urgency and with great intelligence to make their own, for their own needs, large sections of the pre-Christian past.[15]

The preservation through copying and transmission of the very scrolls that take center stage in Greenblatt's story is part of a broader Christian culture whose bones were knit by textuality, whose vast network of sinew, muscle, and nerve were robust discussions of hermeneutical method, and whose lifeblood was the *paideia* (instruction) of Christ.[16] Nor was this learning of Christ somehow a blotting out of Greece or Rome. "In calling Christianity the *paideia* of Christ," writes Werner Jaeger in his classic study, "the intention . . . [was] to make Christianity appear as a continuation of the classical Greek *paideia,* which it would be logical for those who possessed the older one to accept," while nonetheless acknowledging that classical *paideia* was now an instrument of Christ as the center of a new culture.[17] This meant reading and preserving almost

12. Brown, *The World of Late Antiquity.*
13. Brown, *The Rise of Western Christendom*, 23.
14. Brown, *The Rise of Western Christendom*, 28.
15. Brown, *The Rise of Western Christendom*, 8.
16. Young, *Biblical Exegesis and the Formation of Christian Culture.*
17. Jaeger, *Early Christianity and the Greek Paideia*, 12.

everything—even if it was seen as particularly aggressive thorn bush to keep in the garden of Christianity.

Indeed, the culture of reading, preservation, and commentary that Christianity incubated did not just lead to a preservation of the humanitarian arts like literature, but as the historian of science Peter Harrison has recently argued led into the production of what we today consider a scientific culture by providing a way to "read" the world. Traditional arguments that saw in the rise of science the diminution of religion claimed initially that as a more scientific or "literal-empirical" views of nature arose, biblical reading strategies accommodated themselves to this new "clear-eyed" order. Just so, they began to demystify themselves of allegory. This story is already too simplistic (allegory is a much misunderstood category, though we cannot get into it here), but Harrison demonstrates that the order was, in fact, reversed: The culture of reading that Christianity had fostered altered how nature—itself viewed as God's second book alongside and in concert with Scripture—was read. Harrison writes:

> The study of the natural world thus became, as it had been for the discoverers of nature in the twelfth century, a religious activity, albeit in a new sense. Nature no longer comprised a vast array of symbols which pointed to a transcendent realm beyond: instead, the way in which the things of nature were ordered and disposed came to represent a logical premise from which God's wisdom and providence could be inferred. Of equal importance was the emergence of the conviction that God's purposes in the creation could only be realized when the functions of those things originally designed for human use were discovered. Interpreting the book of the creatures became a matter of discerning the intention of its author. In much the same way as the true meaning of a written text came to be identified with the designs of the writer, so legitimate meanings of the book of nature were sought in the purposes for which God had designed its living contents.[18]

While this transition of method would ultimately occur later in the Protestant Reformation and beyond, its impulse to study and understand nature was there from the beginning and is on full display in early Christian texts. St. Ambrose of Milan (d. 387) writes for example in his *Hexameron* (a commentary on the first six days of creation in Genesis) that "We cannot fully know ourselves without first fully knowing the nature of all living creatures." St. Hildegard of Bingen (d. 1179) wrote similarly that "Through [the organs of sense] man looks upon all the creatures knowing them for what they are, distinguishing them, separating them, naming them" and this precisely through the textual warrant that the first man, Adam, did exactly this and was—to speak anachronistically—the first scientist. Indeed, he represented humanity *as* scientist, as a knower and investigator of the world God created. "The turn to the natural world was in some sense a turning away from the sacred page," writes Harrison, "but it could

---

18. Harrison, *The Bible*, 168–69.

hardly be said that it was motivated by a secular, or non-theological impulse. On the contrary, acquisition of knowledge of the order of nature was enjoined on mankind as an integral part of the process of human redemption, and more specifically, a reversal of the losses incurred at the Fall."[19] Indeed, "All creatures were in a sense to be found [by] man, and in another [sense], were to be reunited in him through an orderly knowledge of the natural world." However alien this sense may be to the more hard-nosed secular interpretations of the scientific spirit today, the fact remains that historically speaking for Christianity "to accumulate systematic knowledge of created things was both to restore the knowledge of Adam and approach the very mind of God. Through the acquisition of knowledge came also the redemption of the world, for knowledge was assimilated or incorporated into the human mind, and thus redeemed along with it."[20] The culture of reading and education fostered by Christianity, then, overbrimmed the world with words. And in their overflowing, these words soaked into the seas and stones and stars, allowing them in time to speak back and reveal their secrets.

Monks have often been caricatured as aloof and unworldly (and many no doubt have earned this particular image as dainty spiritualists). Yet, in their retreats to the desert (the "desert" here though, is not just sand, but in ancient and medieval parlance could also mean the infinite marches of forest, the snow-locked roofs of the world, the trackless cobalt of the sea) they understood themselves to be bringing order to the world,[21] through prayer, and patience, and presence, but also through technology and the burden of responsibility to cultivate God's garden that Genesis records is part of what it means to be made in God's image. But as early as the communities and religious settlements of Pachomius (290–346) in Egypt, things like tailoring, metalworking, carpentry, camel and horse driving, plant cultivation, farming, and so on, were all considered part of their spiritual labor.[22] Indeed, in the Rule of St. Benedict, which outlined broad guidelines indicating how Benedictine monks should live and conduct themselves, likewise mingled labor with prayer. "The monastery," proclaimed the rule, "ought if possible to be constituted that all things necessary, such as water, a mill, a garden, and various crafts might be contained in it."[23] Bernard of Clairvaux (1090–1153), a famous theologian spearheading a reform movement within the Benedictine order known as the Cistercians, would later write that the ideal monk was one who mastered "all the skills and jobs of the peasants,"[24] that is, carpentry, masonry, gardening, weaving, metalworking, and animal husbandry, so as to bring stability, that is, to "cosmicize" the world—to bring it to one or another idealized order. The Christian historian and monk Cassiodorus (490–585) could even exult of the marvels

19. These quotes are from Harrison, *The Bible*, 59.

20. Harrison, *The Bible*, 62.

21. Taylor, *A Secular Age*, 335–36.

22. Gies and Gies, *Cathedral, Forge, and Waterwheel*, 9.

23. Quoted in Gies and Gies, *Cathedral, Forge, and Waterwheel*, 48.

24. Quoted in Gies and Gies, *Cathedral, Forge, and Waterwheel*, 9.

of technology that he had gained for the monastery he founded: "cleverly built lamps, sundials and water-clocks, water powered mills and . . . irrigation systems, the Egyptian-invented papyrus . . ." He judged that mechanics were, in fact, "almost Nature's comrade, opening her secrets, changing her manifestations, sporting with miracles"[25] that aided the Christian in their quest to restore God's lost Eden.

What we are seeing with these above examples is a broader change that occurred with the rise of Christianity, namely overcoming the aged disdain for what Aristotle had termed the "banausic" arts involving the utilitarian production of technology (from the Greek *banausikos* meaning "of an artisan" but also simply "unrefined" or "unintellectual"). For the Greeks, science was primarily theoretical, and the use of one's hands in craftsmanship was, while ultimately necessary, beneath the notice of the philosophers. "The industries that earn wages," wrote Aristotle, "degrade the mind." They are in some respect unavoidable, however "to dwell long upon them would be in poor taste."[26] The Romans, on the other hand, inherited a majority of their technology and instead of convenience born from innovation relied upon an unending tide of war spoils and slave labor to oil the gears of empire. The bristling sea of Roman spears and military might camouflage from us the fact that beneath this monumental civilization were technologies born from the Stone, Bronze, and early Iron Ages.[27] As historian Jacques le Goff puts it, "within [its] boundary Rome exploited its empire without creating anything. No technical innovation had occurred since the Hellenistic age." As such, while Rome is often seen as a high point that then descends into the dark of what follows, in truth in regards to the era that was to come the legacy of "Rome both fed and paralyzed it."[28] The church fathers can hardly be faulted for carrying on some of this disdain or simple apathy toward the practical arts that was their cultural inheritance from Greece and Rome. What is more remarkable is how their Christian convictions began to completely deconstruct this prejudice, exulting in the practical arts and the production of technology as part of their spiritual habits to be cultivated.

It was in fact in the uproar of the invasions of Rome that Christians found themselves "by force of circumstance" in the roles of power and thus were turned to as preservers attempting "to fight against violence and ameliorate people's behavior."[29] As such, "As Europe slowly emerged" from the shadow of Rome, "there [also] emerged a new set of values which—for better or worse—regarded the transformation of this world as a sufficient condition for salvation in the next."[30] It should be noted that this quote from historian Brian Stock is all the more striking as in the same essay he aggressively argues that "overall the force of the new religion [that is, Christianity] has

25. Quoted in Gies and Gies, *Cathedral, Forge, and Waterwheel*, 11.

26. Aristotle, *Politics*, 1248b.

27. Gies and Gies, *Cathedral, Forge, and Waterwheel*, 17.

28. Le Goff, *Medieval Civilization*, 3.

29. Le Goff, *Medieval Civilization*, 34.

30. Stock, "Science, Technology, and Economic Progress," 25.

been exaggerated"[31] and in its place he promotes the later, impressive array of Islamic sciences that made their way into the West. Nevertheless, the decline of ancient learning where and when it did happen was hardly the fault of Christians, says Stock, but was fruit born from the Romans' "incapacity to overcome the impediments to human progress through labor saving technology." He continues, saying that "the failure of Greece and Rome to increase productivity through innovation is as notorious" as is commentators like Edward Gibbon attributing said stagnation to the rise of Christianity. Stock concedes therefore that "[Christian] society in the Middle Ages surpassed [the Greeks and Romans] . . . in the success with which periodic waves of new techniques and inventions improved the human condition."[32] Indeed,

> In practical areas, Christians improved upon the aristocratic habits of the classical world. They opened Roman hospitals to the poor, thus extending the benefits of rational medicine and foreshadowing the hospitals of the later Middle Ages, which were among the first institutions of social welfare. In the realm of ideas, the most serious critic of Aristotle after Ptolemy was also a Christian . . .[33]

A list of all innovations in the Middle Ages would take us far too long, and for our story is somewhat beside the point.[34] This era post-Rome but pre-Renaissance often lumped as a thousand-year period where nothing happened (except perhaps bloodshed and disease) turns out to be, rather, "one of the great inventive eras of mankind" as machinery and technology were developed and more importantly put into use "on a scale no civilization had previously known."[35] How was it then that such misconceptions about the Middle Ages occurred? We have already seen in the last chapter many of the more grandiose and sweeping mischaracterizations that distort our perceptions of this long stretch of history. But within those grand misunderstandings, a thousand smaller cuts also bleed our histories dry. To a few of those, we can now turn.

Pictures of the Dark

With the new harness and stirrup technology for horses also came the drawn wagon that could carry heavy cargo across long distances. This was greatly needed, as the increased productivity meant there was more goods that could be shipped and traded. Here, though, we encounter another myth of the Dark Ages: "One of the most misleading claims about the decline of Europe into the Dark Ages concerns the neglect

31. Stock, "Science, Technology, and Economic Progress," 10.

32. Stock, "Science, Technology, and Economic Progress," 23.

33. Stock, "Science, Technology, and Economic Progress," 11.

34. For some lists, see Gies and Gies, *Cathedral, Forge, and Waterwheel*; Gimple, *The Medieval Machine*; Hannam, *The Genesis of Science*; Stark, *The Victory of Reason*, 33–68; Truitt, *Medieval Robots*.

35. Gimple, *The Medieval Machine*, viii; 1.

of the Roman roads—in many places the paving stones were taken up and reused in local constructions. As the roads decayed, the story goes, so did long distance trade, and Europe became an archipelago of isolated and inward-looking communities."[36] Like Gibbon and Petrarch's Roman ruins, the long and broken Roman Roads carry the imagination far beyond what historical evidence can allow. This is, in fact, another peculiar instance where as a matter of course the truth of things is exactly the opposite of the common caricatures of decay. Roman long-distance trade had consisted mostly of luxury items, and was an unproductive and almost purely ornamental aspect of an economy primarily driven by the wealth brought to elites from their estates: "What is apparent," writes Stark, "is that it was not until the [so-called] Dark Ages that Europeans developed the means for long-distance overland transportation of heavy and bulky goods."[37] As such, the Roman roads were principally for the quick transportation of soldiers on foot. In turn, as medieval long-distance trade arose from new technology that supported it, the roads were quickly found to be largely unserviceable for riders or ox-drawn wagons. The latter wagons in Roman times had no brakes, and so were quite the liability for long-distance travel except on extremely flat surfaces.

Being paved with stone, in addition, meant the Roman roads were a hard path to walk, and grew extremely slippery during rain and snow seasons. Most traders found themselves walking to the side of the roads, while nonetheless using them as guides to various destinations. When horses began to be given iron horse shoes in the Middle Ages, the roads became upgraded from dangerous when wet to an active hazard. Incomparably more important for travel than these narrow roads was a Middle Ages abounding in the construction of permanent bridges for river crossing. The historian C. T. Flower called these bridges the "great public work"[38] of the Middle Ages, and one that drastically made up for the typical Roman use of fording rivers or using ferries inadequate to the new heft of Middle Age bulk trade. "Pack animals could navigate even a bad road, but not a river served only by a skiff; or nothing at all."[39] The neglect of the Roman roads, in other words, came not from technological ignorance, or the stupefied laziness of the Middle Ages. Rather, it was precisely because of a booming of their technology. In other words, the Roman roads could not keep up with the demands of new innovations. They were ignored as antiquarian, and so fell through disuse into disrepair.

However, like so many of these myths, the reason for the neglect of the Roman roadways was obscured by historical representations stemming from the nineteenth century. These mistakes, and occasional outright fabrications, reinforced notions of an age of ignorance and wanton neglect. Due to the work of one German scholar named Johann Ginzrot, many of these innovations were illicitly projected backwards

---

36. Stark, *The Victory of Reason*, 48.
37. Stark, *The Victory of Reason*, 50.
38. Quoted in Gies and Gies, *Cathedral, Forge, and Waterwheel*, 70.
39. Gies and Gies, *Cathedral, Forge, and Waterwheel*, 70.

and made out to be innovations of the Greeks and the Romans.[40] The desertion of the Roman roads by those in the Middle Ages was thereby made all the more wanton in the historical imagination of the time, as it suggested that if the roads—now seen in Ginzrot's illustrations as capable of supporting wagons and heavy trade—were neglected, it also meant the Middle Ages lost, or perhaps were apathetic to, these engines of long distance trade. Ginzrot drew his own illustrations and, in great detail, displayed Greek and Roman wagons with very modern looking innovations like a swivel placed into the front axle, allowing sharp turns and precise control, as well as a heavy braking system that would be needed to lug cargo in bulk. Greek and Roman wagons had neither a front axle—and so had to be forced around corners—nor, as we have mentioned, did they have brakes.[41] Both were middle-age inventions, now attributed to the lost glory of Rome. In some sense Ginzrot is only partially to blame for this trickery, as he clearly states—one is tempted to say, confesses—that he has drawn these representations "as I imagine [them]," and this because "I have nowhere found a picture on ancient monuments which would be suitable for this."[42]

Of course, he did not find any such usable examples because no such transportation had existed. Yet, the damage was done, and Ginzrot's beautiful drawings were reproduced hundreds of times in a variety of magazines, dictionaries, journals, and popular literature.[43] Lest this seem a trivial anecdote, we must remind ourselves again how key a piece the "decay of the Roman roads" is to the general story of the Dark Ages. And Ginzrot's illustrations were, as such, a major ballast for this mythmaking. But more to the point, Ginzrot is merely one example in a sea of misrepresentations of the past, as we have been and will continue to uncover in this book. Alone, it is perhaps merely an interesting piece of trivia, an anecdote; together, all of these misrepresentations begin to add up into a giant worldview, where one can conjure for students the lackadaisical attitude of an entire thousand-year period from a simple illustration of a front wagon axle and braking system.

Perhaps the most major casualty of historical misrepresentation that has produced long-lasting shockwaves in our understanding of the Middle Ages is the representation of its medicinal practices. As historian of medicine, science, and religion Gary Ferngren puts it, "erroneous assumptions about early Christian views of science, nature, and medicine have long bedeviled the study of the relationship between Christianity and medicine."[44] Such misinformation reaches a fever pitch when one begins to speak of those tepid middle days bestriding the dividing lines of the Western world. Upon hearing "medieval medicine" one has, perhaps, already pictured plague doctors walking about in their ghastly beaked masks and steampunk goggles, or children

---

40. On this story, see Leighton, *Transport and Communication,* 71.

41. Leighton, *Transport and Communication,* 74–75.

42. Quoted in Leighton, *Transport and Communication,* 121.

43. Leighton, *Transport and Communication,* 71.

44. Ferngren, *Medicine & Health Care in Early Christianity,* 3.

singing "ring around the rosy" as a futile incantation against the plague. And so, already, we stand bamboozled by later inventions projected backward like a movie reel upon the Middle Ages. For neither the beaked mask, nor ring around the rosy, have their actual historical counterparts in the Middle Ages. The plague doctor is an artifact, as it happens, of the seventeenth century; ring around the rosy does not appear until the 1800s.[45]

In fact, the beaked mask was never even a staple of medicinal practice even when it appeared, but was initially a mere concept proposed to King Louis XIII by his chief physician Charles de Lorme in 1619. The beak, designed to be filled with perfume and aromatic herbs, was meant to buffer the doctor from *miasma*—poisoned or corrupted air thought to be a prime vehicle for the transmission of disease. The design was so striking that it caught on—not as doctor's garb, though there are perhaps a few cases where it was used—but as "Dr. Beaky" (yes, that is indeed his name), a sort of ghost story that bemused adults used to scare children. "You believe its just a fable, what is written about Doctor Beaky," reads an inscription from a 1656 engraving depicting horrified children fleeing before the beaked doctor (who also, if one looks closely, has for some unknown reason grown his fingernails into grotesque claws). "He seeks corpses to make a living . . . Ah, believe it, and don't look away from here. . . . His purse is Hell, and gold is the soul he fetches . . ."[46] In reality, while getting sick in the Middle Ages was certainly not a good omen, medieval doctors were not ghouls but learned men and women who had a variety of intricate theories about diseases in general—including the plague—that were conversant with the Greek greats like Hippocrates and, especially, Galen. That these theories all turned out to be wrong or misguided does not suddenly make them irrational. In fact, most were born through careful—if ultimately misplaced—observation and medical theorization. To understand this, we need to start a little further back in time, for Christianity's relationship to medicine is part of a broader transformation of the human personhood of the marginalized.

From the very beginning, Christianity was committed to charity, to the poor, to medicine and its egalitarian access in a way that was unheard of in the ancient world—however haltingly and inconsistently it may have been at executing this vision at times.[47] We have already run into the claim, made by Stock above, that Christians "opened Roman hospitals to the poor, thus extending the benefits of rational medicine and foreshadowing the hospitals of the later Middle Ages, which were among the first institutions of social welfare." More than this though, Christianity brought about the "institutionalization of philanthropy and the creation of establishments to shelter and feed the poor, care for the sick, assist widows and the aged, and raise orphans."[48] This

45. Black, *The Middle Ages*, 213–40.

46. Black, *The Middle Ages*, 226.

47. Holland, *Dominion*, 137–58.

48. Risse, *Mending Bodies, Saving Souls*, 7. For the general history of Christianity's relation to medicine and hospitals up to the Enlightenment, see 69–230.

was, we might remind ourselves, the very definition of "true religion" in the epistle of James (1:27).

As recent scholarship has emphasized, positive attitudes to the personhood and characteristics of children, for example—all children, including orphans and those born outside of wedlock—are historically traceable to Christ's loving attitude to them, and the Christian perpetuation of this ethos. As O. M. Bakke records in his work *When Children Became People: The Birth of Childhood in Early Christianity,*[49] whereas childhood has now for us been soaked in Christian terms that include a sort of wide-eyed innocence and laudable simplicity embodying many qualities that even adults are commended to imitate, the predominant Greco-Roman attitudes from Homer to Cicero classified children precisely and only by frailty, cowardice, unguarded passion, and a total lack of reason.[50] Indeed, in Greek *nepioi* and in Latin *in-fantes* both etymologically reflect the inability of children to communicate as an adult can. As Bakke summarizes, "children tended to be portrayed, along with other weak groups [like slaves and women], as the negative counterfoil to the free, urban male."[51] This was a common trope among both Greeks and Romans. The female, as Aristotle once infamously remarked, is as it were an inadequate male.

As such, the Christian faith and its concomitant endowment of the image of God upon children, women, and slaves along with a hope outside the machine of empire, is not an aside to Christianity and medicine, but reflects the broader social structural changes that Christian agape and charity were bringing into the ambient culture it was now inhabiting as it created some of the first institutions of social welfare. Reflecting this change due to Christianity, Emperor Julian—often called "The Apostate" because he was brought up a Christian but later aggressively renounced this and spent his very brief career as emperor from 361–363 reviving pagan practices[52]—wrote a letter to Arsacius, the pagan high priest of Galatia, complaining of the rampant Christian disposition to charity and care for the poor. Opening his letter by celebrating the quick revival of pagan practices he had initiated as emperor, Julian quickly slipped into a sour tone with Arsacius and notes this revival will simply not be enough. Julian and his high priests must realize, he wrote, that the Atheists (that is, Christians who were called such because of their denial of the pantheon of gods) in their "benevolence to strangers, their care for the graves of the dead, and the pretended holiness of their lives" have done more to spread the cause of Christianity than anything. The pagans must imitate this. As such, Julian orders Arsacius to mimic Christian benevolence and so "in every city establish frequent hostels, so that strangers may profit from our benevolence; I do not mean for our own people only, but for others who are in need of money. . . . I have given directions that 30,000 modi of corn [around 259,000 liters]

49. Bakke, *When Children Became People.*
50. Bakke, *When Children Became People*, 15–55.
51. Bakke, *When Children Became People*, 21.
52. Wilken, *Christians as the Romans Saw Them*, 164–96.

be assigned every year for the whole of Galatia, and 60,000 pints of wine. I order that one-fifth of this be used for the poor who serve as priests, and the remainder be distributed by us to strangers and beggars." And this generosity is precisely because "it is disgraceful that, when no Jew has ever to beg, and the impious Galileans [that is, the Christians] support not only their own poor but ours as well, all men see that our people lack aid from us."[53]

Medicine and the opening of hospitals was, as such, not something merely fortuitous but grew out of the core convictions of the gospel. Just as Christians overcame the prejudice against "banausic" professions, a similar prejudice was overcome in regards to medicine. Even by the first century Cicero, though well read in the medical discoveries of Galen and Hippocrates, was of the opinion that it was a debase profession for Roman nobels. Pliny the Elder similarly was uncomfortable with Romans needing the tutelage from Greeks, and so discounted the profession. It remained a practice occupied mainly by Greek slaves, although having found favor under Julius Caesar the profession did rise in stature.[54] Just a short time after Julian's brief reign as emperor, cities like Edessa especially in the Roman East had "adopted the tenets of the new Christian welfare,"[55] and according to early Church historians like Sozomenus (400–450) by 373 *xenodocheion* or *xenon*—"guest houses" or shelters—had been built for the needy and the sick amidst both famine and plague. Edessa, moreover, was also the locus of "extensive translation projects that sought to render Greek documents, including theological and medical treatises" into Syriac.[56] Indeed, it was the Eastern Christian monks and their *xenon* that supplied the most comprehensive and cutting-edge medical care available.[57] Though the Byzantine *xenons* and their origins in Christian philanthropy have been overlooked or dismissed by most scholars, to write about them is in fact "to write the first chapter in the history of the Hospital itself."[58] Yet, much of this has been lost, or indeed erased, through neglect, misunderstanding, and—as we are about to see once again—blatant misrepresentations rewriting history. The typical attitude in the Enlightenment and beyond is one where "intellectuals have ignored the achievements of medieval charitable institutions and have established a [historiographical] wall between enlightened science and the imperatives of Christian morality."[59]

One reigning assumption among scholars has been that since medicine was based upon the figure of Christ as healer, Christianity must have had a purely miraculous view of healing, and a completely demonological view of illness. Thus Ulrich Mueller

---

53. Julian, *Letter to Arsacius,* quoted in Bowersock, *Julian the Apostate,* 87–88.
54. Miller, *The Birth of the Hospital,* 30–49.
55. Risse, *Mending Bodies,* 72.
56. Risse, *Mending Bodies,* 73.
57. The key study on this remains Miller, *The Birth of the Hospital in the Byzantine Empire.*
58. Miller, *The Birth of the Hospital,* 4–5.
59. Miller, *The Birth of the Hospital,* xi.

can write in his history of medicine that "the dominant view of the New Testament is that demons are the cause of sickness."[60] Certainly, while this "pan-demonological" interpretation as one scholar calls it is not accepted by all New Testament scholars or historians of early Christianity, it is "the currently dominant narrative,"[61] as Gary Ferngren notes. Yet nothing is further from the truth in terms both of the text of the New Testament, or of its historical application. A close reading of the New Testament however demonstrates that only in three cases do illness and demonic possession even overlap. The majority of cases in Scripture readily distinguish Jesus' healing from exorcisms.[62]

There is, moreover, a remarkably naturalist description of the etiology (that is, the origin and cause) of disease. Even in cases of epilepsy—which is often used as a classic example of a natural illness being mistaken for possession—the majority of cases were given purely physiological explanations by Christians, following the earlier diagnoses of the Hippocratic treatise *Sacred Disease*.[63] Yes, demons could in fact cause disease or at least manifest parallel symptoms, but these were incredibly rare even in the New Testament, where the confrontation with demons was believed to be much more numerous than normal due to the peculiar apocalyptic nature of Christ's ministry. And yes, prayer, laying on of hands, and anointing with oil should be utilized for the sick. But these were in concert with the practice of physicians, not in competition with them. In the normal course of things, demons were an unusual and exceptional circumstance, and not one conflated with medicinal diagnoses. As such no antagonism existed between the notion of Jesus' healing and that of normal physicians: "in the New Testament [there is] no condemnation of physicians or medicine, either specific or implied."[64] Just the opposite. Christ's compassion toward the suffering fueled an early and aggressive Christian interest in "secular" medicinal treatises and practices, as we already alluded to above. And yet, to summon Charles Freeman's *Closing of the Western Mind* from the dustbin, he jeers that the rise of Christianity led to "the rejection of a scientific approach to medicine."[65] This is echoed as well in Richard Gordon's 1993 *The Alarming History of Medicine* where, because of Christianity, "anatomy was dead, and medicine was stillborn. . . . [Christianity] scuppered healing for fifteen centuries."[66]

To explain this failure of understanding, we may turn to a series of misrepresentations that arose—you guessed it—in the nineteenth century, which reinforced the notion of absurd supernatural and magical cures abounding from Christian

---

60. Quoted in Ferngren, *Medicine & Health Care*, 42.

61. Ferngren, *Medicine & Health Care*, 42.

62. See Ferngren's analysis, *Medicine & Health Care*, 42–63.

63. Ferngren, *Medicine & Health Care*, 57.

64. Ferngren, *Medicine & Health Care*, 48.

65. Freeman, *The Closing of the Western Mind*, 320.

66. Quoted in Black, *The Middle Ages*, 173.

irrationalism. To be sure, such cures no doubt existed. There were all sorts of magic, folk remedies, and incantations that strike us as hopelessly fantastical and muddled. And, of course, such things still exist today. Yet we tend to fascinate in upon the uncanny in the Middle Ages as representative. An assessment of Charles Singer, the author of *Anglo-Saxon Magic and Medicine* summarizes this when he writes, "Surveying the mass of folly and credulity that makes up Anglo-Saxon leechdoms, it may be asked: is there any rational element here? The answer is, of course, very little."[67] Singer is, with this opinion, generally representative of the literature of the last century. Yet as recent scholarship has unearthed, the main reason for these opinions can be "traced to a negative bias set in place by a very few nineteenth- and early-twentieth-century scholars"[68] who were the few who bothered to write on the neglected topic of medieval medicine, and in fact actively set out to reinforce the negative stereotypes that surround the genre.

Before one can turn to these literary representations of medicine, they intersect in an unexpected manner with the assumption of the frequency and nature of witch hunts in the Middle Ages. For wise women were often, so it goes, the only source of cures for what ailed you. The acrid smoke of witches burning at the stake, stinging the eyes, and choking the sky, would from nearly any popular book on the Middle Ages be the uncanny image left in the mind. Black-robed inquisitors would also line this picture, dour faces looking out upon the unwashed masses to find their next hapless victim. The accusations would be false, of course; yet the torturous confessions wrung from broken lips would be bricks to line yet more of the foundations of what amounted to the church's paranoiac police state enforcing orthodoxy. For the tale of medicine in particular a double service is rendered by this picture: the only medicine truly available at the time was the arcana of the witches (sometimes involving herbs known by wise women; sometimes the invocation of elves or demons), but even this odd compendium was quashed by the arm of the church. It is also a fabricated image.

On the one hand, recent studies into the Inquisition have revealed it to be a much different, much more complex thing than is rendered in jokes like "no one expects the Spanish Inquisition." Indeed, the Spanish Inquisition itself has been revealed to be drastically different than its frenzied caricatures.[69] To be sure, much of this history (especially from our vantage point) is unhappy, and some of the anti-Inquisition propaganda has been so successful even for scholars it is laborious "to separate fact from fiction."[70] But much of the litany of atrocities (not all, of course, these are people we are speaking of, after all) have been shown to be fabrications of Protestant polemics against Catholicism, especially to the extent such could be coopted to gain rhetorical advantages in the polemics between nations, such as the "Black Legend" of Spanish

---

67. Quoted in Arsdall, "Rehabilitating Medieval Medicine," 135.
68. Arsdall, "Rehabilitating Medieval Medicine," 135.
69. See especially Kamen, *The Spanish Inquisition*; Rawlings, *The Spanish Inquisition*.
70. Kamen, *The Spanish Inquisition*, 374.

exploration and totalitarianism that fed easily into notions of the Spanish Inquisition wrought by their many English, Dutch, and French opponents. In other words, like so many things in this story "The Inquisition" is an evolving mythology.[71] Curiously, not only have the judgments of the Inquisition been shown to be more tolerant than court decisions of the different states, using torture far less than state courts. Their methods of due process led directly into modern legal systems along with Church canon law. Indeed, the use of torture *declined* in the West, because inquisitors "themselves were skeptical of the efficacy and validity of torture as a method of conviction."[72] Much of this has been revealed through new access to Catholic and state archives. Nonetheless, the Inquisition became one of the primary sites where the initial use of history as a weapon came to be clearly perceived by both Protestants and Catholics.

Protestants, wanting to display the paranoid authority of the Catholics, exaggerated (and sometimes even fabricated) the executions we have come to associate with the movement (which, it should be noted, was never one thing, but a multitude that was dependent upon different factors like time, place, and personality for its local temperaments). Moreover, on the Catholic side, many new histories were written by zealous apologists who turned all of Catholic history (ironically enough) into a history of the Inquisition as a move made in its own defense. These were histories of the victory of truth triumphant in the midst of being assaulted by error on all sides.[73] In the Enlightenment, once again, much of the polemical origins (and mitigations that come from such beginnings) were lost; one hardly needs to use much imagination to reflect upon how such histories could easily be wielded uncritically and turned to the polemical advantage of the *philosophes*.[74]

None of this quite expresses the level of mythology that inquisitorial witch hunts in the Middle Ages have reached. But just so, the reveal is equally shocking: there were no Church-led witch hunts in the Middle Ages. None. We can pause here for a second as the perplexity works itself out. Though it goes beyond the scope of what we can deal with here, the mass-scale witch hunts were actually a feature of *modernity*,[75] and championed by many of its most "enlightened" scientific figures. In the Middle Ages, there on occasion may have been a local execution by uproarious town folk. But the official position of the Catholic Church at the time was that witches did not exist, and so women could not be punished for claims from themselves or others that they were, in fact, witches.[76] Indeed, in many instances the Inquisition was instituted precisely to discourage mob justice and enact equity. In the nineteenth century histories

71. See the work of Peters, *Inquisition*, esp. 122–295.

72. Rawlings, *Spanish Inquisition*, 33.

73. Peters, *Inquisition*; Kamen, *Spanish Inquisition*, 92–118, 374–94.

74. Peters, *Inquisition*, 155–88.

75. Trevor-Roper, "The European Witch Craze."

76. For this suprising position on witches in the middle ages, see: Decker, *Witchcraft & The Papacy*, ch. 1.

were written indicating that covens of witches were in fact the early invention of the Inquisition to justify the torments they were inflicting on many who were simply wise women attempting to provide herbal and spiritual cures to their local communities. Due to the investigating of the historian Norman Cohen, however, it was discovered toward the end of the twentieth century that the documents used to create these histories were in fact a quite ubiquitous fabrication invented in the nineteenth century to condemn the Inquisition and perpetuate the sharp rise in anti-clerical sentiment that we have been exploring from yet another angle. It didn't hurt that it would improve the notoriety of the forger and make a little coin in the process.[77] "There is no doubt about it," Cohen pauses to remark at one point. "There never was an inquisitorial witch hunt at Toulouse or at Carcassonne. Not only the famous trial of 1335 but the whole saga was invented . . . We are faced with a spectacular historical hoax."[78] Ironically, however, when Pius IX declared himself infallible, one of his most academically esteemed opponents whom we have mentioned—the famous Catholic historian Ignaz von Dollinger—used the spurious texts as another point of evidence to marshal against papal infallibility.[79] From there, it was but a short hop into the work of Dollinger's acquaintance and fellow anti-infallibilist, John William Draper, and onward into the many fabulations of the warfare thesis that were to follow.

For this chapter, the point of these fabricated witch hunts was not just that they were another pillar supporting the notion of the Dark Ages, or the more specific tyranny of the Inquisition. They also incorporated deep imagery of medieval medicine as, at its best, little else than the crockery of these wise women witches, a miscellany of wizarding and occultism (that, of course, was still a bit too advanced for the besotted Catholics trying to stamp it out). Contemporary sleuthing has revealed a man named Oswald Cockayne as a particular pronounced vehicle for this misrepresentation.[80] Cockayne was the first translator and editor of three medical texts written in Old English before the year 1000—which he published in 1864 as *Leechdoms, Wortcunning, and Starcraft of Early England*. This is, as the reader no doubt already apprised, hardly a title to convince one of any "serious" medicine contained therein. One of the three translated works—the *Lacnunga*, or *Cures*—contains the bulk of esoteric or nonrational folkloric cures that we might expect given a certain image of the Middle Ages—charms, prayers, incantations, even—on occasion—declaring tiny elves to be at the root of some ailments (this last one was particularly memorable to Victorian audiences, as one might expect). Despite these eccentricities, such make up a mere pittance of the total material contained in the *Lacnunga*, not to mention medieval medicine as a whole. Yet, of course, "it is *this* work [and this subsection of the work]

---

77. Cohn, *Europe's Inner Demons*, 181–201.

78. Cohn, *Europe's Inner Demons*, 186.

79. Cohn, *Europe's Inner Demons*, 194.

80. I am here indebted to Arsdall, "Rehabilitating Medieval Medicine," in this and the following paragraph, esp. 136–37.

that has received the most attention, and is often held up as typical of medieval medical writings and practices."

To compound this prejudicial distortion, Cockayne "much like . . . folklorists of the nineteenth century" went into these texts searching not for medicine, but saw them "only as sources for pre-Christian German charms and superstition, pointing out, without any type of proof, where certain terms of uses originated, and suggesting the kind of fanciful etymologies that were popular in the nineteenth century." So, Cockayne decided to provide for his audience translated prose filtered through an intentionally "quaint and antiquated translation style" and crafted his translation to be intentionally fantastic. For example, he used the word "leechdom" (which did not exist in his then-modern English) to translate the Old English *laece* ("leech"), which as it happens, was not a term representing the small parasitic creature but merely the Old English word for "physician." It, of course, made Cockayne's Victorian readers clutch their pearls in discomfort (and perhaps macabre fascination) as they pictured patients riddled with actual leeches, and this is what Cockayne was hoping.

As it goes, however, the creature that bears the name was only eventually named leech after the long-established practices of the *laece* such as bloodletting and surgery, not the other way around. Leeches are still, in fact, used on occasion in legitimate medicine, so the prejudice conjured by them is perhaps itself beside the point. Regardless, this was not the only curiosity of Cockayne's dissimulation. In addition, instead of the respectable and thoroughly mundane Old English phrase "knowledge of medical plants" Cockayne decided that substituting the shorthand "Wortcunning" (again, not a word before then) every time the more mundane phrase appeared was, for some reason, a good idea. It certainly was better suited to conjure images of witchcraft and superstition. To add a cherry on top, every time the *Lacnunga* spoke of women, Cockayne would translate this from Old English into Latin, and leave it this way, giving his Victorian readers the impression that the Latin was in fact original to the text. This particular decision was meant to imply to Cockayne's audience, it seems, that whenever medievals had to speak of women and their bits, such taboos were to be quarantined from common speech via the arcane tongue of Church theologians.

As with so many of the misrepresentations we have run into, Cockayne's shenanigans would perhaps be mildly amusing were it not for the fact that scholarship on medieval medicine was almost entirely shaped by his work for the next several generations. In 1952, J. H. G. Grattan wrote his *Anglo-Saxon Magic and Medicine*, which was reliant on Cockayne's translation of the *Lacnunga*. In 1962 Charles Singer's *A Short History of Medicine* was likewise dependent upon Cockayne, and believed "medieval medicine was the last stage of a process that has [thankfully] left no legitimate successor, a final pathological disintegration of the great system of Greek medical thought."[81] Indeed, even a textbook aimed for children grades three through seven entitled *A Totally Gross History of Medieval Europe* summons the *Lacnunga* and notes those icky

81. Arsdall, "Rehabilitating Medieval Medicine," 137.

Christians in medieval times thought all disease was due to either elves, demons, or God's punishment.[82]

Cockayne's deceptions were further exacerbated by two other medicinal myths that developed around the time he was translating his three texts. These were the notions that the medieval church prohibited the dissection of bodies, and so further slowed the progress of medical and anatomical knowledge. The second related to supposed Christian prohibitions regarding the use of chloroform as an anesthetic for women during childbirth. This came about, so went the tale, because theologians believed that in alleviating a woman's birthing pains, they were thereby subverting God's curse upon women after the fall of Adam and Eve. But we shall return to this canard in a moment.

It turns out the first notion about the medieval Church prohibiting dissection was created by Andrew Dickson White. There is, perhaps, no surprise at this revelation. Indeed, the myth fits particularly well with the notion that the Renaissance—where anatomy and dissection did indeed become widespread—was a heroic turn away from the dogmatic queasiness of the church toward science. It also, interestingly enough, fit with Kant's notion summarizing Enlightenment as *Sapere Aude!* or "Dare to think [for yourself]." For, etymologically, the word autopsy comes from the Greek *autopsia*, namely, "to see for one's self." Thus, in performing dissection one was not only undermining dogma, but one was conjuring a wordplay embodying the Enlightenment ideal of free thought. White could, through this single myth, rally several layers of anti-church periodization and propaganda that had already been settling into the hearts and minds of Europe and America for quite some time. And so, White writes:

> From the outset [sixteenth-century anatomist] Vesalius proved himself a master. In the search for real knowledge he risked the most terrible dangers, and especially the charge of sacrilege, founded upon the teachings of the church for ages. . . . Through this sacred conventionalism, Vesalius broke without fear; despite ecclesiastical censure, great opposition in his own profession, and popular fury, he studied his science by the only method that could give useful results.[83]

White thought he could cite, for example, a papal bull *Of detestable cruelty* (*Detestande feritatis*) by Pope Boniface VIII in 1299–1300. The bull, says White, threatened with excommunication and persecution any who practiced the vile art of dissection. It turns out, however, that White was not a careful reader, and this citation is one of many of his fake footnotes leading us astray in our historical wayfaring. White also cited a church decree entitled "The Church Abhors the Shedding of Blood" (*Ecclesia abhorret a sanguine*) of 1248. Neither of these, it turns out, says what White thought they did. The earlier "The Church Abhors the Shedding of Blood," as Katharine Park points out,

82. Black, *The Middle Ages*, 173–74.
83. White, *Warfare*, 2:50.

has the little defect of not existing. It is, in fact—as Charles H. Talbot demonstrated in 1967—a "literary ghost" invented by "an inept French historian"[84] and then passed on to be popularized by White. Park clarifies that while there were indeed church declarations forbidding clergy—not doctors per se—from performing surgeries that involved cauterization or cutting, these declarations were meant to avoid clergy with varying competencies in medicine from being tempted to put people's lives at risk for monetary reasons. In other words, as Hannam puts it, Church law prohibited "priests from moonlighting as physicians."[85] They said nothing whatsoever about dissection as such.

On the other hand, the bull *Of detestable cruelty* does exist, so we can celebrate that low hurdle being cleared. But instead of dissection, it is in fact forbidding a funerary practice that arose toward the end of the thirteenth century, which "involved cutting up the corpse and boiling the flesh off the bones, in order to make it easier to transport for distant burial."[86] This was a procedure that was popularized among European Crusaders in the holy land, as it aided in transporting their dead. Its origins also appear to be peculiar to Germans—and following them, the English and the French—especially among the aristocracy. As Park has argued, the general disgust among Italians for the practice of their German counterparts is really the only reason we have for why Boniface VIII (himself an Italian) had for writing *Of detestable cruelty*.[87] He cites no Scripture or church tradition, but merely references words of disgust like "horror" and "abomination." Some scholars, says Park, argue therefore that Boniface merely put into writing a general Italian prejudice against Germanic funerary rites. However, she counters that this is not the case, as Italians were clearly performing autopsies by that time. Whatever his reasoning, though, Boniface's bull clearly did not forbid dissections for the purpose of medical investigation.

Quite ironically, dismemberment of the body post-mortem was, it turns out, a Greek and Roman taboo, not a Christian one. To speak of the rise of autopsy and dissection is therefore not to call our attention to the revival of Greek science after the Christians lost it. It is in fact to point out that it was the Christians who removed the aura of the ritual uncleanness of corpses. "Dismemberment," says Park, "had long since been domesticated by the Christian cult of relics," that is, dismantling the holy dead in the belief that the merest hints of them in whatever location could still be a channel of grace for whomever drew near.[88] Christian doctors thus practiced in an environment primed to make them feel at ease with dissection of the dead. Now, as a matter of course dissections happened very infrequently—if at all—prior to the thirteenth century. Yet, as medical dissection increased in frequency, we do not detect

---

84. Park, "Myth 5," 46–47.

85. Hannam, *The Genesis of Science*, 103.

86. Park, "Myth 5," 47.

87. Park, "The Life of the Corpse."

88. Park, "The Life of the Corpse," 115.

the faintest hint of dissent at the practice—*except* in cases where a few overzealous anatomists hired graverobbers to find more subjects. This was, indeed, not allowed by the church. Moreover, White's reference to an event where the anatomist Vesalius was censured by the Inquisition for dissection, upon investigation, turns out to be a denunciation of Vesalius because he accidentally cut someone open who—unhappily—was still quite alive.[89] As such, White's entire list of evidence on this count falls apart.

White was similarly a force behind the idea that the church and its theologians were rallying against the newly invented chloroform anesthesia used in childbirth, pioneered by James Young Simpson in 1846. While obviously not a myth about the Middle Ages, it easily coalesced with the other myths being paraded about by White and others at the end of the nineteenth century. Over time it has become it difficult to distinguish what was a protest of the time, and what was, supposedly, more perennial to the obstinate posture of Christian dogma. As we mentioned, the alleged reasoning behind this religious protest was that labor pains were given to women by God as part of the consequences of the fall. So, White: "From pulpit after pulpit Simpson's use of chloroform was denounced as impious and contrary to Holy Writ; texts were cited abundantly, the ordinary declaration being that to use chloroform was 'to avoid one part of the primeval curse of woman.'"[90] Draper records a similar objection.[91]

Admittedly, by 1847 Simpson did write a pamphlet directed to doctors, but entitled "An Answer to Religious Objections Advanced Against the Employment of Anesthetic Agents in Midwifery and Surgery." So, on the surface it appears such religious objections did, for a time, exist. However, several things must be noted. The comprehensive work of A. D. Farr on this issue has discovered only the barest hint of evidence for "either theological opposition to anesthesia from the institutional churches, or any widely held (or express) opposition from [religious] individuals."[92] Farr thus concludes that Simpson's tract is, in fact, was "written to forestall objections which, in the event, did not arise."[93] The publication of Simpson's pamphlet has subsequently, says Farr, been misinterpreted as an entrenched reaction to religious opposition by White, and others. But religious reactions were actually enthusiastic, and "theologians and clergy from Presbyterian, Anglican, and dissenting churches [that is, Christians who had separated from the Church of England] sent overwhelmingly positive responses to Simpson for his 1847 pamphlet."[94] It seems they were thus quite open to correction on these matters, even if there was a smattering of initial dissent. Simpson himself noted

89. Park, "Myth 5," 47.
90. White, *Warfare*, 2:63.
91. Draper, *History of the Conflict*, 318–19.
92. Farr, "Religious Opposition to Obstetric Anesthesia," 166.
93. Farr, "Early Opposition to Obstetric Anesthesia," 906.
94. Schoepflin, "Myth 14," 124.

that by 1848 he "no longer met with and objections on this point, for the religious, like other forms of opposition to chloroform, have ceased among us."[95]

The notion of "other forms of opposition" is important. For as Farr notes, it was in fact the medical community at large—and not the religious—that was initially most hostile to the use of chloroform, and this for thoroughly medical reasons. In part, such objections arose from a completely different evaluation of the use of pain in medicine than we have today. Whereas we tend to see pain as something entirely negative to be quarantined, in the reigning Victorian practices of the day, many theorized that pain was in fact the body's great safeguard.[96] If pains were eliminated, some doctors hypothesized, it would in turn eliminate the body's healing response. Others argued that if pain was eliminated in childbirth, it would in fact dull or even obstruct the natural course of the birth, harming both the child and the mother. Not without cause, others insisted ether could produce psychotic episodes that would—understandably—interfere with labor. A fairly intense caution arose rightly indicating that no one knew what effect the "etherization" of the mother would have on the baby after birth.[97] In other words, Simpson encountered the normal, healthy course of science that was, perhaps on occasion, embroidered with fairly shallow scriptural objections that ceased nearly immediately after his publication.[98]

Again, however, these revelations were too late. Ginzrot, Cockayne, White, and others had produced enough ostensible evidence that the narratives of Petrarch, Gibbon, Burckhardt, Michelet, and the general Protestant prejudice against the Catholic Middle Ages seemed more secure than ever.

An Age of Too Much Science?

The Middle Ages—or the Dark Ages—as we have seen are something of a mobile category, changeable, moldable, often applicable to anything one simply does not like.[99] Perhaps nothing gives evidence to this than the curious change of fortunes that science and the Middle Ages took in the latter half of the twentieth century. While "Science" was seen as pure and clear in its azure heights as the final and highest virtue that humanity could achieve, opponents of Christianity were quick to claim that in the Christian past there was not enough science, perhaps none, and indeed even an active suppression of its practice. This changed in the mid-twentieth century as questions about the sciences and their position as as a chief virtue of humankind began to germinate after the two World Wars, the bombing of Hiroshima and Nagasaki, and an increasingly suspicious eye turned toward an ever-widening human ecological

---

95. Quoted in Schoepflin, "Myth 14," 124.

96. See Farr, "Early Opposition," 898–900.

97. Farr, "Early Opposition," 900.

98. Farr, "Early Opposition," 906–7.

99. Davis, "The Sense of an Epoch," 41.

footprint. This new uneasiness—and on occasion outright hostility—to science and technology could co-opt the then still-tender sprouts of scholarship rediscovering science in medieval Christianity and gesticulate wildly that, far from too little, there was in fact too much of the newly villainized science in the Middle Ages![100]

Lynn Townsend White, Jr. (though himself a Christian of a fairly liberal persuasion) can be seen on both sides of this interesting sea change. Early on, White was one of the most important historians doing the groundbreaking in terms of rediscovering previously overlooked or misunderstood medieval technology. In 1940 he wrote an essay entitled "Technology and Invention in the Middle Ages," in which he records his famous line: The Middle Ages' "chief glory . . . was not its cathedrals or its epics or its scholasticism: it was the building for the first time in history of a complex civilization that did not rest upon the backs of sweating slaves . . . but primarily on" technological advances that had been born from Western theology's "activist" traditions. Christians undid the Greek reticence toward craftsmanship and practical technology, brought to bear the notion of God as universal Creator through his Word and Wisdom, and "the implicit assumption of infinite worth of even the most degraded human personality," which therefore meant a complete abhorrence and "repugnance of subjecting any man to monotonous drudgery [of unaided labor]."[101]

Whether or not he harbored reservations about technology from the beginning, by 1967 White embodied the alarm of the time, but used his particular expertise to give it historical context:

> The victory of Christianity over paganism was the greatest psychic revolution in the history of our culture. . . . Our daily habits of action, for example, are dominated by an implicit faith in perpetual progress which was unknown either to Greco-Roman antiquity or the Orient. It is rooted in, and is indefensible without, Judeo-Christian teleology. . . . We continue today to live, as we have lived for about 1700 years, very largely in a context of Christian axioms. . . . Christianity inherited from Judaism not only the concept of time as nonrepetitive and linear, but also a striking story of creation. God planned [all of creation] explicitly for man's benefit and rule . . . Man shares, in great measure, God's transcendence of nature. Christianity, in absolute contrast to ancient paganism and Asia's religions . . . not only established a dualism of man and nature but also insisted that it is God's will that man exploit nature [to] his proper ends.[102]

Precisely because its theology and practice so intrinsically guided and collected the production of technology and science in the west, concludes White, Christianity "bears a huge burden of guilt for environmental deterioration" and that "we shall

---

100. Brague, *The Kingdom of Man*, 19, also notes this humorous valence switch.

101. White, Jr., "Technology and Invention."

102. White, Jr., "The Historical Roots of Our Ecological Crisis," 1205; relevant too is the now classic work of Merchant, *The Death of Nature*.

continue to have a worsening ecological crisis until we reject the Christian axiom that nature has no reason for existence except to serve man."[103] It is often overlooked by those who wield this latter essay gleefully as another weapon against Christianity, that White cites the Franciscan order, and in particular St. Francis, as embodying a Christian ideology that would counter this abuse. Christianity is, as the saying goes, a broad church.

Regardless, it is clear enough that White's ultimate argument against technology and science underwritten by a particular Christian theology. It should be noted that, even if we take White at face value here, it confirms and does not suspend the judgment of Christianity's positive—albeit complex—relationship to the production of knowledge of the natural world and technology in the Middle Ages, the scholarship surrounding which White himself was a father figure.[104] Nonetheless, this new myth of the Christian Middle Ages—indeed Christianity in the West as a whole—is concerning. White's essay on history has become "almost a sacred text for modern ecologists,"[105] and in 1989 the notion had become so prevalent that even *Time* magazine ran an article arguing that the Bible had bequeathed to Western civilization our impending technological devastation through the notion that we may use the earth however we see fit.[106] We have gone from viewing this period of history as a long and barren Dark Ages, to one full of machines and wheels and metal burning red and bright with the leaping fires of dominion and deforestation. It is in fact in the 1960s as well that J. R. R. Tolkien found his neo-medieval masterpiece *The Lord of the Rings* (originally published between 1954–1956) central to the libertarian and environmental movements, much to his puzzlement. Yet, his vision of a medieval struggle between virtuous craft in the hobbits of the Shire, and the threatening industry of Isengard, allowed popular imagination to easily assume such a rapacious transition occurred in the Christian Middle Ages.

White in some sense hangs the entire Christian ethos of "domination" upon the single invention of the heavy plow in Europe. This singular agricultural advance embodied and further catalyzed an ethos where "once man had been part of nature; now he became her exploiter."[107] It is worth noting that White cites no particular instances of abuse, deforestation, pollution, corruption, despoliation, or the like, but is arguing at a broader theoretical level that the Christian ethos that produced these medieval inventions also has its inevitable historical terminus in the corruption of the planet once that technology has reached a sufficient capacity. In other words, White has taken several of the now commonplace facts that he himself helped establish about the medieval world—again, their technological advances and theological orientation to a study of

103. White, Jr., "The Historical Roots," 1207.

104. See S. Worthen, "The Influence of Lynn White Jr.'s *Medieval Technology and Social Change*."

105. Quoted in Sampson, *6 Modern Myths*, 72.

106. Sampson, *6 Modern Myths*, 72.

107. White, Jr., *Medieval Technology and Social Change*, 56.

nature—and, despite some hand-waving about "Christianity [being] a complex faith" so that "its consequences differ in differing contexts,"[108] White merely restates these discoveries, soaked now in the tone of righteous anger permeating 1960s eco-culture. These Middle Age ingenuities, theories, and practices are therefore baptized with a violent, even catastrophic necessity that was never theirs.

A subtle but increasingly prevalent tale told is the notion of a pre-civilized humanity living in a sacred equilibrium with nature—an equilibrium broken by the advent not just of society, but because of domineering ideologies such as those supposedly contained in that damnable Christianity. The Christian notion of being made in the image of God, given license by the Almighty Himself to do as we see fit, is embroidered with a somewhat Baroque mythology as it is contrasted to an almost entirely fictitious picture of pre-civilized societies frolicking in Edenic equilibrium with mother earth. Such a picture of paleolithic, or perhaps merely pagan, humanity is sometimes referred to as "Dark Green Religion." Undoubtedly there are instances where such scenarios of harmony are more or less true. On the whole, however, the notion of the "noble savage" in this sense is a romanticized myth that represents a reworking of the notion of Eden, translated now into a language that coats itself with the thin veneer of archaeological respectability. Such wistful memories of a life that society has denied us, like the angelic flaming swords guarding Eden, not only goes far beyond the large archaeological silence on the matter in terms of interpretation. More importantly, the bright pasts of "Dark Green" religious historiographies are deeply contradicted by what evidence we do have.[109] Deforestation, massive brushfires to control animal migration, tribal warfare, even large-scale animal depopulation are all part and parcel of the various stories of neolithic humanity recounted by the long memories of stone.

On the other hand, as scholars like James Barr have pointed out, the word *dominion* in the Bible—the Hebrew verb *rada*—is a particularly soft expression, and was, for example, used to describe King Solomon's peaceful and wise rule. It does not mean a rapacious and unlimited use on the part of humankind, who are encouraged to learn from all the creatures and plants (Job 12:7–10) by listening to them. In fact, such reworking of domination is pictured in Scripture as the cause and by-product of sin (Isa 24:5–6), where not even warfare justified deforestation (Deut 20:19–20). On the other hand, "subdue"—*kabash*—is used in a limited agricultural sense referring to the working and tilling of the earth.[110] Scripture—though its opinions on creation cover a remarkably wide range[111]—hardly recommends humanity to take an aggressive and uncaring posture toward nature—which is represented in Genesis as God's garden, after all, and humanity its chosen caretakers (see also Deut 25:4).

---

108. White, Jr., "The Historical Roots," 1206.

109. LeBlanc, *Constant Battles*; Provan, *Convenient Myths*.

110. Barr, "The Ecological Controversy and the Old Testament," quote at 22.

111. Perhaps no recent study has captured that variety like Brown, *The Seven Pillars of Creation*.

But the meaning of a text hardly accounts for its later use and reception. Did early and medieval Christians in fact use it to justify wanton pillaging of natural resources? The answer will unsurprisingly be: no. "For the first fifteen hundred years of the Christian era," writes Peter Harrison, "there is little in the history . . . to support White [Jr.'s] major contentions."[112] While the number of vastly different approaches to nature makes it ambiguous to speak of anything like "the medieval view," it nonetheless would in fact be difficult for the early and medieval Christian imagination to comprehend—let alone produce—the notion that things so vast as rivers, forests, oceans, animal populations, and the like could be dominated by man, diminutive as he was surrounded by God's vast creation. Such domination was either a distant dream, or a threatening nightmare. Medieval theologians' interpretation of the "Creation and the Fall revealed a God-ordained world dominated by human beings, whose role in respect to nature, however, was not exploitation but stewardship and cooperation."[113] The key term used here was not "dominion" but "cultivation." Nature was seen not just as a tableau of material resources like trees and bears and mountains, but very often as embodying also a deep spiritual component that could only be discerned by the careful and prayerful interpreter. "Nature, in this scheme of things, was to be known in order to determine its moral and spiritual meanings," along with its use for man, "and so not . . . be materially exploited."[114] The major course of technological advancement is pictured by early theologians not in brute domination, not in strip mining, not, in a word, in exploitation—but cultivation, a beatifying, a bringing of new life.[115]

Of course, even theologians are not always theologians. Even more so for less interested, even if no less pious, practitioners of religion. And there were indeed problems that arose with the new scale of operations, as is inevitable. But, "What is lacking in [White's argument] . . . is [therefore] the identification of a *religiously* motivated ideology of exploitation."[116] Indeed, White in his *Medieval Technology and Social Change* can write that all of the improvements we spoke of above should be seen in terms of "the emergence of a conscious and generalized lust for natural energy and its application for human purposes."[117] There is next to nothing to commend this interpretation, and it seems much simpler, as Harrison counters, to say that medieval increases here were simply a component of the fact that more humans need more food and more clothing, and hence more efficient ways to get both.

Seeing nature as a somewhat helpless victim before the dominance of humankind is in fact thoroughly modern. While this does not absolve Christians who participated and even enacted this changing understanding, it is nonetheless important

112. Harrison, "Subduing the Earth," 90.

113. Gies and Gies, *Cathedral, Forge, and Waterwheel*, 7.

114. Harrison, "Subduing the Earth," 91.

115. Sampson, *6 Modern Myths*, 71–91.

116. Harrison, "Subduing the Earth," 95.

117. White, Jr., *Medieval Technology*, 129.

to record these shifts and transitions in theological thought. It is far too easy to treat history as merely the outworking of a few mostly unchanging ideas, and, like White, seeing domination as in some sense the perennial destiny of Christianity in the West. Ironically, White represents this rapacious domination as the sad victory of Christianity over paganism. In fact, it was the reincorporation of certain strands of Greek and Roman paganism that changed the traditional Christian interpretation of dominion pictured as a dialogue with nature.[118] Ernst Cassirer for example notes how in the Renaissance the typical Christian notion of humanity made in God's image—which as we saw above meant primarily that we were caretakers of God's creation—because of the rising fascination with classical texts, combined with the Greek myth of Prometheus stealing fire from the gods. "Man arrives at his determination, fulfills his own being, only by using this basic and primary power" to dominate, he says.[119] This combined with the rising fascination with the mystical melting pot known as Hermeticism, supposedly founded by the Egyptian Hermes Trismegistus (Hermes the Thrice-Great), but shown to be a much later, non-Egyptian body of texts. Nonetheless, one of the aspects of this mysticism is the idea that man fulfills his imaging of God in the enactment of power and mastery over all.[120] The notoriously syncretic Hermetic corpus thus absorbed the Genesis command to have dominion into itself, changing the very meaning, and providing a parallel to the rising notion of Promethianism. Hermeticism also found common cause with a strand of Christian theology known as Voluntarism, which so elevated the Will of God as to make it apparently unrelated to goodness, mercy, or restraint.[121] As Cameron Wybrow has demonstrated, this cocktail of Promethean and Hermetic glosses on Christian theology made its way into Frances Bacon (whom we might recall is usually named "The Father of Modern Science"), where Bacon now saw the Christian task to restore Eden as one that fundamentally involved exerting mankind's power over nature through technology and experiment.[122] Eventually, to summarize an increasingly complex story, with the secularization of scientific knowledge, the Christian elements reining in the Promethean ones were expunged, and so the (still residually theological) will-to-power of mankind found itself rechristened as scientific and was often freed from any lingering, overt religious restraint. Without too much tongue in cheek, wanton dominion or ecological abuse can be called something of a Christian heresy, but hardly an inevitable by-product of its essence.

118. Brague, *The Kingdom of Man*, 17–25.

119. Cassirer, *The Individual and the Cosmos*, 95.

120. Trinkaus, *In Our Image and Likeness*, 1:485–86.

121. Gillespie, *Nihilism Before Nietzsche*.

122. Wybrow, *The Bible, Baconianism, and Mastery Over Nature*.

Conclusion

Of course, heroes and villains exist. But, once again we discover in the stories related to the warfare of science and Christianity clear villains and heroes seem maddeningly absent, slippery, hard to pin down. But, since our histories until recently have been systematically pruned of the elements secular culture often finds most alien—the theological, the mystical, the alchemical, the mythical—we have remained unaware how the many changing and swirling currents of these modes of thought have played into our contemporary "scientific" notions of the world, for better and for worse. When these themes and their alterations are ignored, it becomes notoriously easy to find a line through history where Christian theology is understood as one, mostly homogenous thing that terminates its inevitable destiny in some disaster, be it the Dark Ages or the environmental crises currently raging. Things are not so easy. In fact, as we have seen, Christianity in a fashion opposite to many expectations shaped many of the values unthinkingly used today as weapons *against* Christianity, like care for the environment, technological innovation, care for the poor, the orphan, the widow. Without complexifying our histories in this way—taking in the strange twists and turns that theology and other currently unfashionable ideas have played—we simply begin to import our own assumptions about what must have been, what terms must have meant, what Christians must have done.

—Chapter Ten—

Closed Worlds and Infinite Universes
Copernicus, Galileo, and the Church

Oh, and night: there is night, when a wind full of infinite space
Gnaws at our faces. . . . Fling the emptiness out of your arms
Into the spaces we breathe; perhaps the birds
will feel the expanded air with more passionate flying. . . .

Appearance ceaselessly rises . . . like dew from the morning grass
What is ours floats into air, like steam from a dish
Of hot food. O' smile, where are you going? O' upturned glance:
New warm receding wave on the sea of the heart . . .
Alas, but that is what we are. *Does the infinite space*
We dissolve into, taste of us then?

—Rilke, *Duino Elegies*

On April 12, 1633, Galileo was condemned by the Inquisition (also known as the "Holy Office") for "holding as true the false doctrine taught by some that the sun is the center of the world." If one were to make the trek to the Villa Medici, a gorgeous abode that sits bright-eyed on Rome's Pincian Hill in Italy, next to it one can read engraved in a column "it was here that Galileo was kept prisoner by the Holy Office, being guilty of having seen that the earth moves around the sun."[1] The implication is, of course, that Galileo saw the truth, and so was condemned. Galileo to many is a name more or less equated with scientific genius, much as Leonardo or Newton. Moreover, he is also representative—much as the Scopes Trial is for Americans—of science suffering at the hands of a backward Christian religion. The situation is ripe for drama. Indeed, the twentieth-century German playwright Berthold Brecht took this quite literally and so

1. See Finocchiaro, *The Trial of Galileo*, 135.

put it on stage in his 1938 *Life of Galileo*. When the actor Richard Griffiths, who played the part of Galileo, reviewed a then-recent biography of the man himself, he wrote his now oft-quoted phrase: "[B]y stifling the truth, which was there for anyone to see, the church destroyed its credibility with science."[2]

Indeed, when the scientist Alice Dreger visited Italy with her mother to gaze upon several sacred objects of scientific history—Galileo's telescopes—she recounts a very humorous (and somewhat disturbing) encounter with an object of another kind: Galileo's mummified middle finger. Like the relic of a saint it stood in its repellant glamor beneath glass, and she looked down tentatively at her English guide book, hoping it might explain this morbid curiosity. A century after his death, explained the book, Galileo had been exhumed to be reburied in a grander gravesite. During the transition, an unnamed devotee cut off Galileo's middle digit, and another man named Tommaso Perelli provided a shrine with the engraving "This is the finger, belonging to the illustrious hand that ran through the skies, pointing at the immense spaces, singling out new stars, offering to the senses a marvelous apparatus of crafted glass, and with wise daring they could reach where neither Enceladus nor Tiphaeus [both giants in Greek mythology] ever reached." Dreger notes, of course, that the middle finger does not mean for Italians what it does for Americans and for the English,

> But the more I thought about it—about Galileo's contentious nature, his belief in the righteousness of science, his ego, his burning knowledge that he and Copernicus were *right*, and especially about what the Church put him through—the more amusing the middle finger thrust skyward seemed. . . . Eventually I couldn't stand it anymore, I burst out laughing, dropping the tour brochure on the floor. I picked it up and found the docent giving me a rather severe look. But I couldn't help myself, and started laughing uncontrollably again.[3]

And so goes the typical pictures of Galileo: his middle finger was on the pulse of cutting-edge science, locked and loaded as well to gesture wildly at the Church, and perhaps even antagonize God above. Nonetheless, a major question poses itself: why Galileo? Why condemn him? If, as it was just represented, the issue was truly about heliocentrism—the idea that the sun, and not the earth, was at the center of our solar system—why did the church not persecute Copernicus, whose work *On the Revolution of the Spheres* clearly did challenge the Aristotelianism of many in the church, and by the time of Galileo's condemnation had been circulating for nearly ninety years? The answer has to do with the fact that Galileo's condemnation was in fact a mess of grandstanding personalities, political intrigues, and indeed an impressive sense of bad timing. But, again to no one's surprise, it has taken the form we are so familiar with—that of Galileo persecuted for his science—because of retellings that occurred

2. Griffiths, "Very, Very Frightening."
3. Dreger, *Galileo's Middle Finger*, 17–18.

in the Enlightenment, which were magnified and mythologized even further in the debates surrounding the professionalization of science and the reception of Darwin.

Dust in an Infinite Sky

What, then, is man in the midst of two infinities? Nothing in comparison to the universe; infinite in comparison to the atomic.... The eternal silence of these infinite spaces terrifies me.

—Pascal, *Pensées*

The "Why Galileo?" question becomes even more curious when we realize that it is the rank and file of astronomy 101 textbooks, popular historical reviews, public opinion, and even a good deal of dictionaries and encyclopedias that Nicolaus Copernicus, by discovering that the sun and not the earth was in the center of our solar system, displaced and reduced the dignity of humanity as traditionally held in the Church. Surely, if this was the case, the Church should have been up in arms about Copernicus and those following him well before Galileo? But they were not. This did not stop such a story from spreading. Not only were we not the center, or so goes the story—Copernicus discovered that there was no such thing as a center. The horizon of the sky opened suddenly, and we were sent into stupor to find ourselves adrift like dust in an infinite sky, from nowhere to nowhere, amounting to nothing. No one more succinctly and famously embodies this understanding than Sigmund Freud in his *Introductory Lectures on Psycho-Analysis*. It perhaps seems odd that a figure in astronomy should appear in a text on psychology, and yet, Freud claims, Copernicus is one of three men who are in essence responsible for contemporary anxiety. Science, he alleged, had inflicted upon humanity "two great outrages upon its naïve self-love." The first was Copernicus, who caused humanity to clutch its collective pearls in horror "when it realized that our earth was not the center of the universe, but only a tiny speck in a world-system of magnitude hardly conceivable." The second shock was Charles Darwin, demonstrating that we were not crowned in glory but sired from apes, who in turn found parentage in an oozy little pond somewhere in the depths of an alarmingly deep time. A third shock in this same lineage was Freud himself, showing that our waking life was little more than flotsam washing up on the shores of our consciousness from the unfathomable depths of irrational desires and urges.[4]

Orienting all of world history around an event or three is perhaps at this point hardly remarkable, given how often we have already seen it done. Yet, we must not underplay how influential Freud's schema of the world and its history through these three great humiliations has been. Showing up in a text on psychology is perhaps the least startling place Copernicus has been invited to appear—indeed an entire

---

4. Freud, *The Complete Psychological Works*, 240–41.

cosmology, and entire metaphysics, has been conjured from the man, and from how Freud linked him to Darwin, and to himself. "We should revel in our newfound [lowly] status," writes biologist Stephen Jay Gould in reference to the humiliating trifecta of Copernicus, Darwin, and Freud. "[We need to cherish the idea] that we must construct meaning for ourselves."[5] In one sense there is nothing wrong with turning this into a metaphysics—the problem is that no one seems to acknowledge that is what they are doing. Rather, Freud's history and the vision of the world attached to it is taken as a plain truth spoken by redoubtable advocates for objective science. Carl Sagan, taking his cue directly from Freud, spread this historical periodization far and wide by declaring that Copernicus was the first of several "Great Demotions . . . to human pride."[6] This was a signal change that occurred on the far end of that great gap of the Dark Ages we have seen in detail. But Sagan's considerable artistry was just warming up.

In what is perhaps one of the most famous pictures ever taken, Sagan embroidered this history with the image of the earth as a "pale blue dot." Taken by the probe Voyager 1 as it turned back to face the earth while it was flying out on its lonely interstellar journey, what it captured was a remarkable picture of the earth, just about the size of a dust mote. Hanging there in a pale ray of cosmic light, warming itself against the immense dark surrounding it, one almost hears Sagan's steady, learned voice tell us "that's here, that's home, that's us." Beautiful, and too true. But behind it lurks a deep meaninglessness that continues to ring long after the wizened timber of Sagan's voice, assuring us that it will all be okay, has dissolved into that same unroofed sky of his moralizing. "Once upon a time," wrote Friedrich Nietzsche, "in some out of the way corner of that universe which is dispersed into numberless twinkling solar systems, there was a star upon which clever beasts invented knowing." What a tale! Yet it is not the ultimately moralizing one that Sagan presented. "That [invention of knowing] was the most arrogant and mendacious minute of 'world history' but nevertheless, it was only a minute. After nature had drawn a few breaths, the star cooled and congealed, and the clever beasts had to die."[7] Sagan's consolation for cooler, scientific heads to prevail is easily deconstructed into Nietzsche's morose alternative, which itself relies upon Copernicus. "Has the self-belittlement of man, his will to self-belittlement, not progressed irresistibly since Copernicus?" said Nietzsche. "Alas, the faith in the dignity and uniqueness of man, in his irreplaceability in the great chain of being, is a thing of the past—he has become an animal, literally and without reservation or qualification, he who was, according to his old faith, almost God . . . Since Copernicus man seems to have got himself on an inclined plane—now he is slipping faster and faster from the center into—what? Into nothingness? Very well!"[8]

5. Gould, *Life's Grandeur*, 18.

6. Sagan, *Pale Blue Dot*, 26.

7. Nietzsche, "On Truth and Lie in an Unmoral Sense," 79.

8. Nietzsche, *Genealogy of Morals*, 155–56.

Something is off about all of this for any with even a passing familiarity with the history of theology, whether it be Freud's or Sagan's or Nietzsche's variation. More on that shortly. Nonetheless, reinforcing Freud's notion of the church's supposedly immense fear of the newly infinite sky, the story of the burning of Giordano Bruno (1548–1600) for his allegiance to Copernicanism like all of these myths, easily combines with this Copernican myth, and slithers in. We must be clear in calling Bruno's death a myth: one can certainly not conjure from the various details an edifying story involving Bruno's sad fate. It is tragic no matter how one spins its ultimate reasons. Bruno wanted to die a martyr for his theologies, and the Church was more than happy to oblige. Nonetheless, again in the nineteenth century the figure of Bruno provided a convenient place for the rapidly increasing anti-clericalism to develop an intricate mythology surrounding not just the Inquisition, but also of course the relationship of the Church to science.[9] The truth is that Bruno's was an almost forgotten case until nineteenth century anti-clericalism found in it a parallel to Galileo and so another weapon with which to needle the Church by creating a mostly fabricated martyrology of science.[10] Indeed, this martyrology tradition led directly into John William Draper's use of Galileo in his narrative of conflict, so that "the work of Huxley, Draper, and White added the last chapters to the myths of inquisition, building on earlier myths and modernizing them to take contemporary concerns into account."[11] It is, for example, incorrect to cite his claim that an infinite number of worlds exist and are—perhaps more scandalously—populated, as the "scientific," catalyst that placed torch to pyre. Here, another thread that will unravel the whole myth of warfare appears. We spoke earlier of how theology has been deleted from the historical record, and here is plays an especially prominent role. "Even a cursory reading [of Bruno's work] *The Ash Wednesday Supper* suggests that any interpretation of Bruno as the first martyr of the Copernican cause fails to do justice to the complexity of his life and thought,"[12] as one scholar has put it. Bruno slighted the virgin Mary, publicly questioned the divinity of Christ, and converted to Protestantism. These were, it goes without saying, dangerous things to be a chorus-leader on the local street corner about. The reconsideration of such theological matters—while in no way softening the sad nature of how the church dealt with its feelings about heresy, or Bruno's own apocalyptic way of dealing with those who irritated him—nonetheless does mean, as one expert on the topic has recently pronounced, that "the legend that Bruno was prosecuted as a philosophical [or scientific] thinker, [or] was burned for his daring views on innumerable worlds or the motion of the earth, can no longer stand."[13]

9. Peters, *Inquisition*, 242–54.

10. Peters, *Inquisition*, 246.

11. Peters, *Inquisition*, 253–54.

12. Harries, *Infinity and Perspective*, 243

13. Yates, *Giordano Bruno*.

As far as the new science went, it could even be said that Bruno in his own way was a chief opponent. While his philosophical musings may in the most severely abstract formality appear similar to today's multiverse speculations, he had "a well-known and clearly expressed distaste for the new mathematics, which he saw as a schematic abstraction attempting to imprison . . . matter into static formulae of universal validity." Bruno was not part of the Scientific Revolution, but was a mystic who reasoned through "visualization, through images and symbols, rather than through direct observation."[14] This certainly does not mitigate his possible contributions to what in hindsight was labeled a "revolution," but noting that he was a "scientist" as the term came to be seen is deeply misguided. In many ways Bruno's mystical visions of the cosmos are actively non-Copernican, in that Copernicus in no way advocated for an infinite universe. He merely inverted the Ptolemaic-Aristotelian consensus of earth's centrality, now placing the sun where our homely earth once stood. The outer rim in the Ptolemaic system was a somewhat literal roof of star-salt made from the fifth element, or quintessence. In other words, for Copernicus, the rim of the universe was firmly fixed and not of an infinite depth. Where Copernicus differed from the earlier system, for example, was where in the Ptolemaic layout the lid of stars rotated every twenty-four hours, Copernicus and others noted with some bemusement that this was a pretty inefficient way to run things. Since we could create the same appearance of the heavens if it was assumed that only the earth, and not the whole universe, was that which rotated, Copernicus made the decision that his choice spoke more readily of that elegance the Creator always used to create. Regardless, the edge of the universe remained as it was. Actually, "the Copernican cosmos might . . . seem even *more* bounded than the Ptolemaic cosmos [where the earth is at the center of the universe, because for Copernicus] the stars are truly fixed in the heavens [now that the earth's rotation accounts for their apparent movement]."[15]

This was hardly uncommon among the luminaries of the time. Leonardo da Vinci, for example, in his famous image of the human in his *Vitruvian Man*, represents the human in the shape of a Greek *chi*, that is, a cross (of Christ), while also depicting man as bound by a circle that as far back as Plato defined the limit of the world. There were, to be sure, hints of an infinite universe in Copernicus's revision, just as the universe even in its most geocentric vision was by no means considered "cozy." The vastly influential Roman senator and Christian philosopher Boethius (c. 477–524) noted as much in his famous *Consolations of Philosophy*. Quite plainly he states that we "have learned from astronomical proofs that the whole earth compared to the universe is not greater than [an infinitely small] point. That is, compared to the sphere of the heavens, it may be thought of having no size at all. . . . the space of man hardly even deserves the name 'infinitesimal.'"[16] Nor was Boethius alone on this point.

14. Gatti, *Giordano Bruno and Renaissance Science*, 3.

15. Rubenstein, *Worlds Without End*, 90.

16. Boethius, *The Consolations of Philosophy*, II.vii.

It takes a naïve reading of the classics to come to the conclusion that they thought their universes small, or akin to some terrarium just because it was seen through the lens of the human as microcosm.[17] In a manner hardly even conceivable for us anymore, the frictionless twirl of the heavens seen from here below was often considered the best, most pure, more guileless model for ethics, reality, and meaning. The result was "something like the cosmologization of history," where here below was given infinitely more meaning to the extent it could imitate the sky above.[18] Man as a microcosm—embodying in himself all of the elements that made up the earth, the sky, the animals, and reason—was, to be sure, the meeting place and horizon of all things. But this was humanity's importance precisely because of their task *within* the grandeur and scope of the world. The medieval cosmos, as C. S. Lewis reminds us, "was anthropoperipheral" and not "anthropocentric" in the way many contemporary polemics attempt to level at it.[19] For the ancients, Christian or otherwise, it is enough to say that "to live on earth does not correspond to our most profound aspirations."[20] Regardless, for Bruno, his version of infinite worlds was resolutely and thoroughly theological. In his speculation Bruno conscripts no less than Lucretius—who, in the last two chapters, we saw Greenblatt laud as changing medieval Europe—but ironically here Lucretius is only lionized insofar as he serves Bruno's God. The universe was infinite because, as infinite, God was like a pregnant mother bound to birth infinite infinities. The outrage was not the infinity of worlds, per se, but that God was described by Bruno as being under obligation (even by God's own nature) to create them. It was a perceived assault upon God's freedom, and not infinite worlds, that was the touchy subject.

Ultimately, however heterodox many of Bruno's formulations may have been, in terms of the infinite worlds of the infinite universe he was following the profound and orthodox Nicolas of Cusa (1401–1464) who was, in fact, a Cardinal of the Catholic Church. The non-centrality of earth (and so, humanity) was anticipated before Copernicus by Cusa, and is one of his forgotten influences.[21] In his classic *On Learned Ignorance*, Cusa supposed that, "it always appears to every observer, whether on the earth, the sun, or another star, that one is . . . at an immovable center of things and that all else is being moved."[22] Indeed, such reasoning occurred even earlier in thinkers like Nicole Oresme and Jean Buridan, with whom Duhem's story was so often centered. In what appears to be a startlingly early anticipation of Einstein's special theory of relativity, Cusa set the earth into a relative orbit around other celestial bodies, themselves navigating relative causeways. And despite the scientific ring this will have to modern ears, it had an irreducible theological component to it. Because God

17. Brague, *The Wisdom of the World*, 85–105.

18. Brague, *The Wisdom of the World*, 98.

19. Lewis, *The Discarded Image*, 58.

20. Brague, *The Wisdom of the World*, 93.

21. Harries, *Infinity and Perspective*.

22. Cusa, *On Learned Ignorance*, 2.12.162.

is infinite, nothing could be in the center of his creation; things spill on in an endless series of perspectives as each is infinitely close and simultaneously infinitely distant from God. The outer edges of the world peer back at us, thinking they are the center; while we would for them etch the eerie periphery of the known with our dim and distant light. This created, perspectival infinity is nothing other than an expression of the nature and character of the infinite, Triune Creator God.[23]

As such, the displacement of the earth from the center of the universe was hardly an assault upon the theological concept of humanity by a scientific one, but was in fact initially based upon theological reasoning. So too, far from a meaningless universe, or the demotion of man, "the cosmic system [of Copernicus] was a product of divine design and order." Galileo even insisted that to censor heliocentrism "would be but to censure a hundred passages of holy Scripture which teach us that the glory and greatness of the Almighty God are marvelously discerned in all his works and divinely read in the open book of the heavens."[24] In a particularly humorous moment, the glory of God seen through heliocentrism was because, quite the opposite of many accounts today, the center in many traditions was far from the best location. For Aristotle, for example, instead of prime real estate the center was a place of heaviness, where the refuse of the world naturally and necessarily accumulated. For Dante, the center of the universe was, in fact, quite literally hell itself. And again, this was necessarily so. The unusual historical discovery (or reminder) of late was the fact that for many in the Copernican debates, far from Freud's demotion, "heliocentrism was seen as 'exalting' the position of humankind in the universe . . . and conversely placing the divinely associated sun into the central yet tainted location."[25] One's reaction thus varied accordingly. The deep irony of Freud's claim is that of the initial complaints surrounding Copernicus and Galileo was, in fact, that by moving the earth to where the sun used to be, we were in fact demoting the sun and exalting the earth! And that is where Copernicanism had even taken hold. Even a quarter century after Copernicus published, a thinker like Michel de Montaigne perpetuated a more ancient line and wrote that "we are lodged here in the dirt and the filth of the world, nailed and riveted to the worst and deadest part of the universe, the lowest story of the house, and most remote from the heavenly arch."[26] Galileo, conversely, noted that "as for the earth, we seek to ennoble and to perfect it, when we strive to make it like the celestial bodies and, as it were, place it in heaven from whence your philosophers [the Aristotelians] have banished it."[27]

In the end, so successful was Copernicus's "strenuous revaluing and refurbishing of the center [in which he placed the sun]" along with the exaltation of humanity by

23. Lai, "Nicholas of Cusa and the Finite Universe."

24. Glacken, *Traces*, 376.

25. Danielson, "Myth 6," 54.

26. Quoted in Danielson, "Copernican Cliché," 1031.

27. Galileo, *Dialogue*, 42.

making us participants in the heavenly spheres, "that we have ever since been blinded to how Copernicus' predecessors truly viewed the central location."[28] This leads, in fact to a very curious historical rewriting:

> Once the center was seen as being occupied by the royal Sun, that location *did* appear to be a very special place. Thus we anachronistically read the physical center's post-Copernican excellence back into the pre-Copernican world picture—and so turn it upside down. [In this manner] the Copernican cliché is in some respects more than just an innocent confusion. Rather, it functions as a self-congratulatory story that materialist modernism recites to itself as a means of displacing its own hubris onto what it likes to call the "Dark Ages." When [Bernard le Bovier de Fontenelle (1657–1757)] and his successors [e.g. the *philosophes*] tell the tale [which it seems they also invented], it is clear that they are making no disinterested point; they make no secret of the fact that they are "extremely pleased" with the demotion they read into the accomplishments of Copernicus [because it is a direct attack on their newly invented straw man of church history].[29]

Moreover, Freud's combination of this new story with Darwin and then ultimately himself, was likewise no accident. Nor was Freud the first to make this equation between the two great men. After Darwin's death in 1883 Ernst Haeckel, a biologist who can for convenience be seen as Thomas Huxley's German counterpart, became after his wife died perhaps the most influential advocate for a thoroughly anti-Christian materialist interpretation of Darwinism.[30] His wife's passing left him with the deep burden of a feeling that all meaning had fled the world, to be replaced only by tragedy. One is reminded of the similar death of Michelet's wife, and his reevaluation of the Middle Ages once held so dear, now descending into an evening bordered by no dawn. Like his contemporaries Goethe and Humboldt, Haeckel's science was "transported by deep currents of aesthetic inspiration," and was profoundly affected by the tides of a movement known as German Romanticism. For the Romantics, nature was a display of the attributes of God, a God who was now in hiding. Tragedy was in some sense a metaphysical principle; it resided within the heart of the Hidden God—or The Absolute—and endlessly repeated itself here below.[31] The death of God was, of course, a deeply Christian theme. As early as Melito of Sardis in his homily on Passover, expressions like "God is murdered" were pious phrases of hymnody.[32] In Romanticism, instead, the death of God metastasized into a cosmic principle, and likewise was imprinted deeply upon their interpretations of nature and society. The materialist implications of evolution that Haeckel championed often echoed with this "speculative land

---

28. Danielson, "Copernican Cliché," 1032.
29. Danielson, "Copernican Cliché," 1033.
30. This is the thesis argued expertly and at length by Richards, *The Tragic Sense of Life*.
31. Krell, *The Tragic Absolute*, e.g. 104–209.
32. See Jüngel, *God as the Mystery*, 43–104.

of gothic dreams." And because of his great personal tragedy, evolution was his lance to thrust into the heart of the world, but also the backbone of his theological campaign to display his new and profound sense of the great calamity of things.[33]

More than Huxley—more even than Darwin himself—Haeckel nearly single-handedly created the "Darwin industry" as we know it today. "The controversial implications of evolutionary theory for human life—for man's nature, for ethics, and for religion—would not have the same urgency they still hold today had Haeckel not written."[34] Indeed, had Haeckel never lived—or perhaps had his wife never died in the manner and time she did—one historian, through an intensely detailed studied of his life and thought, can claim "certain non-essential aspects of modern evolutionary theory, namely its materialistic and anti-religious features" would have hardly found the prevalence they did. Haeckel's own version of "Darwinian theory would have lost its markedly hostile features, and these features would not have bled over to the face turned toward the public." Nonetheless, after Haeckel's wife passed, he took up his crusade "and the cultural representations of evolutionary doctrine took a different cast: evolutionary theory became popularly understood as materialistic and a-theistic, if not atheistic. . . . It was Haeckel's formulations that . . . created the texture of modern evolutionary theory as a cultural product."[35]

More to the point, Haeckel was one of the first to connect the newly minted Copernican cliché to Darwin in order to detail his materialist vision: Copernicus's heliocentrism "led to the most decisive revolution in the thought of the world of his day" he says, because it destroyed the assumed place of man in nature. Similarly, one thinks, notes Haeckel, of the "theory of Darwin and the extraordinary stir that it has occasioned. . . . [Darwin] puts in the place of a conscious creative force, building and arranging the organic bodies of animals and plants on a designed plan, a series of natural forces working blindly . . . without aim, without design." Elsewhere, Haeckel could put it simply that Copernicanism and Darwinism corrected the two great errors of the book of Genesis regarding the cosmic centrality of man, and his separation from the lower animals. As one scholar who unearthed this legacy of Haeckel for the Copernican cliché has noted:

> Whether or not Freud was informed by Haeckel . . . is beside the point. What is important to recognize is that the . . . evolutionist [was] articulating a connection between Copernicus and Darwin that was quite similar to the observation made by Freud [and was most likely the first to note] . . . that Darwin continued a process of decentering humanity that was begun by Copernicus and modern astronomy.[36]

---

33. On romanticism, see Abrams, *Natural Supernaturalism*, esp. 197–252.
34. Richards, *The Tragic Sense of Life*, 8.
35. Richards, *The Tragic Sense of Life*, 13–16.
36. Hesketh, "From Copernicus to Darwin to You," 195.

Haeckel's tragic world gained a history. This story of Haeckel's, like so many myths, played quite well with its fellow myths. Visions of the universe as a grand machine began to combine with Haeckel's horrifying Copernican revolution. The cosmos was soulless, running without any consideration of humankind, thrumming and humming and ceaseless as our fathers and mothers in unending time broke upon its myriad wheels and were stretched open upon its springs. And so, the original meaning of a "mechanical universe" began to be twisted to the ends of this new cosmic centerlessness, what is often called the seventeenth-century "death of nature" as cosmos became machine.[37] The notion of the cosmos as a machine is frequently wedded too to the idea of a broader transition in worldview tied to the evacuation of meaning from the world, as well as the joblessness of God in the face of the newly understood regular operations of nature, as was mentioned in chapter 4.[38] Yet this was not so, at least not initially. "The paradox is that among those seventeenth-century scholars who did most to usher in the mechanical metaphors were those who felt that, in so doing, the were enriching rather than emasculating conceptions of divine activity."[39] Moreover, "mechanism" or to be "mechanical" did not have the same connotation as it does today of being lifeless or inert. As late as the Middle Ages (where, it should be noted, robots and machines were actually everywhere) machines could evoke notions of spirituality, enchantment, vitality. The notion that a purely material entity must be inert did not arise until Reformation polemics with Catholicism. Indeed, "the need to establish that human beings [or the cosmos at large] are not machines cannot have had the same urgency in 1500 or 1600 as it had in 1900 [during the evolutionary debates]."[40] Nonetheless, as time commenced hindsight made it clear just how easy it was to turn the mechanical universe into a Godless one, both in philosophy and in the retroactive tracings of the historian. "Proponents of this new idea that the material world was intrinsically inert and passive," writes Jessica Riskin, "inaugurated an intense and world-transforming conflict with a persisting older tradition in which matter and mechanism remained active and vital, and in which automata represented spirit in every corporeal guise available and life at its liveliest."[41] The famous Dutch physicist Eduard Jan Dijksterhuis (1892–1965) for example made it the central point of transition in his famous *The Mechanization of the World Picture*, as did E. A. Burtt (though in a tone of lament) in his *The Metaphysical Foundations of Modern Science*.[42] Showing a deep reliance upon Duhem's work—especially in Dijksterhuis's reflections—the mechanical world was less sinister than it would soon become, being a development of the medieval *machina mundi* (machine of the world) tradition using analogies like

37. Brooke, *Science and Religion*, 117.

38. Snobelen, "The Myth of the Clockwork Universe."

39. Brooke, *Science and Religion*, 118. See also Osler, *Divine Will and the Mechanical Philosophy*.

40. Riskin, *The Restless Clock*, 42.

41. Riskin, *The Restless Clock*, 42–43.

42. H. F. Cohen, *The Scientific Revolution*, 59–72 for Dijksterhuis, 88–96 for Burtt.

the great clock of Strasbourg to envision God's creation.[43] The argument is not that this therefore was a felicitous decision to picture God and the world; but rather, that it was initially seen as positively theological. It does not mark a moment of transition into Godlessness until it was renarrated as such, often at the behest of Laplace as we have already seen with the secularization of Newton and the disappearance of God from the universe. The key date for both Dijksterhuis and Burtt in their well-meaning narratives is 1543: the date that Copernicus published *de Revolutionibus*. A moment many could then take as a rightful ally with Haeckel's story.[44] Once again we run into the paradox that pervades the heart of this book: "it was not physics [or science] that produced the nihilistic clockwork universe, but philosophy," and the reworked histories that carried such philosophies, especially surrounding the professionalization debates over science and the reception of Darwinism.[45]

The problem was again that in the late nineteenth century up through the late twentieth, these stories without any note of their highly specific origins migrated into textbooks everywhere as part of the universal record of scientific advance. Indeed, after the Harvard astronomer Cecilia Helena Payne-Gaposchkin (1900–1979) inserted the cliché of impoverished centerlessness into her *Introduction to Astronomy* (1954), the myth received another new life and a legion of others followed suit.[46] Even Peter Bowler—a historian of evolutionary theory who has spearheaded many of the revisions that play into our story—opens his own (extremely good) introduction to the history of evolution by noting "for historians of science, the 'Darwinian Revolution' has always ranked alongside the 'Copernican revolution,' as an episode in which a new scientific theory symbolized a wholesale change in cultural values."[47] Without the overt polemic, then, Bowler proceeds in broad outlines to repeat Haeckel and Freud's framework for the course of history—and this amidst the same few paragraphs where Bowler also goes out of his way to debunk the warfare thesis! It survives unscathed by Bowler's many keen insights about striking an even-keeled approach to the history of science and religion. Freud's framework is as such, one imagines, made stronger and driven even deeper by that which did not kill it.[48]

43. Bartlett, *Natural and the Supernatural*, 35–70.

44. H. F. Cohen, *The Scientific Revolution*, 68.

45. Josephson-Storm, *The Myth of Disenchantment*, 74; Brooke, *Science and Religion*, 118–19.

46. Keas, *Unbelievable*, 99.

47. Bowler, *Evolution*, 1–3.

48. The so-called "Copernican diminution" was assaulted head on at a conference in Cracow, Poland on the 500th anniversary of Copernicus' birth. At "Symposium 63" Brandon Carter presented a paper titled "Large Number Coincidences and the Anthropic Principle in Cosmology," which would end up being a bombshell reorienting twentieth century discussions on the so-called "Anthropic Principles" that appear to be embedded into the structure of the universe. Instead of unimportant, Carter argued that the observational location of humans revealed a key fact: that the universe is such that observers are possible. Here for the first time "large number coincidences" or "Fine-tuning" were pointed out, referring to the fact that should certain minute, nearly infinitesimal alterations happen to a myriad of conditions, life would have been impossible (see Witham, *By Design*, 37–56; Davies, *The*

Unread Books and Meridian Lines

When it became apparent that the church had not been up in arms about Copernicus, and that the notion of the world's demotion was not what many thought it was, confusion arose. Perhaps, some thought, no one had read Copernicus's work? If such was the case in making this work known, Galileo would of course have been the first to have been persecuted for it. It was a neat concession that saved some of the original antagonism between church and science so dear to so many. Though it was it a slightly different context—and he could not prove it—Arthur Koestler had originally suggested just as much in his bestselling 1959 history of astronomy, *The Sleepwalkers*. The question was taken seriously enough that, in the preparation for the 500th anniversary of Copernicus's birth in 1973, no less than the astronomer Owen Gingerich and the historian of science Jerry Ravetz began to ask themselves about it. To investigate such a claim, it was thought, one could look at the handwritten marginal notes of all first- and second-edition copies of Copernicus' *On the Revolution of the Spheres* to try and decipher the journeys between owners and commentators these copies had undertaken. If one could do this, a larger picture of the network of scholars would emerge. Who had read it, when, and what did they think of it? For Gingerich in particular, a wonderous journey began with these questions, one that would take

---

Goldilocks Enigma; Leslie, *Universes*, 25–56; Rees, *Just Six Numbers*; McGrath, *A Fine-Tuned Universe*, 109–126; Spitzer, *New Proofs*, 47–74). In 1986 John Barrow and Frank Tipler released their now-classic *The Anthropic Cosmological Principle*, which caused the design argument to "come roaring back" by bringing the idea outside of obscure journal articles into broader publication (Rubenstein, *Worlds Without End*, 14) and caused human importance to suddenly regain its position. Ironically, while many took these findings to be the undoing of the Copernican principle (as its historical caricature usually represented it), these new developments are historically more in line with Copernicus' idea of how God must have created the universe for the sake of humanity (Harries, *Infinity and Perspective*, 119–123). Ironically, just as Copernicus was invoked in the Darwin debates, the renewal of design via the anthropic principles of a fine-tuned universe had a similarly "Darwinian" attempt to circumvent the need for God. That is, the odds of achieving the constants we need for life are so astronomical as to make the world appear as "a put up job" as the physicist Fred Hoyle once said. If there were an infinite number of universes, however, then of course one among them would have the constants set as we see them in ours. "Just as Darwin and Wallace explained how the apparently miraculous design of living forms could appear without intervention by a supreme being, the multiverse concepts can explain the fine-tuning of physical law without the need for a benevolent creator who made the universe for our benefit." (Rubenstein, *Worlds Without End*, 17, quoting Stephen Hawking). As physicist Bernard Carr put the matter, "If you don't want God, you'd better have the multiverse" (Rubenstein, *Worlds Without End*, 17). Of course, just as natural selection does not per se eliminate God, neither does selection amped up to the level of the multiverse. Regardless, the Darwinian and Copernican moments in time are intertwined in the historiographical stories of the development of our understanding of the universe, just as it was with Haeckel. "For better or worse," writes Rubenstein, "multiple-world cosmologies consistently rearrange the boundaries between and among philosophy, theology, astronomy, and physics," thus creating, even in moments explicitly attempting to circumvent theology, "the possibility of new relationships across these boundaries" (Rubenstein, *Worlds Without End*, 18). So too, in the same spirit, we might riff on Rubenstein and state that "for better or worse, Copernican cosmologies consistently rearrange the bigger picture taking place among Darwinian historiographies, and vice-versa, creating the possibility of new relationships across historical boundaries."

him nearly forty years to complete. Much as with Duhem, at the beginning, Gingerich recalls, "I could scarcely have imagined the scope of the search."[49]

In his wonderful *The Book No One Read*, Gingerich—now the Senior Astronomer Emeritus at the Smithsonian Astrophysical Observatory and research professor in the history of science and astronomy at Harvard—recounts his life-spanning quest with great charm and detail. The search involved well over 500 copies, quite literally scattered across the globe in libraries, museums, and private collections—some even behind the then nearly impenetrable iron curtain of the Soviets. What he discovered was that Koestler had been wrong, and with gusto. Some of the discoveries along the way were mere novelties. Giordano Bruno's copy, for example, contained "a bold signature" but "no evidence that he had actually read the book," which in general supported the fact, gleaned from his own writings, that he "seemed at best to be rather ill informed about Copernicus' ideas."[50] There were meatier considerations afoot as well. After reviewing the marginal notes of every extant first and second edition, not only was Gingerich able to come to the conclusion that the book was well-known among scholars of astronomy. What was revealed as well was a huge network of astronomers both religious and otherwise, all of whom engaged with the work with varying degrees of detail. Indeed, even the "copies in Lyons, Clermont-Ferrand, and Bouges had all been own by clerical libraries," and amidst the many notations left by scholars, "not one had seen the censor's hand, even in a Catholic country."[51] Even when *On the Revolutions* was asked to be censored (because of Galileo), Gingerich notes the censorship was both minor—and mostly ignored. As we will see in a moment, one of the issues was whether Copernicus was teaching a convenient mathematical hypothesis (which no one had a problem with), or whether he demanded that this was a physically true statement of the solar system—which at the time did not have enough proof. The censorship request that was caused by Galileo only applied to bits that demanded it be a physically true theory. The rest remained untouched.

Indeed, humorously enough, as part of the decree for censorship it was demanded that the book of Copernicus be "entirely preserved" except for the few passages suggesting the theory demanded the physical truth of its hypothetical layout because it helped predict Church feasts and, most importantly, Easter. Tracing down all extant copies, Gingerich noted a "remarkably complete picture" of censorship emerged. While a majority of copies in Italy had followed through with the minimal censorship, elsewhere censorship was almost nonexistent, with areas like Spain and Portugal ignoring the Vatican decree entirely. In some instances, "the ink cancellation [for censorship] was so slight," they almost appear to "highlight the passages in question." One wonders if a few cheeky theologians were getting their laughs at the expense of authority by doing it this way. The Jesuits, for example, felt Copernican censorship was

49. Gingerich, *The Book Nobody Read*, 27.

50. Gingerich, *The Book Nobody Read*, 64.

51. Gingerich, *The Book Nobody Read*, 91.

a Dominican affair, and so either ignored it or, on occasion, subverted the Dominican efforts at enforcing the censorship rule. As the historian Rivka Feldhay has explored in detail, in fact, one of the many overlooked contexts that complexifies the Galileo affair is how it came about as it was ground and fed through internal church politics between the Dominicans and the Jesuits. This was not "science against the Church," but rather a struggle of different understandings of "science within the Church."[52] Beyond the Jesuits, most Catholics considered it "a local Italian imbroglio" and not only ignored the calls for censorship, but continued circulating both Copernicus *and* Galileo.[53]

An unexpected curiosity of Gingrich's quest was that he began to notice "the general lack of [predictive] improvement made when the Copernican system replaced the old Ptolemaic system."[54] Tracing yet another Victorian-era myth, Gingerich discovered the false nature of the commonly stated opinion in the latter half of the nineteenth century that Copernicus was inspired to set pen to paper because the old Ptolemaic system had accrued a legion of clumsy additions attempting to string along its quickly fading explanatory power. Such notions were meant to magnify all the more Copernicus's achievement, increasing the emphasis upon the uniqueness of his genius and the singularity of his triumph against a background of confusion and darkness. Even revisionist scholars like Kuhn not only maintained this piece of trivia, but even made it a centerpiece for his notion that scientific "crises" spur on paradigmatic changes to what he also termed "normal science" in any given period.[55] It also proved untrue. When looking at ephemerides tables (giving the predicted locations of stellar objects at regular intervals), Gingerich concluded that "the errors reach approximately the same magnitude before and after Copernicus."[56] Moreover, the Copernican system is not actually simpler than the Ptolemaic by most standards.[57] The curious conclusion reached by all this was that Copernicus's system was not the response to a scientific crisis. It was rather Copernicus's attempt at a new aesthetic and indeed, theological, vision of the universe.

> There is no particular astronomical reason why the heliocentric cosmology could not have been defended centuries earlier, and it is in fact shocking that Copernicus, with the accumulated experience of fourteen more centuries, did not come up with a substantial advance in predictive technique over the well-honed mechanisms of Ptolemy. The debased positivism that has so thoroughly penetrated our philosophical framework urges us to look to data as the

52. Feldhay, *Galileo and the Church.*

53. Gingerich, "The Censorship of *De Revolutionibus*," 282–84.

54. Gingerich, *The Book Nobody Read*, 116.

55. Kuhn, *The Structure of Scientific Revolutions*, 67.

56. Gingerich, "Crisis vs. Aesthetic," 194.

57. Gingerich, "Crisis vs. Aesthetic," 197.

foundation of scientific theory, but Copernicus' radical cosmology came forth not from new observations, but from insight.[58]

Despite how startling this might seem, many others have since come to similar conclusions. Thomas Kuhn, for example, noted that "available observational tests provided no basis for a choice" between the Copernican and Ptolemaic theories.[59] To be sure, there were many new observations in Galileo's time that highly qualified and deconstructed portions of the Ptolemaic system. But they did not necessarily recommend themselves in any automatic fashion to the new, heliocentric model. Indeed, the Ptolemaic system was often flexible enough to withstand and incorporate these new observations into itself. Heliocentrism was, rather, an elegant act of insight. And Copernicus's new insight was, in a large respect, a theological one that incorporated mathematical reflection. Whereas in the Ptolemaic system, each movement of the cosmos was independent, for Copernicus we should expect all things to be part of a unified system precisely because of the unified Creator. "Nor have they [the Ptolemaists] been able [from their system] thereby to discern or deduce the *principle thing*," says Copernicus, "namely the design of the universe [by God] and the fixed symmetry of its parts." With the Ptolemaic system, it is as if "one were to gather various hands, feet, head, and other members, each part excellently drawn, but not related to a single body. And since they in no way match each other, the result would be a monster rather than a man." What struck Copernicus was "a new cosmological vision, a grand aesthetic view of the structure of the universe," that in his mind reflected the symmetry and ubiquity one would expect from the infinite, omnipotent God.[60] Yet, perhaps more than any other discipline, given how central it has been in constructing an image of the "Scientific Revolution," astronomy as a whole was carefully culled of its robust theological dimensions and is a prime example of the theme from our second chapter, in particular with the work of Mach.

> So many writers on the so-called "Scientific Revolution" (a historians' term coined around the mid-twentieth century), especially if they are coming from an anti-religious direction, cherry-pick from the sixteenth- and seventeenth-century discoveries only those which suit their "triumph of science" agenda. . . . To develop a balanced understanding . . . however, we must weigh in other factors and treat them with respect, and not shy away in embarrassment.[61]

Gingerich's efforts revealed a complex world where Copernicus and Galileo were incorporated in a variety of ways into the lengthy pedigree of church astronomy among Christians. They were not singularities insulated from the church, nor were they "scientific" in contrast to other models like those of the Jesuits. This has not stopped

58. Gingerich, "Crisis vs. Aesthetic," 199–200.

59. Kuhn, *The Structure of Scientific Revolutions*, 75–76; cf. Chapman, *Stargazers*, 84.

60. Gingerich, "Crisis vs. Aesthetic," 199.

61. Chapman, *Stargazers*, 86 87.

many from invoking the gap of the "Dark Ages" that we saw in the last two chapters, of course. The menagerie of examples is filled with the sadly not so exotic species of *academics who should have known better*, like Charles Freeman in his popular book *The Closing of the Western Mind: The Rise of Faith and the Fall of Reason*. Invoking something akin to Carl Sagan's infamous chart detailing the thousand-year lacuna that Christians set upon humankind's collective learning, Freeman laments the decline of something he quite fancifully labels the "Greek empirical tradition"[62] and the general subservience of reason to the ascendant shadow of faith leering over Western history. He continues: for more than a thousand years after the "last recorded astronomical observation in the ancient Greek world [he says by Proclus (c. 410–485)]" such studies then lay dormant "until Copernicus [1473–1543]."[63] Indeed, "by the fifth century, not only has rational thought been suppressed [by Christianity], but there has been a substitution for it of 'mystery, magic, and authority.'"[64]

Claims such as Freeman's have been around for a while, and many arose at similar time periods to combine with other instances of the legendarium we have been covering in this book. Again and again the rush and welter of myths reinforced an ever-increasing network of interconnected ideas and create a bulwark against this or that factual debunking. Regardless, one does have to wonder at the sort of blindness that made possible some of the more outlandish claims like those of Freeman's. Astronomy did not stand still in any sense, and theology and science continued to be interwoven in complex ways amidst a potpourri of allegiances, parties, and theories.[65] While Gingerich was still undertaking his journey, another scholar, John Heilbron, was making similar discoveries within Catholic cathedrals themselves. "I can say I literally tripped over the subject," says Heilbron to journalist Geoff Manaugh in an interview with *Atlas Obscura*. As he continued the interview, he began to laugh. The lesson, he said, "is that in a church you need to look down as well as up."[66] And his discovery? A thin whisper of bronze running through any number of cathedrals called a meridian line. To step over the line is not merely to pass over an odd sliver in the flooring, but as one eighteenth-century commentator noted, it marks "an epoch in the history of the renewal of the sciences." Indeed, the story of this delicate strand, as Heilbron records in his book *Sun in the Church* recording his lost discoveries in a manner reminiscent of Duhem, ties together "many fields now usually held apart." Not just

62. Freeman, *The Closing of the Western Mind*, 307–23.

63. Freeman, *The Closing of the Western Mind*, 322.

64. Freeman, *The Closing of the Western Mind*, xix.

65. For this claim, see especially: Scotti, *Galileo Revisited*, 127–174; Graney, *Setting Aside All Authority*; Freely, *Before Galileo*; Nebelsick, *Circles of God: Theology and Science from the Greeks to Copernicus*.

66. Manaugh, "Why Catholics Built Secret Astronomical Features Into Churches."

theology, but "architecture, astronomy, ecclesiastical and civil history, mathematics, and philosophy."[67]

Beneath the wandering sun making its practiced course westward sits the half-orbed dome of the *Santa Maria degli Angeli* (Saint Mary of the Angels), for example, a grand basilica constructed in the sixteenth century by Michaelangelo Buanarroti. (Yes, that Michaelangelo). The sunlight, having made its eight-minute journey through the blackened void of space approaches Saint Mary of the Angels and effortlessly falls through an opening in the ceiling painted with rays resembling the sun. Passing through as a thin column, a small disc of white and yellow light then opens upon the polished marbled floor as an exhalation of the morning where the day stretches its limbs and welcomes itself into this place of worship. A menagerie of subtle colors embedded in the marble in turn greet this small breath of light, and stir as if alive. "Shafts from the sun fall through the dome and windows of the cathedral to make puddles of light on the marble floor. Shadows cast by the obelisk in the square outside serve as a gigantic solar clock. At sunset, rays shining through the stained-glass window over the western altar dramatize the presence, and indicates the aptness, of the sun in the church."[68] As the seconds pass, the trembling column of day will slowly moves in a straight line, creeping steadily toward solar noon. There, it will intersect with the meridian, an inlaid track of thin brass running through the floor.

One can easily miss it, distracted by the grandeur of the building that surrounds and enframes and pronounces the glory of God. But as Scripture notes—where God's strength is made perfect in our weakness—the line's frail bronze whisper through the floorboards may nonetheless stand out: far from being in symmetry with the construction of the building, the track obeys its own, apparently arbitrary course cutting through all architectural sense and heading where it will. Like Christ, its kingdom is not of this world. In other cathedrals such as the Basilica de San Petronio in Bologna, Italy, such lines run their apparently random race for a length of almost 220 feet. To deepen the mystery of the headstrong little trail, on either side signs of the zodiac appear. But this is no callback to paganism. Rather, it is an astronomical calendar that recalls for many an unknown past: cathedrals as solar observatories. And much like the Israelites plundered Egyptian gold to use for the purposes of worshipping, not the Egyptian coterie of divinities, but YHWH, so here the band of the celestial zodiac has been marshalled for a new purpose: to calculate the resurrection of the Son of God, and more precisely locate the date of Easter.

Yet, in a theme that has frequented these pages, even to historians this is something of a forgotten segment of history. Heilbron's discovery, in parallel with Gingerich and many others we have spoken of, began to unweave the long war of Christianity and science. More than that, it reveals that the opposite of war was more often the case: "The Roman Catholic Church gave more financial and social support to the study of

67. Heilbron, *Sun in the Church*, 5.
68. Heilbron, *Sun in the Church*, 21.

astronomy for over six centuries, from the recovery of ancient learning during the Middle Ages into the Enlightenment," Heilbron writes. This in turn was more "than any other, probably all other, institutions" put together.[69]

Scripture and Science

Yet, so many stories have come down to us indicating theologians hated the heavens, and cowered before their infinity. These stories and images come in all shapes and sizes. One thinks of the accusation of the churchmen Cesare Cremonini, who would not even acquiesce to look into Galileo's telescope to view the roof of nature "there for all to see," as the typical accusations go. Or, for example, the courageous tale of the mathematician Pierre Simon de la Place telling Napolean Bonaparte he had no need of God as a hypothesis for his theory regarding solar system formation. Or, yet another, the ludicrous example of Pope Calixtus III (1378–1458) who is condemned in the collective memory of our textbooks for excommunicating Halley's comet from the Catholic Church after seeing it as an ill omen. Every time a notable comet jaunts through the sky in recent memory, one news agency or another brings up this humorous canard. One is reminded of a moment in *The Simpsons* when Grandpa Abe Simpson—a character known for his senility, irrelevance, and irrational anger—is lampooned by the local Springfield newspaper with a picture of him shaking his fist at the sky under the caption "Old Man Yells At Cloud." The tale of Calixtus is not true (in fact, it was invented by Pierre Simon de la Place drawing on some very questionable historical documents),[70] and though it is certainly less subtle than some of the myths we have been (un)covering, it too takes its place in the menagerie of examples of warfare. Even such easily debunked examples reinforce a grand story of church buffoonery and scientific heroism simply by force of repetition among several familiar names, and a few others unfamiliar like "*Laplace . . . Draper . . . White . . . The Scientific American* [all use this apocryphal tale]."[71]

What is created is a very neat answer to the "why Galileo?" question through utilizing a Scripture vs. science picture of history. Bertrand Russell, in his immensely popular *History of Western Philosophy*,[72] for example, placed a very anti-Copernican phrase in the mouth of Calvin, and thus reinforced the idea that in general the church was against the scientific revolution of heliocentrism because of its theological and scriptural commitments[73]: "Who will venture to place the authority of Copernicus above that of the Holy Spirit?" said this Calvin. Edward Rosen, who is an expert on the reception of

69. Heilbron, *Sun in the Church*, 3.

70. Rigge, "An Historical Examination," who summarizes a lengthier forty-page document released by the Vatican that year debunking the myth.

71. Rigge, "An Historical Examination."

72. Russell, *A History of Western Philosophy*, 528.

73. Rosen, "Calvin's Attitude toward Copernicus."

Copernican theory and has something of a friendly rivalry going with Gingerich for who knows more about Copernicus and Galileo, was quite interested in this juicy quote. After all, Calvin's opinion had previously been recorded in his commentary on Genesis that the story of creation did not compete with astronomy, for Scripture is not the revelation of science but of God who, in His infinite wisdom and mercy communicates truths in terms accommodated to the unlettered man.[74] Rosen was as such quite intrigued. Now, Russell's book was originally based on lectures, and so Russell up front apologizes and says that he has not tracked all his sources down for citation. This is what happened here. As Russell did note, however, he was "quite indebted" to Andrew Dickson White. Given what we have already covered in this book, a certain level of foreboding at this has no doubt already set in for the reader.

Rosen turned to White's *Warfare* next to find the source for this quotable Calvin. When he compared how a similar quote appeared in a later work from White, to the one in Russell, Rosen concluded, "We shall feel fully justified that it was from White, not Calvin, that Russell took the anti-Copernican exclamation that interests us."[75] Further sleuthing revealed that White himself got the quote from an Anglican canon named F. W. Farrer (at one time a chaplain to Queen Victoria), but from there as happens quite often in these strange tales, the trail again goes cold and abruptly vanishes. Farrer was typically dependent upon his normally capacious memory, and so did not himself cite where he got the quote in Calvin's works. In fact, as his journey continued, Rosen came to the conclusion that not only did Calvin never utter this sentence, looking over the entirety of Calvin's works, Calvin had most likely never even heard of Copernicus. "What was Calvin's attitude toward Copernicus? Never having heard of him, Calvin had no attitude toward Copernicus."[76] And yet, in addition to Russell, Rosen notes nine other prominent historians had used this phantom quote to illustrate general religious attitudes toward Copernicus, or even to push the warfare agenda.[77] Even the groundbreaking work of Kuhn followed White here, as well as trusting White's word on the general Lutheran response to Copernicanism where Kuhn's quotations are, like Russell's, taken directly from White.[78] It is no accident that in an essay entitled, "Shadow History in Philosophy," where the philosopher Richard A. Watson argued that "the shadows of great philosophers . . . constitute a shadow history of philosophy more influential than philosophy itself,"[79] Russell occurs as his main example of such a shadow caster.

74. Oberman, "Reformation and Revolution."

75. Rosen, "Calvin's Attitude Toward Copernicus," 163.

76. Rosen, "Calvin's Attitude Toward Copernicus," 171.

77. It turns out Calvin said something that on the surface could be take as vaguely similar. But the mitigating circumstances around this qualify it as directly anti-Copernican. Cf. R. White, "Calvin and Copernicus," and Kaiser, "Calvin, Copernicus, and Castellio."

78. Barker, "Lutheran Responses to Copernicus," 63.

79. Watson, "Shadow History," 95.

Calvin's phantom is nonetheless similar to a real remark given by Luther at a Table Talk in 1539, several years before the release of Copernicus's masterwork, *De Revolutionibus*. Word had been spreading initially about some of Copernicus's theories with the publication of the slender *Commentariolus*, which had been circulating since 1514; and Luther was recorded to have said:

> He who wants to be clever must agree with nothing that others esteem. He must do something of his own. This is what that fellow does who wishes to turn the whole of astronomy upside-down. Even in these things that are thrown into disorder, I believe Holy Scriptures, for Joshua commanded the sun to stand still, not the earth.[80]

Case closed, right? This seems to be a clear instance of the idea that Scripture was a vehicle used against scientific advance. However, Luther's quote cannot really be regarded as an informed dissent. First of all, it was not recorded or authorized by Luther himself, but by someone present at the Table Talk who did not publish these remarks until after Luther's death. Indeed, it isn't entirely clear that Copernicus is specifically in view, and not some other astronomer's second-hand (and potentially second-rate) reworking of Copernican ideas. Moreover, this statement attributed to Luther was hardly influential as an opinion among the Reformers. In fact, Luther's colleagues at Wittenberg, like Erasmus Reinhold, not only eventually promoted exuberant study of Copernicus, another colleague at Wittenberg, Georg Joachim Rheticus, was the very man responsible for convincing Copernicus to publish in the first place![81] This is a fact made all the more remarkable when we remember these are Lutherans fond of a Catholic. While White and others love to cite Luther's brother in arms in the Reformation, Melanchthon, as being a fellow conspirator of Luther's against Copernicus, they fail to note that whatever Melanchthon's disagreements with Copernicus, he in his own words "[began] to love and admire Copernicus all the more"[82] as he substituted Copernican calculations in the place of Ptolemaic ones (if not wholesale adopting Copernicus's work as a physical model of the solar system).[83]

Indeed, comparisons can even be made between Galileo's view of the Scriptures and those of Cardinal Bellarmine, the man who had the task of admonishing Galileo on behalf of the Inquisition. One would assume, given the historical caricature, that Galileo would perhaps take a loose, allegorical, or accommodationist approach to account for progressing "science," while Bellarmine would be a hard-nosed literalist opposed to anything that smacks of revisionism. Yet, if we hold up representative quotes side by side without attribution, it is difficult to tell who is who:

80. Luther, *Luther's Works* vol. 54, 358–59, as cited in Barker, "Lutheran Respones to Copernicus," 63.

81. Gingerich, "The Censorship of *De Revelutionibus*," 269–85.

82. Melanchthon, *Corpus reformatorum*, 11:839, as cited in Westman, "The Melanchthon Circle," 167.

83. On this, see Barker, "The Role of Religion in Lutheran Responses to Copernicus."

1.) I say that if there were a true demonstration that the sun is at the center of the world and the earth in the third heaven, and that the sun does not circle the earth but the earth circles the sun, then one would have to proceed with great care in explaining the Scriptures that appear contrary, and say rather that we do not understand them, than that what is demonstrated [in science] is false.

2.) In the learned books of worldly authors are contained some propositions about nature which are truly demonstrated, and others which are simply taught; in regard to the former [that which is demonstrated] the task of wise theologians is to show that they are not contrary to Holy Scripture; as for the latter (which are taught but not demonstrated with necessity), if they contain anything contrary to the Holy Writ, then they must be considered indubitably false and must be demonstrated such by every possible means.

Both statements, in good Augustinian fashion,[84] appear to say that if heliocentrism could be demonstrated, then Christians would have to be cautious about understanding the meaning of Scripture in those passages that appear to teach geocentrism. On this they are united. The second statement appears, however, slightly more reactionary: propositions taught but not demonstrated with *certainty* that conflict with Scripture must be considered false, and "demonstrated [false] by any means necessary." This, too, is Augustinian language. But here comes the tricky part undermining our usual historical narratives about the church using Scripture against science: Galileo's is actually the second, more conservative-sounding statement, while Bellarmine's is the first, more cautiously "progressive."[85] A lot turned on what each thought was a sound scientific demonstration, and indeed whether scientific demonstration was "functionalist" or "realist."[86] More on that presently.

Since alternative readings of Scripture were seen to hinge upon the success of scientific demonstration, it therefore becomes imperative to ask: didn't Galileo prove his position? The consensus of recent scholarship is quite startling: Galileo *did not* prove it—even by the standards he set for himself. And so we come to a crucial point that is all too often underemphasized—especially by those eager to turn these scenarios into straightforward cases of Scripture being used against science—the opponents of Copernicus, even through Galileo's day, were not critiquing it primarily religiously, but scientifically.[87] Even after *De Revolutionibus* was printed and circulating among scholars through Europe, there was no real reason to assume it had proven heliocentrism as

84. McMullin, "Galileo on Science and Scripture." For some subtle differences between Augustine and Galileo, see Reeves, "Augustine and Galileo."

85. Cf. Bellarmine, "Cardinal R. Bellarmine to P. Foscarini, 12 April 1615," 68. For Galileo's statement, cf. Galileo, "Letter to the Grand Duchess Christina (1615)," 101–2. I owe the idea of comparing these two quotes to Brooke and Cantor, *Reconstructing Nature*, 24.

86. Howell, *God's Two Books*. Cf. also Feldhay, *Galileo and the Church*, 201–292.

87. Graney, *Setting Aside All Authority*.

a physically true theory. Though it will sound odd to modern ears, a large piece of the struggle was how much Copernicus and Galileo's position relied upon mathematics, a discipline which was, in the Aristotelian system, in a rather subordinate position. That mathematics should tell one the physical layout of a system was rather scandalous. So now, not only has Galileo not justified why his telescope should be trusted from a theoretical standpoint, neither has he given reason to upend the heretofore time-tested principle that mathematics was subordinate to physical theory and followed after it. Galileo's "proofs" would have (and did) read as so much audacious speculation ornamented by a bravado that tested the patience of many.[88]

Let us recall then the general agreement between Galileo and Bellarmine that we were met with above regarding good hermeneutical practice: only if a sound demonstration of heliocentrism was on offer would we be led to a cautious rethinking of passages that appear to teach otherwise. Luther—and anyone else—would not have immediate cause to change their scriptural interpretations as such—and this based upon Galileo's own insistence on how to read the sacred text. It is as such not a clear case of "Scripture vs. science" at all, but rather the apparently geocentric passages of Scripture corroborating a widely held, predictively successful Ptolemaic theory in the face of an as-yet-unverified rival.

Moreover, on top of the scientific debates still ongoing about Copernicanism, an anonymous preface entitled "Letter to the Reader" (which Johannes Kepler later discerned had been written by the Lutheran Andreas Osiander) indicates that what is contained in *De Revolutionibus* is a clever mathematical theory that aids predicting stellar motion, but not a physical theory about the actual physical layout of the solar system. Pierre Duhem notes that Osiander was not contriving an *ad hoc* device to make Copernicanism palatable; rather such a "functionalist" understanding of astronomy had a venerable pedigree which Osiander was following.[89] Robert Westman argues at length about this particular disciplinary distinctive, noting Osiander was trying to keep the peace, not between science and religion, but between the proper ordering of the humanities.[90] Thus, this functionalist gloss to Copernicus was a double boon for Osiander's academy at Nuremberg, in that it both preserved geocentrism and avoided the internecine debates between Aristotelians and theologians about how to organize knowledge into a total package:

> Concerning hypotheses, I have always thought of them, not as articles of faith,
> but as a foundation for calculation so that, even if they are false, they carefully

---

88. Galileo's eventual alienation of the Jesuits was here pivotal, as they were one of the primary advocates for the elevation of mathematics in the face of the more traditionalist Dominicans. See: Dear, *Discipline and Experience*, 32–62; Feldhay, *Galileo and the Church*, for the general Jesuit and Dominican clashes as they were relevant for the Galielo affair.

89. Duhem, *To Save the Phenomena*.

90. Westman, "The Astronomer's Role in the Sixteenth Century."

exhibit only the appearance of the motion. . . . In this way, you may pacify the Aristotelians and the theologians that you fear contradicting.[91]

We are thus in the peculiar position to acknowledge that historically speaking even those who completely accepted Copernicus's system may not have felt obliged to change their reading of Scripture. On the other hand, those who did preserve their reading of Scripture against Copernicus—perhaps even Luther—were to reiterate not doing so by using Scripture as a straightforward trump card against science, but in support of a longstanding, predictively successful alternate scientific model. As such, there are at least two conditions that even a brief account such as this must keep in mind: first, the status of Copernicanism as either functionalist or realist, and second the demonstration of its accuracy either way. The pope himself agreed on the conditions of this situation, and even used the functionalist interpretation of Copernicus as a cautionary remark against his friend, Galileo:

> Let us grant you that all of your demonstrations are sound and that it is entirely possible for things to stand as you say. . . . [Yet] you will have to concede to us that God can, conceivably, have arranged things in an entirely different manner, while yet bringing about the effects that we see. And if this possibility exists, which might still preserve in their literal truth the sayings of Scripture, it is not for us mortals to try to force those holy words to mean what to us, from here, may appear to be the situation.[92]

This no doubt pricked Galileo's pride even more deeply because the astronomer Tycho Brahe had actually come up with one such alternative physical arrangement. In what has been dubbed "geoheliocentrism," Brahe imagined the earth in the center of the universe being orbited by the sun as did Ptolemy. The twist was that the rest of the planets in turn orbited the sun, not the earth. Constantly infuriating for Galileo, observationally and mathematically it was not possible at the time to differentiate between the Tychonic and Galilean models. In part, this was also responsible for the fallout that Galileo had with the Jesuits, who supported Galileo throughout his career but were more interested in Brahe because he both preserved the phenomena and allowed one to retain geocentrism. Heliocentrism as a scientific hypothesis for Galileo was really more of a cluster of aesthetic intuitions and anti-Aristotelian arguments held together by assertion and the force of Galileo's wit. It never rose to the level of proof: "So Galileo had arguments rather than proof," as David Lindberg puts it.[93] But, even by his own standards, it was proof—not mere probabilistic arguments—that Galileo needed before scriptural interpretation would be reconsidered. And yet, starting at least with Voltaire, the condemnation was "caused by Galileo's *demonstration*" to which Voltaire also added Catholic "ignorance and hatred," of such a demonstration

91. Quoted in Howell, *God's Two Books*, 46.

92. Quoted in Santillana, *The Crime of Galileo*, 166.

93. Lindberg, "Galileo, The Church, and the Cosmos," 43.

to the root of the whole affair.[94] In stark contrast to this polemicizing counter memory, it remains quite interesting to note that in October of 1741 the Inquisition in fact authorized a project intent on publishing all of Galileo's works (even the banned parts) should the editor's acquiesce to revise the work "to make it 'hypothetical'" that is, functionalist in interpretation![95]

To make matters of proof worse, not everyone trusted Galileo's telescopes to do the heavy lifting he claimed they could. "We must divest ourselves of the gift of hindsight when reviewing this circumstance," writes Oxford historian Allan Chapman.[96]

> Since classical times, the *true* instruments of the astronomer were graduated, angle measuring instruments based on the properties of a 360-degree circle. Astronomical truth lay in measurement and geometry: it did *not* lie in strangely magnified images of familiar bodies like Jupiter or Saturn, sometimes complete with image distortions and false color fringes. . . . Light being bent and distorted down a tube [was considered] an absurdity [and without further justification] clearly full of false information that no wise philosopher could be expected to give credence to what appeared through the eyepiece.[97]

Empiricism and observation are so commonsense today that the theoretical maps that go with them in order to make appearances understandable, separable, combinable, or even appear at all are forgotten. But such theoretical justifications would prove to be indispensable and necessary for things like the telescope to be operable (and such theoretical trappings are still in play today, however reflexive they have become). Think of a ball falling from a tower. Since it lands at the base of the tower, empirically all things being equal this would in fact be evidence for a stable earth. But all things are not equal. What Galileo and others had to do was to convince their audiences that there were other theoretical frames of reference (like inertia) that needed to be taken into account to understand what the ball at the foot of the tower was truly saying. The same goes for the observations made through the telescope. And precisely here problems arose.

"[Galileo] offers no theoretical reasons why the telescope should be expected to give a true picture of the sky. . . . Nor does the initial experience with the telescope provide such reasons." When we think of Galileo's telescope, we may be imparting images of our own modern giant observatories or even modern personal telescopes, but this simply was not the case either in terms of proficiency, or even the theoretical justifications for why such knowledge should be trusted in the first place. "The first telescopic observations of the sky are indistinct, indeterminate, contradictory, and in conflict with what everyone can see with his unaided eyes. And the only theory

---

94. Finocchiaro, *Retrying Galileo*, 116, 119.

95. Finocchiaro, *Retrying Galileo*, 127.

96. Chapman, *Stargazers*, 160.

97. Chapman, *Stargazers*, 160. Italics in the original. Perhaps one of the most important works to come out regarding the philosophical and even political conditions behind experimentalism and the use of instruments, see Shapin and Schafer, *Leviathan and the Air Pump*.

that could have helped to separate telescopic illusions from veridical phenomena was refuted by simple tests."[98] Indeed, many professional astronomers across Europe, *with aid of Galileo's personal telescope*, could not duplicate his findings. For example, Johannes Kepler's pupil, with the amusing name of Horky, wrote on the twenty-fourth and twenty-fifth of April, 1610 that Galileo had taken his own personal telescope to the house of the astronomer Magini of Bologna to demonstrate to twenty-four professors in attendance. "I never slept on the 24th or 25th of April," writes an almost breathless Horky:

> But I tested the instrument of Galileo in a thousand ways, both on things here below and on those above. *Below it works wonderfully; in the heavens it deceives* . . . as some fixed stars are seen as double. I have as witnesses most excellent men and noble doctors . . . and all have admitted the instrument to deceive . . . This silenced Galileo and on the 26th he sadly left quite early in the morning . . . not even thanking Magini for the splendid meal.[99]

Magini himself, writing a letter to Kepler, added this regarding the occasion at his abode: "He [Galileo] has achieved nothing; for more than twenty learned men were present yet nobody has seen the new planets distinctly; he [Galileo] will hardly be able to keep them."[100] Then, Kepler himself—the great champion of heliocentrism!—sent a letter to Galileo that must have stung deeply: "I do not want to hide it from you that quite a few Italians have sent letters to Prague asserting that they could not see those stars [the moons of Jupiter] with your own telescope."[101] Later, in his *Optics*, Kepler in fact uses *naked eye observations* against Galileo's telescope: "It seemed [as I looked through Galileo's telescope] that something seemed to be missing on the outermost periphery [of the moon],"[102] as the telescope seemed to warp and smooth the outermost edges of observation. Chapman confirms from his own experience how difficult Galilean telescopes are to work with: "The ray of light emerging from a Galilean eyepiece is very narrow, and making it fall, sharply focused upon one's retina can be quite a job, until one gets the hang of it. In 1610–1611, that was only the beginning: making logical *sense* of what you saw was another matter entirely."[103]

Conclusions: The Condemnation of Galileo

Unfortunately, as Galileo had to navigate the complex dynamics of patronage mechanics and court politics to maintain both his reputation and his paychecks, he also

---

98. Feyerabend, *Against Method*, 99, 121.

99. Quoted in Feyerabend, *Against Method*, 85. Emphasis added.

100. Quoted in Feyerabend, *Against Method*, 85.

101. Quoted in Feyerabend, *Against Method*, 85.

102. Quoted in Feyerabend, *Against Method*, 89.

103. Chapman, *Stargazers*, 161.

had to become increasingly vocal—and not less so—about the *physical* truth of his system so tied to his name and notoriety. Far from "science vs. religion," Galileo's particular self-fashioning and his clash and eventual condemnation by the Church cannot be understood outside of this context.[104] Galileo's feud was particularly bad timing as well. "If Copernicus' book had been published either one hundred years earlier, or one hundred years later, the Galileo affair would probably not have happened," as one historian, Richard Blackwell, has aptly pointed out. "But, in fact, it was published in 1543, when the Reformation was in full bloom, and the [Catholic] Counter Reformation was just beginning. Hence it was by 1616 all of the actors and cultural forces were in place for the drama of the Galileo affair to begin."[105]

Moreover, while ostensibly Galileo, Bellarmine, and the pope could at least at one time be said to have agreed on the relevant hermeneutical principles in regard to Scripture, they could not agree on the application of those rules. And this is so precisely because they did not believe Galileo had demonstrated his claims. The pope was surely not anti-science or in any sense a dunce—Galileo originally being quite excited that his friend, a man of learning and letters, had been promoted to the papal throne. Nonetheless the pope's skepticism regarding the physical layout of Galileo's system was "abhorrent to him."[106] Of course, Galileo could believe little else than positive things about his demonstrations for the sake of his own financial survival and reputation. Even the midst of disagreement, however, the pope would later reflect that "it was never our intention [to prohibit Copernicanism]; and if [it] had been left to us, that decree . . . would not even have been made."[107] The Inquisition, however, had been on alert, and supposedly prohibited Galileo from teaching the physical truth of heliocentrism. Galileo took this to mean (perhaps with a bit of wishful thinking) that he could teach heliocentrism as a hypothesis. Yet, it appears that Galileo did not make this concession in his *Dialogue*. Moreover, to promote his point, Galileo made the blunder of putting a few choice points the pope had personally made to Galileo (and so their source was unmistakeable) into the mouth of the character Simplicio, in essence making the pope play the role of the fool. This, to say the least, did not sit well. Having previously been friends with Maffeo Barbarini before he became Urban VIII could only make this series of choices seem like a deep betrayal of confidence. Not only was the pope infuriated by the Simplicio character, he was outraged that Galileo had not informed him of the previous standing order from the Inquisition. Altogether, with the continuing teaching of heliocentrism, hiding the standing order, and mocking the pope publicly, to Urban VIII it appeared Galileo had intentionally betrayed his friendship in the name of his pride at least three times (one recalls Peter

---

104. See Biagioli, *Galileo Courtier,* esp. 313–53, for the recontextualization of his trial.

105. Blackwell, "Galileo Galilei," 108.

106. Scotti, *Galileo,* 270.

107. Quoted in Brooke and Cantor, *Reconstructing Nature,* 109.

thrice betraying Jesus the Galilean; here, in something of a reversal Galilei has thrice betrayed Peter). Amidst everything, a personal squall could only add fuel to the fire.

As John Heilbron and Richard Westfall have put it each in their own way, Galileo simply could not let go of his own ego, and was one of the chief architects of his own disaster.[108] On the other side of things, both the pope and the Catholic Church as they were feeling the intense pressures of the Reformation, and having grown more centralized, more bureaucratic, more totalitarian, were deeply to blame as well. While the whole affair hardly reduces to anything remotely resembling "science vs. religion," this is not magically to exonerate the parties from their actions. An immovable object here met an incorrigible force, and everything around them paid the price.[109] The historian David Lindberg reminds us that, strictly speaking, ideas do not and cannot clash. Only people can. Nowhere, perhaps, is this more true than with Galileo and Urban VIII.[110] We must unfortunately for the sake of space bypass the ins-and-outs of the Inquisition's handling of Galileo, and the questions of whether his trial was mishandled or was (however lamentably) following protocol. Some of the papers stolen by Napoleon in fact reveal that things were more considered than the hype of myth would lead us to believe.[111] Given the myths moving about, "although the condemnation" recovered by the trial documents rediscovered in Paris hardly painted the Catholic Church in favorable light, the truth was in fact much "less discreditable than [the myths] about the trial."[112] Such details are of course all relevant to the continuing point, yet for now it has been made: in the end, while his silence can hardly be justified in modern eyes, it is also surely not a case of the warfare thesis, which is enacted more by those who sought to shape the legacy of Galileo's trial rather than the event itself, as the work of Maurice Finocchiaro has so meticulously displayed.[113] Clever metaphors may die hard, but they can, in fact, die. Yet, as we saw with Haeckel, the Galileo affair not only served nicely as yet another polemical accoutrement to the X-Club's portfolio of Church interference. In many instances the mythic form we today recognize regarding the place and importance of Copernicus and Galileo itself evolved right along with the Darwinian debates, was of a piece with them, and was nurtured "in the nineteenth century" by the "implicit assumptions about the nature of science and the nature of religion," along with the explicit "representations of a total clash of ideology and values" that Huxley, Draper, White, and so many others took care to manicure to their specifications.[114]

---

108. Heilbron, *Galileo*, vii; Westfall, *Essays*, 33.

109. Scotti, *Galileo*, 269.

110. Lindberg, "Galileo, the Church, and the Cosmos," 34.

111. For this story, see Chadwick, *Catholicism and History*, 14–45.

112. Chadwick, *Catholicism and History*, 26.

113. Finocchiaro, *Retrying Galileo*.

114. Feldhay, *Galileo and the Church*, 3–4.

—Chapter Eleven—

American Memory

The Scopes Trial, Creationism, and a Public Stage for War

Sadly, . . . the language of wonder has become riddled with the rhetoric of adversity. In their fight against "soulless science" creationists champion a view of creation so narrow that it is decidedly unbliblical. At the other extreme, certain scientists construe faith in God as the enemy of scientific progress, and human well-being. As one might expect, misunderstandings and distortions abound as each side reduces the other to laughable caricatures. Illiteracy, both scientific and biblical, reigns.

—William P. Brown, *The Seven Pillars of Creation*

FOSSILS HAVE ALWAYS HAD a peculiar place in the national imagination of America. Indeed, the Mastodon, whose bones were originally discovered in 1705 along the Hudson river, were deeply intertwined with the emerging national ethos of America. Thinking the bones as belonging to a massive predator (possibly still alive), Thomas Jefferson not only sent Lewis and Clark out to try and find it lumbering somewhere within the Louisiana Purchase, he and others also saw this American monster as reinforcing the great need for seizing, controlling, and civilizing this new world as elaborated in the quasi-religious and theological doctrine of manifest destiny. As (apparently) both a spectacular predator and incredibly ancient, the Mastodon was to be used by Jefferson against European rhetoric of the infancy and impotency of the new American project. The artistic renderings of the primordial and savage biosphere of the "American *Incognitum*" as it was called, were not just popular renderings to aid in communicating scientific discoveries, but were shot through with religious and political meaning. Bones were never just bones in America, but were oddly enough often that to which the ligaments of warring national scientific, religious, and political ideologies attached themselves.[1] An entire chapter of American ideology could be written from the perspective of the fossils that always strangely went hand in hand with them.

Indeed, in one of those intriguing coincidences of history, Darwin's *The Origin of Species* arrived in America just weeks after John Brown was publicly executed for his attacks on the armory of Harper's Ferry to gain supplies for an abolitionist revolution. Not far behind Darwin's book, another transatlantic passenger as well were the first Gorilla specimens making their way to the new world. As such despite Darwin quite explicitly leaving human evolution out of his theory until his later, second book *The Descent of Man*, primate descent, religion, and abolition were all inextricably linked in the American mind—both through the intrinsic overlap of their ideas, and the sheer happenstance of serendipity.[2] This was especially so in the South, as the science of evolution and the controversies surrounding Darwin's mechanism of natural selection in particular became entangled with Southern values during the Civil War in a culture amidst a literal and theoretical battle to continue to define and defend its own identity.[3]

Given the charge that bones and ancestry and apes all had already in the American imagination, one can see how it is not too much of a stretch for the warfare thesis to convince us all that it appeared that every path through history, every battle it supposedly fought, all lead into America in 1925 to a little town called Dayton, Tennessee, where "the trial of the [twentieth] century" was to occur precisely over creation and evolution.[4] There, a young substitute teacher named John Thomas Scopes was found guilty for violating a newly implemented law put into effect by then Tennessee governor Austin Peay in March of that year, which forbade evolution from being taught in the classroom. More specifically, it targeted any teaching that described the descent of man from lower species. As the story goes, Scopes was a courageous teacher—perhaps even a martyr, likened to men such as Galileo and Socrates—who wanted little else than to encourage free inquiry in his students and instruct them in the wonders of recent scientific discoveries. Already it began to link up to the broader faith vs. reason, religion vs. science histories. To that end, so continues the story, Scopes defied the law and carried out his convictions. After Scopes was arrested, the famous lawyer Clarence Darrow came to defend him at the behest of the American Civil Liberties Union (ACLU), while a thrice-failed Democratic presidential candidate, William Jennings Bryan, would advocate on behalf of Tennessee and the anti-evolution law. Not only did all eyes turn to Dayton, a conspiracy of all sides ensured that the whole thing

---

1. Semonin, *American Monster.*
2. Fuller, *The Book that Changed America*, 79.
3. Hampton, *Storm of Words.*
4. Larson, "Law and Society."

slowly but surely poured into the mold of a religion vs. science memory. From there, it was said, fundamentalism was dealt a mortal blow.[5]

But, of all places, why Dayton? Given the universality implied in the warfare thesis, the answer would inevitably seem to be that this could have happened anywhere, and Dayton is simply where it did occur as a matter of fact. Yet the local and contextual conditions surrounding Dayton reveal that this is not the case. Even before the trial had started, the "Why Dayton?" question was on everyone's mind. In fact, in it lay a key element to how a grandiose "faith vs. science" narrative was distilled out of a peculiarly odd blend of big personalities, idiosyncratic beliefs, and a trial that seemed to completely forget why it was originally convened mere moments after it got underway. Tensions between urban and rural areas throughout the United States was a major theme in the early twentieth century, and Dayton was no different. In order to capitalize on the trial and potentially staunch the bleeding of their population (which had dwindled to a mere eighteen-hundred by 1925) the citizens of Dayton created pamphlets advertising Dayton as the quintessential American town. More than this, as Edward Larson and other historians have noted, a major part of the reason many town members even decided to host the trial was precisely as a means to publicize the town and save it from extinction.[6] "Why Dayton—of *all* places" the pamphlets read, "of all places, *why not Dayton?*" It was a clever maneuver. Turning their indistinctness into an icon of Americana, Dayton was no longer a backwater location in Tennessee, but rather it was everywhere—or, rather, anywhere, in America. "*This is America,*" read another line in the pamphlet. But at the same time, this "all-American" ploy made anywhere in America seem Daytonian.[7] Rather than an event that bore all the markings of its particular geographically and temporally bound contexts, the Scopes trial began to grow into a timeless and essential struggle that could have happened at anytime and anywhere in America. The trial, as advertised, was rooted in "generic America themes," and was a clash between religion and science that was "epic and essential."[8] And over its course, one sociologist out of North Carolina estimated that the trial was discussed by "some 2,310 daily newspapers in this country, some 13,267 weeklies, about 3,613 monthlies, no less than 392 quarterlies, with perhaps another five hundred including bi-monthlies and semi-monthlies, tri-weeklies, and odd types."[9] In addition, for the first time ever, radio entered into a court of law for the first ever criminal trial broadcast to the American public at large. The message, in other words, clearly got out. Evolution already intersected with any number of issues

---

5. As Larson, *Summer for the Gods*, 236 notes, as with so much of what has transpired in this book so too our image of Scopes as a massive blow to Fundamentalism in the early twentieth century is a result of textbook representation.

6. Larson, *Summer*, 93–95.

7. Shapiro, *Trying Biology*, 9.

8. Shapiro, *Trying Biology*, 10.

9. Numbers, *Darwinism Comes to America*, 78–79.

near and dear to the American heart. But even if it hadn't, there is little doubt that amidst those particularly torrid summer days, a legend would have been born. And that was only the beginning.

How to Build a War

What happened next is slapstick comedy . . .

—Michael Roberts, *Evangelicals And Science*[10]

For those familiar with the Scopes trial, as well as the clash between creationism and evolutionism, the notion that a revisionary history can even be undertaken will seem quite absurd. And yet, the Scopes trial can not only be disassociated from the warfare thesis generally, even when looking at the trial itself the notion that it was fundamentally about "science vs. religion" blurs and falls apart. The memory of a clash of science vs. religion emerged in part because of the rhetoric and intentions of all of those involved, which were in fact far more complicated than the simple warfare model can handle. The "genre" of a trial is of course already fundamentally oppositional, and so ready-made for easy binary opponents to clash. However, even the inevitable antagonism of a trial did not make the titanic presence of the Scopes trial inevitable. In fact, just a year before in Nebraska, another similar case occurred for nearly identical reasons. But the case was just as quickly whisked out of memory. If evolution and creation were the two explosive elements needed to mix and produce outrage, one would have certainly expected it in Nebraska where all the needed factors were present. But nothing of the sort occurred. To explain Scopes, then, a broader view is required, one that discards both the warfare thesis and the inevitable couching of the main actors in terms of science and of religion. From a larger perspective both the specific planning of the participants of the Scopes case, as well as the manufacturing of mass-distributed recollections of the trial have deeply affected its significance in American memory, which goes far, far beyond anything that the trial itself achieved.[11]

Without any surprise, much of this has been due to the continuing legacy of White and Draper. "During the years leading up to the Scopes trial, [the reaction to White and Draper] inspired an outpouring of academic books, articles, and essays discussing the conflict between science and religion, with an increasing emphasis on the seemingly pivotal issue of Darwinism."[12] Indeed, if the infamous discovery of the Piltdown Man fossil was not explosive enough for those debating about human origins, in 1922 one of the experts, Arthur Keith, wrote of Darwin and Huxley that "they made it possible for us men of today to pursue our studies without persecution—without

10. M. Roberts, *Evangelicals and Science*, 144.

11. See Waggoner, "Historiography," for an excellent summary on the changing historical literature.

12. Larson, *Summer*, 22.

being the subject to the contumely Church dignitaries."[13] Ironically, the Piltdown specimen turned out to be a sensational counterfeit, just like the history of science that Keith had attached to it. But it hardly mattered. "It is obvious in retrospect," writes one historian, "that the meaning of 'evolution' [by] the 1920's went far beyond the developmental theories of Darwin or Spencer. The word became a symbol for everything that was wrong with the nation in that decade."[14] Like a tune that just would not get out of America's collective mind, the warfare thesis kept on playing and nearly any popular scientific finding was soon asked to dance to its beat. Indeed, the lawyer in charge of Scopes's defense, Clarence Darrow, had been reared on the warfare thesis as a child due to his zealously anti-clerical father who read Draper, Huxley, and Darwin with the devotion of a monk at morning prayer.[15]

With the deconstruction of the warfare narrative in mind, many historians have come to some general conclusions about the Scopes trial and its broader relationships. In terms of the ambient history Scopes has been seen as perpetuating mass deconversions supposedly due to Darwinism or the rise of science in the Victorian period; but these have been discovered as rhetorical magnifications (and sometimes fabrications) for effect. Conversions of secularists (often even among their leaders) to Christianity in this period were similarly always systematically played down, or even outright ignored. A picture of Christians horrified *en masse* by Darwinism, pragmatism, and a host of secular movements was enhanced by this artificial contrast and provided an exaggerated background for those presenting a particular image of Scopes.[16] The truth was that the secularist and free-thought movements were as harrowed by the continuing spiritual power and intellectual viability of Christianity as Christians were by atheism.[17] The historian Timothy Larsen notes rather wryly that when his colleagues "say that most thinking Victorians experienced a crisis of faith, what they really mean is that they ought to have."[18] Moreover, we need to constantly remember that the scientific consensus on Darwinism was quite different than today—even with widespread acceptance there were still large concerns about how trait inheritance and other particularities worked. Treating it as a settled fact opposed only by those

13. Quoted in Larson, *Summer*, 22.

14. Quotation from Lienesch, *In the Beginning*, 85.

15. Larson, *Summer*, 22.

16. Larsen, *Crisis of Doubt*, 239–53.

17. This is the elegant major assertion of Larsen, *Crisis of Doubt*, e.g. 241: "Both as a percentage of the entire leadership, and in terms of the place of respect they held within the leadership pool, the reconverts [to Christianity away from secularism] represents a significant reality in the story of the nineteenth-century Secularist movement. Despite the way that the narrative of the Victorian loss of faith has loomed so large in the existing literature, the Secularist movement lost a far greater percentage of its top leadership to reconversion [to Christianity] than the Christian ministry lost due to a crisis of faith. Moreover, much of the literature of the angst in church circles that is chronicled in the loss-of-faith literature has its mirror-image in the Secularist movement."

18. Larsen, *Crisis of Doubt*, 245.

emboldened by religious audacity is far from the truth.[19] Nonetheless, it remains true that everyone was talking about mass deconversions and the "atheist threat" at that time. This was mostly idle gossip and speculation rather than a reflection of experience, more "a literary than a historical theme." To be sure, "In a religious age, doubt loomed large as the bugbear of faith," but only insofar as it was a creature lurking "out there," and spoken of as one might of the monster in some fairy tale.[20] As far as reality on the ground went, at the elite level, for example, it seems the encounter with Darwinism did little to change any affiliation with religion. Indeed, while how Christians engaged with science might have been characterized by a certain "orthodox idiosyncrasy of the mind," they were no obscurantists in the nineteenth century, and dealt robustly with the works of Comte, Mill, Darwin, Huxley, and Spencer.[21] In a statistical study of members of the National Academy of Sciences—who ranged from atheist, to agnostic, to Christians of all stripes—Ronald Numbers notes "I have found no evidence in either biographical or autobiographical accounts to suggest that a single one of these men severed his religious ties as a direct result of his encounter with Darwinism." By and large, Numbers continues, "Catholic naturalists in the Academy remained Catholic, the Presbyterians remained Presbyterian, and the agnostics remained agnostic."[22]

Moreover, as the historian Aileen Fyfe has shown, far from using the rhetoric of warfare which was often nearly exclusive to the elite circles of Huxley and the X-Club, most conservative Christians especially in Britain through the use of the Religious Tract Society actively worked against the warfare thesis to display the harmony of science and faith. By distributing science pamphlets that were cheap, and communicated scientific discoveries in a clear way understandable to the public, they demonstrated that it was not particular scientific discoveries themselves that concerned evangelicals but rather "they were worried about what they regarded as the distorting manner in which those discoveries were presented to the reading public."[23] Such popular presentations have been actively overlooked ironically because of the professionalization campaigns affected by Huxley and at large, and because Scopes and later Creationism arising in the mid-twentieth century have in turn coopted earlier history as stand-in representatives of conservative Christian thought.[24] Not being part of the elite generation of knowledge, the publications Fyfe examines were not sought as reflective of broader attitudes.[25] The truth is, however, on closer inspection nearly everyone—the

---

19. Bowler, *The Eclipse of Darwinism*; Bowler, *The Non-Darwinian Revolution*.

20. Larsen, *Crisis of Doubt*, 11.

21. Hampton, *Storm of Words*, 63.

22. Numbers, *Darwinism Comes to America*, 41.

23. Fyfe, *Science and Salvation*, 4.

24. Numbers, *Darwinism Comes to America*, 130.

25. This created the circularity mentioned previously, where the assumed lack of evangelical interaction with the sciences led scholars to be lackadaisical in searching for such sources. Because of

X-Club included—agreed that ultimately science and religion were in harmony—from the staunchest fundamentalist to the most zealous evolutionist.[26] Like the professionalization debates, so too in America in the early twentieth century the issue was more over what counted as religion and what counted as science—and hence what also counted as their interaction. But the posturing of both sides in the Scopes trial, as well as the memories later crafted of such things, led to the conflict becoming one where "the defense came to be seen as proponents of biological science and opponents of faith in the truth of the Bible. The prosecutors came to be seen as defending scripture and opposing evolution. As a result, the conflict over the boundaries of the concepts called 'science' and 'religion' (within an overarching belief in their ultimate harmony) became [represented as] a conflict *between* religion and science."[27]

Far from the embodiment of inevitable forces clashing, it is much rather the case that "the Scopes trial took place because several of its participants connived to ensure that it would happen. These different groups each saw the trial as a useful way to achieve goals that had little to do with the guilt or innocence of John Scopes."[28] Because the life of the warfare narrative had in some sense already begun to shape how the memory of science and religion at large functioned, instead of a trial over whether Scopes broke the law, or even if the law was just, it kept ballooning into a war over such massive things as "Science" and "Religion," where evolution is little else than the creation narrative for atheists.[29] What was originally a trial about academic freedom spiraled out of control as innumerable hands began reworking its colors and hues. No longer about Scopes, it represented a canvas to paint in the terms that each party brought to it and sought to effect. So, in the end "the trial consisted of these different parties fighting over the meaning of the thing they had together created."[30]

Indeed, "one of the reasons the Scopes trial was so significant was precisely because it was a good show."[31] A circus-like atmosphere began to gather around the trial and throughout the small streets of Dayton. Rather like suspicious-looking storm clouds, the crowds were thunderheads threatening at any moment to explode—not with lightning, but with headlines, snappy article leads, spooked reporters flocking to and fro like a murmuration of bloodthirsty starlings. "Cranks and Freaks Flock

---

this they found none, thus reinforcing the suspicion that they did not exist. See: Fyfe, *Science and Salvation*, 6.

26. We must remember that those like Huxley (who fashioned himself as a new Luther enacting a New Reformation) were against theology, and not religion per se. Huxley and other agnostics even lauded scripture. "What made us freethinkers?" asked one, "Why, reading the Bible!" See: Lightman, *The Origins of Agnosticism*, 120–125.

27. Shapiro, *Trying Biology*, 98.

28. Shapiro, *Trying Biology*, 91.

29. Giberson and Yerxa, *Species of Origin*, 58.

30. Shapiro, *Trying Biology*, 91.

31. Lienesch, *In the Beginning*, 141.

to Dayton," as one headline read.[32] And while the trial certainly attracted "hucksters and proselytizers,"[33] spontaneously creating their own miniature carnivals circling in orbit around the courthouse, it was only by a collective sleight of hand of the over 200 reporters who had travelled to Dayton that they themselves were excluded from being represented as colluding in the feeding frenzy. What made the Scopes trial "a spectacle was not the court case that inaugurated it," writes Adam Shapiro, "but the nature of its audience." Not only was the "spatial boundary between court and spectators blurred," but even more importantly "the separation between participants' roles as creators and consumers also became increasingly untenable."[34]

The journalists not only recorded events or shaped them to terms they thought suitable for quick paper sales. They also orchestrated the expectations of the crowd in Dayton by presenting things in terms they thought the multitudes would respond to most. On the other side, the crowds in turn internalized the feedback of the journalists and, wanting a good fight, began themselves to embody and emulate and enforce the increasingly clear identities of different sides in the hopes of a straightforward clash. This was, it seems, a war that everyone wanted. Far from embodying a clash of science vs. religion, in reality it was a scripted wrestling match operating under the direction of some, and taking a few hapless others along for a very particular ride. Indeed, in many senses this was recorded in the earliest opinions of the trial, where from afar "the early analyses of the Scopes trial reflect the belief that the trial was best understood as an episode, more a media event, than any sort of critical watershed."[35] But we are getting ahead of ourselves. The point to keep in mind as we move forward is simply that "to accept that a universalized science-religion conflict made the Scopes trial inevitable is to accept the explanation put forward by the trial's participants, who of course accounted for the trial in ways that justified and valorized their own involvement."[36] Such justifications quickly replicated themselves as overnight Scopes became "the world's most famous court trial."[37] There is little denying that aspects of science and religion made themselves felt here. But that the trial occurred solely or even primarily because of science and religion, or that it represented a long history of their conflict, cannot be maintained. Rather, "the trial had origins in debates over American education," that, especially initially had little to do with science or religion. The Scopes Trial, and the struggles of evolution and creationism were part of the broader emerging genre of the "evolutionary epic," and treating Darwinism and evolution in religious terms that substituted for the now defunct cosmogenic myths of

32. "Cranks and Freaks Flock to Dayton," *New York Times*, July 11, 1925.

33. Larson, *Summer*, 139–44.

34. Shapiro, *Trying Biology*, 93.

35. Waggoner, "Historiography," 161.

36. Shapiro, "The Scopes Trial Beyond Science and Religion," 201.

37. Darrow and Bryan, *The World's Most Famous Court Trial*. This book is a complete stenographic record of the trial, for those interested in the first-hand account.

yore. "As a narrative format, the evolutionary epic provided the opportunity to create evolutionary heroes who could capture the imagination of readers, as had Odysseus and Achilles in Homer's classical epics," writes Bernard Lightman. "Both forward-looking evolutionary theorists, locked in battle with religious bigotry, and as clever and courageous animals, fighting with their natural enemies in a hostile environment, [they] could be portrayed as heroes engaged in an epic war."[38]

Trials and Tribulations

John Scopes had taught at the Rhea County Central High School for nearly a year when he was asked to substitute over the course of a few days for the regular biology teacher, who had fallen ill. This substitution happened to coincide with the newly minted Butler bill passed by the newly reelected governor, John Peay. Several local business leaders, including the chief architect of the Scopes trial, George Rappleyea, invited Scopes out to the town drugstore to ask him a few questions. Rappleyea, the manager of the Cumberland Coal and Irony company, similarly spearheaded the approach to Scopes where they asked the twenty-four-year-old teacher if he had covered evolution in the course of his substitute work. What was curious, as we will see in more detail, is that looking into the cause of Rappleyea's disgruntlement that led him to becoming one of the point men contacting Scopes was not the long history of fundamentalist anti-evolutionism that finally pushed him over the edge. There was no such long past at any rate, as we will see momentarily, and certainly Dayton was not particularly notable as a hotbed of activity with—well, nearly anything. A Methodist whose minister in fact put his mind to rest regarding the question of evolution, Rappleyea—much as both Darwin and Haeckel had been—found himself pushed into a more extreme position on evolution not because the theory demanded it, but because Rappleyea was traumatized by that age-old haunter of men: the problem of evil.

Whereas Darwin's emphasis on randomness and waste accelerated after the death of his daughter Annie, while Haeckel's anti-theistic take on evolution began its crazed and tragic quest after his wife had passed,[39] Rappleyea was tormented by the funeral of a local six-year-old boy. During the ceremony, the poor mother was inconsolable. Between sobs she could only manage to find the wherewithal to repeat the phrase "Oh! If I only knew he was with Jesus!" The response of the minister, in earshot of all

---

38. Lightman, "The Evolution of the Evolutionary Epic," 221. Cf. Ruse, *Darwinism as Religion*, 82–114.

39. See: Keynes, *Darwin, His Daughter, & Human Evolution*; Richards, *The Tragic Sense of Life*. Neither work is claiming, of course, that had these tragedies not occurred that randomness and waste would have miraculously been absent from a now benign and demur evolutionary theory. What did change, however, was the interpretive stress, extent, and centrality that waste and tragedy had as interpretive keys for contingency (rather than, say, spontaneity, freedom, play, joy, and the like). After the tragic death of Darwin's daughter and Heckel's wife, the stress on the negative valence of contingency in evolution increased exponentially.

in attendance, was nothing short of savage. Noting the boy had never been baptized, the minister replied that "there was no doubt, at this moment, that he is in the flames of Hell." Even the thin tissue of control the mother had maintained was now abandoned, hope had been dashed, and the Victorian war with the doctrine of hell seemed suddenly to rise unbidden into this sad scene. Rappleyea's blood boiled over, and a furious response to the minister hardly eased his anger. A few days later, still fuming, "I [Rappleyea] heard that the . . . Fundamentalists, had passed that anti-evolution law, and I made up my mind I'd show the world [their true colors]."[40] It was thus that the Scopes trial found a major moment of its birth having nothing at all to do with evolution per se, but with Christians (or, at least one) behaving badly.[41]

On the other side of things, Scopes's response to Rappleyea's questions was hardly one that glimmered with the trappings of destiny. Far from answering Rappleyea whether he had taught evolution with feet firmly planted and shoulders broad to answer a clarion call from history itself, in reality Scopes was embarrassed to find that he could not remember. "To tell the truth," he later wrote, "I had no idea."[42] In some respects this did not matter at all. As we mentioned above, the trial was hardly a trial. Far from a legal case investigating the nature of Scope's violation "it was a public performance; a stage on which William Jennings Bryan and Clarence Darrow—two of the most celebrated and controversial figures of the era" would posture and stalk in the oppressive July heat.[43] Scopes was hardly alone in his confusion. While the defense had early on opted to admit Scope's guilt from the outset so that they could explore its constitutionality, the irony was that the issue whether he violated the law at all was far from clear. The Butler Act—passed by a vote of seventy-one to five in the Tennessee House of Representatives—far from outlawing the teaching of evolution generally, specified that the illegal teaching consisted in promoting opinions on *human* origins from lower animals. It was unlawful for state-supported schools "to teach any theory that denies the story of the Divine Creation *of man* as taught in the Bible, and to teach instead that man has descended from a lower order of animals."[44] That's it. The text in question, Hunter's *Civic Biology*, much like the early Darwin himself in *Origin of Species*, avoided that particular issue. As such, Scopes could have indeed taught evolution—even quite thoroughly—and not have violated the specific parameters that had been set as guards over Tennessee's youth. Things become even more curious when we realize that no less an individual than Governor Peay, when signing the act into reality,

40. Quoted in Erdozain, *The Soul of Doubt*, 174.

41. It is perhaps worth noting that beyond the death of his daughter, Darwin's loss of faith came, too, by way of hell and his utter disgust at the notion that his unbelieving father was being wracked by unending flame. We can not make too much of anecdotes alone, but there does seem to be enough evidence to support the claim that beneath the so-called war of science and religion the true battle was against the traditional notions of hell, salvation, and punishment.

42. Lienesch, *In the Beginning*, 139.

43. Lienesch, *In the Beginning*, 139.

44. Taken from Numbers, *Darwinism Comes to America*, 77.

commented that to his mind after review the law did not affect any textbook currently in circulation in the Tennessee, including Hunter's.[45]

Peay and his team had indeed had ample opportunity to review *Civic Biology,* which, instead of an artifact embodying what was surely an inevitable clash between science and religion, was ironically a textbook there by a good deal of happenstance. In late 1923 a deadline was approaching for the renewal of textbooks that also happened to coincide with Governor Peay's bid for reelection. Having run on a platform of reforming and improving the Tennessee school system, the costs families would have to shoulder in adopting new textbooks—which rapidly bloated due to post-World War I printing inflation—would be disastrous for the tax increases Peay proposed to form the later foundations for school reform. Peay as such proposed to defer the adoption of new textbooks and simply buy older books still covered by their previous contract. In this case, it was Hunter's *Civic Biology.*[46] Rather than representing any given standardized work at the time, the 1914 textbook was less cautiously worded than many that came after it as publishers realized invoking controversy with incautious wording was quite bad for business. While later textbooks still taught evolution, and contained much of the same information as Hunter's book (with relevant scientific updates as time went on), Hunter's book in terms of rhetoric and presentation was something of a bull in the china shop of public education in the South. Indeed, the whole Scopes episode might have resolved without confrontation "had Tennessee not been using a book that pre-dated the entire controversy."[47]

What was at issue in fact was the notion of compulsory education, which became a minefield for public discourse to wade through especially in the long fall-out following the Civil War. "Religion initially intersected the school antievolution movement, not because evolution disproved Genesis," which, surprisingly, was not a topic at issue until later, "but because the issue of compulsory public schooling was frequently seen in religious terms, especially in the South." Books like Hunter's were fought against precisely because they became flags of an invasion. "Control of the schools was viewed by many Tennessee parents as a matter of cultural religious identity. . . . To reform the culture, as progressives tried to do in the early twentieth century, would mean a symbolic rejection of the ways of the fathers."[48] Intensifying this wariness even further was the fact—recovered in the early 2000s through some historical sleuthing—that Hunter's biology text in fact had two different editions: one meant for high schoolers living in urban areas, and the other rhetorically geared toward education occurring amongst rural students. Truth is often stranger than fiction, as they say, but this peculiar duality in Hunter's texts amped an already contentious atmosphere when it was realized that the urban texts were in fact being used (mistakenly or not) and distributed in rural

45. Shapiro, "The Scopes Trial Beyond Science and Religion," 201.
46. Shapiro, "The Scopes Trial Beyond Science and Religion."
47. Shapiro, "The Scopes Trial Beyond Science and Religion," 202.
48. Shapiro, *Trying Biology,* 88.

school districts. As such, on top of the perceived threats coming by way of compulsory education, it was felt Hunter's text and others like it were training rural children to abandon their upbringing and head to the city. This was not some ornamentation to how biology was being presented, but thoroughly saturated the appearance of biological information. It hardly takes a diligent reader of *Civic Biology* to find its heavy bias toward city life and the opportunities such affords as opposed to the backwater woods and farmland.[49] The conclusion that must be drawn from this is that the anti-evolution law and its supporters were initially attacking evolution not because of its conflict with Scripture, but because of its allegiance with compulsory education and the broader philosophical and ethical divide this represented in Southern culture.[50] To be sure, scriptural arguments soon followed, but these were, strangely enough, garlands strung upon the much more central trunk of philosophical differences regarding compulsory public education.

Indeed, so far from a religion vs. science clash, the first approach of defense strategized by Darrow and the ACLU was in effect to argue that evolution and Scripture were *not* incompatible, but had been held in various harmonies by a large array of thinkers. Therefore, the anti-evolution law was tacitly sanctioning a single interpretation of Scripture, demonstrating that one religion—evolutionary versions of Christianity—were being discriminated against by another religion—anti-evolutionary Christianity. The Tennessee anti-evolution law could then be shown to violate Tennessee's own "religious preference" clause that prohibited the state from taking any action that would give preference for or show discrimination against any religion. Such arguments were not to be. To be sure, later the trial would degenerate into the burlesque of taking Bryan's personal opinions on Scripture not only as the witness of an expert theologian engaging with science, but as representative of conservative Christians at large. But that turn was, yet again, hardly representative of an essential and titanic clash. Rather, Darrow's attempted humiliation of Bryan was only made possible because the first line of defense involving *harmony* was thwarted by the "plump, round-faced, and middle aged" Judge John Tate Raulston's ruling on Friday July 17 that such an argument—and the expert witnesses needed to establish it—were irrelevant to the case at hand regarding Scope's guilt or innocence. This would prove to be an ironic dismissal, for it would prove to be one of the last times the specific notion that originally animated the trial would be referenced. Scopes himself recalls that "technical point being argued was forgotten and was not, to my knowledge, mentioned after the first five minutes of Bryan's speech."[51] The trial was already off the rails, and both defense and prosecution were wrestling to lay down tracks as fast as they could to channel where it would go from here. The point being that, Darrow's ruthless

---

49. Shapiro, "The Scopes Trial Beyond Science and Religion," 205.

50. Moran, *American Genesis*, 47–71.

51. Scopes, "Reflections—Forty Years After," 25–26.

interrogation of Bryan that has become synonymous with the trial itself, was a strategy made possible only because a very contingent sequence of events preceded it.

William Jennings Bryan was also a prime reason for the change in tactics. To be sure, the judge had ruled the initial strategy irrelevant to the case at hand. But Bryan, hearing a month before the trial that the defense's plan was to concede the legal guilt of Scope's violation while arguing that teaching evolution was morally innocent, began shifting his own case to answer in kind by focusing against the *moral* qualities of evolution. In this regard Bryan's case was actually much stronger than most records allow us to recall. In America, as was mentioned, most received their understanding of evolution and Darwin not from the man himself but from Herbert Spencer, among others. It is no small thing that Spencer was also the inventor of the phrase "survival of the fittest." While of course this notion is flexible (the fittest very well could be the friendliest and most cooperative given the right circumstances),[52] it generally turned on the understanding that it represented the strong overcoming the weak, even in terms of the artificial management of society. Such notions—sometime termed "eugenics" after the concept coined by Darwin's cousin Francis Galton—were in fact prevalent in Hunter's *Civic Biology*, and many including Bryan were unhappy to say the least that these topics were being taught to youth. The chapter on evolution in *Civic Biology*, for example, explained with regards to the mentally ill, the disabled, criminals, and even epileptics that:

> If such people were lower animals, we would probably kill them off to prevent them from spreading. Humanity would not allow this, but we do have the remedy of separating the sexes in asylums and other places and in various ways preventing intermarriage and the possibility of perpetuating such a low and degenerate race.[53]

While this will ring especially offensive in today's contemporary egalitarian atmosphere, the existential threat this posed will nonetheless still be lost on us. That is because this was not some fringe opinion, representing isolated cabals of individuals perhaps possessed of a few radical notions. Rather, in Bryan's day, because of the Spencerian and Galtonian hues that Darwinism was presented and received in, many associated Darwinism with "a survival-of-the-fittest mentality," that justified even the worst excesses of "laissez-faire capitalism, imperialism, and militarism."[54] In this same spirit Hunter's textbook justified as a thought experiment the idea that the handicapped, those with mental ailments, or moral deviancies "take from society and give nothing back." To be sure, on occasion this was put in terms of saving others from harm, but at the end of the day the justification was primarily pecuniary: these people are expensive and require socialized care that is a direct interference with the

52. Nowak and Coakley, *Evolution, Games, and God.*
53. Quoted in Moritz, *Science and Religion*, 33.
54. Larson, *Summer*, 27.

normal operations of the invisible hand of the market. Bryan's horror was not that of a backwater moralist, but of someone who in the political world often came face to face with avatars who represented the radical end points of such a chain of reason. As a Democrat, Bryan had built his presidential campaign bids on the premise of fighting the excesses of capitalism and militarism, and so was readily predisposed to be allergic to those who would make them out to be an inevitable and essential core to evolutionary theory, even nature itself.[55] More than this, Bryan did not argue against evolution because in points of fact it disagreed with a literal interpretation of Scripture, but rather because in Bryan's mind it clashed "with the belief that there were truths beyond science and human reason."[56] It seems clear that this is attacking evolution as a particularly constructed worldview—one that has escaped even the visions of Huxley to enter to the heart of America—rather than evolution per se.

To many, of course, this is mere hairsplitting. For what else could Darwinism lead to if not a harsh view of nature "red in tooth and claw"? It is, as they say, a dog-eat-dog world, and Darwin was able to turn this into a powerful theory of biology. Surely, then, Bryan was correct to equate Darwinism with this particular moral expression? The historical record shows a great deal of complexity here, and it is not too much of a stretch to say that the morality one can extract from Darwinism will be of a piece with whatever worldview is using Darwinism as its avatar. To be sure, Darwinism was often as controversial for its biology as it was for what many saw as its potentially disastrous theological implications. But as we already saw in chapter 4, what Darwinism *meant* was more often than not at the heart of these early struggles. On the one hand, the morbid vision of the world that many saw in Darwin actually was in large part either inherited or at least paralleled by Darwin's reliance upon a natural theological theodicy and economic model produced by Thomas Malthus and the darker, more violent bits of Adam Smith.[57] The winnowing of the weak because of competition over limited resources was for Malthus's particularly grim view of things exactly what one would expect from God's providence. It was, moreover, one of the direct inspirations (or at least confirmations) of the principle of natural selection for Darwin, who explicitly drew on both Malthus and Smith.

On the other hand, as Randall Fuller has artfully recorded in his *The Book That Changed America*, a surprising part of the rapid acceptance of *Origin of Species* in America was because Christians saw in it a means to combat racism and promote abolitionism through the Darwinian notion of the unity of the human race.[58] In this way, curiously enough, Darwinism for a time found itself in a moral alliance with those who believed in the world's first couple, Adam and Eve, in order to emphasize the

55. Davis, "Science and Religious Fundamentalism," 254–55.

56. Cited in Shapiro, *Trying Biology*, 103. Cf. Lienesch, *In the Beginning*, 71–72.

57. Oslington, *Political Economy as Natural Theology*, 60–79; Depew and Weber, *Darwinism Evolving*, 113–140; Boer and Petterson, *Idols of Nations*, 131–166.

58. Fuller, *The Book that Changed America*, x.

equality of all races with one another. Indeed, so eager was the acceptance of Darwin's work for this reason that many of these same figures evinced a good deal of worry after they rushed to accept it for abolitionist purposes, when they came to realize how Darwin was being utilized in manners akin to the X-Club. Regardless, at the time both scientific and theological arguments had also been used to justify slavery. In relation to Adam, for example, the historian David Livingstone has done a remarkable job of tracing the strange and sometimes esoteric currents of speculation in Christian circles that Adam was not the first man, nor Eve the first woman.[59]

Spurred on by questions like where all the people populating the cities Cain was so afraid of in Genesis chapter 4 came from, lines of pre-Adamites for many came to populate the distant world of the past. For some, this was an apologetic move that made sense of the deep past populated by the likes of the Neanderthal (which had first been discovered in 1856). But more often than not it was turned to more nefarious ends.[60] Fascinating and largely lost to time, such arguments had begun intersecting with scientific theories of polygenism of the time, which hypothesized that different types of humans were different species entirely, and that some—such as the Caucasians—had the right of domination because of a natural superiority. Yet Darwin himself saw the ontological unity of the race which his theory described as of a piece with the Christian faith he grew up in, "which had a decidedly evangelical cast."[61] It is no accident that Edward Tylor, who we met in chapter four, continued his Quaker heritage and believed in the fundamental unity of the race implied by pedigree from Adam and Eve as well as the *imago Dei*. Indeed, scientific monogenists were more often than not Christians who held to the theory not just for scientific reasons, but because this stance also "preserved the integrity of scripture."[62] Darwin's grandfather Josiah Wedgewood, in fact, was a supporter of Thomas Clarkson who, along with William Wilberforce, were some of the earliest English abolitionists for just such reasons. In an ironic twist on the religion vs. science narrative, the polygenist position was held more often than not by vocally anti-Christian anthropologists, and so if one can say it was monogenetic religion opposing polygenetic science, it appears this was for the betterment of both science and morality.

Regardless of what one might think regarding Bryan's knowledge of evolution as a scientific theory, he was responding to a clear and present danger in terms of how evolution was being applied as a worldview. The attack upon evolution as a worldview becomes even clearer when an unexpected name begins showing up again and again in the trial transcripts: that of the philosopher Friedrich Nietzsche. The looming presence of the mustachioed provocateur has not been investigated at much length, where

---

59. Livingstone, *Adam's Ancestors*.
60. Livingstone, *Adam's Ancestors*, 80–108.
61. Desmond and Moore, *Darwin's Sacred Cause*, 58.
62. Livingstone, *Adam's Ancestors*, 20–23.

it has been noticed at all.[63] Yet, Nietzsche's *Genealogy of Morals*—containing Nietzsche's own mature engagement with the sciences—somewhat prophetically touched upon many of the issues at stake in the overheated Tennessee courtroom. Ironically, it was precisely because of the figure of Nietzsche that Darrow and Bryan (at least indirectly) had come to loggerheads just the year before. In turn at the Scopes trial, Nietzsche became a convenient symbolic figure for Bryan to point toward yet more ethical horrors wrought by evolutionary theory.

During the afternoon of May 21, 1924, two University of Chicago students, Richard Loeb and Nathan F. Leopold, Jr., picked up a rented car. With them they also brought a chisel, a ransom note, and in their hearts they "brought with them a Nietzsche created in their own image." Stopping along the way, they picked up a completely unsuspecting fourteen-year-old Bobby Franks—whom they had met at random—and with chilling savagery murdered him with the chisel before dropping his naked and mutilated body in a nearby marsh.[64] Papers, scientists, and religious figures all quickly began pumping out commentary on this grizzly scene. Scientific experts, criminologists, and Freudian psychiatrists all offered their own scientific takes on what could cause two wealthy, well-educated boys to commit such an atrocity with no prior signs of violence or maladjustment. Religious leaders like Billy Sunday commented upon their godlessness. Bryan wrote particularly damning reflections on the moral hollowness of our culture. And yet, all of these rang hollow in turn. It was left up to Leopold and Loeb's defense attorney to make a stronger case: none other than Clarence Darrow.

The problem with the myriad of explanations were that they merely covered symptoms of the root, confessed by the boys themselves: they fancied themselves Nietzschean supermen who, in a world that lacked value, would create their own values and live in accord with them.[65] Not even Darrow's silver tongue could get rid of that fact. Instead, Darrow went on a very peculiar offensive that would ultimately come back to haunt him some at the Scopes trial. Not able to exonerate the two college students from their crime, he could attempt to soften the charges just enough that they might narrowly avoid the death penalty. As such, Darrow meant to spread the blame around as liberally as he could. He turned to extol the power of Nietzsche's rhetoric, how widespread and popular it was, how even Nietzsche may have been a victim of his own philosophy as he went insane at the end of his life. No one was safe from Darrow's finger pointing. Even the library at the University of Chicago, the publishers and the distributors, the professors of philosophy who continued to teach Nietzsche, all in a sense bore some blame. "It is hardly fair," Darrow opined at one point of his impassioned talk, "to hang a nineteen-year-old boy for the philosophy that was taught him at university." While, to be sure, the boys applied Nietzsche to themselves in ways no

63. Konoval, "Nietzsche and the Scopes Trial."

64. Ratner-Rosenhagen, *American Nietzsche*, 144.

65. Konoval, "Nietzsche and the Scopes Trial," 553.

others had, they did so within parameters Nietzsche himself set, or so Darrow argued. While Darrow painted himself as something of an expert on Nietzsche—having read everything the man had written—he remained vague about the actual content of Nietzsche's superman. What he did say was that "Nietzsche believed that some time the superman would be born, that evolution was working toward the supermen. . . . He wrote that one book, 'Beyond Good and Evil,' . . . a treatise holding that the intelligent man is beyond good and evil; that the laws for good and the laws for evil do not apply to those who approach the supermen."[66]

After Darrow gave what is still considered one of the best speeches of his career, the judge was moved to mercy. The defendants avoided the death penalty, though Loeb and Leopold were sentenced to life plus ninety-nine years. For Darrow, this was a major victory considering the no-win conditions he had entered into. Apart from a surge of negative press that this defense bought him, Darrow's own arguments would come back to haunt him in the Scopes trial. Precisely because Darrow had based his defense of Loeb and Leopold upon the effectiveness of Nietzsche's arguments, as well as how Darrow had connected Nietzsche's superman to evolution (and the general connection that the American mind had already made between Darwin and Nietzsche at that point), this gave Bryan everything he needed during the Scopes trial. On the Thursday of the Scopes Trial, Bryan continued his attack on the morality of Darwinism by recalling the murder committed by Loeb and Leopold:

> These people come in from the outside of the state, and force upon the people of this state, and upon the children of the taxpayers of this state, a doctrine that refutes . . . their belief in God . . . It is this doctrine that gives us Nietzsche, the only great author who tried to carry this to its logical conclusion, and we have the testimony of my distinguished friend from Chicago [Darrow] in his speech in the Loeb and Leopold case that 50,000 volumes had been written about Nietzsche, and [that] he is the greatest philosopher in the last hundred years, and have him pleading because Leopold read Nietzsche and adopted Nietzsche's philosophy of the superman, that he is not responsible for taking of human life . . . That is the doctrine, my friends, that they have tried to bring into existence [which] they commence in the high schools with their foundation in the evolutionary theory, and the statement of that distinguished lawyer [Darrow] that this is more read than any other in a hundred years, and the statement of that distinguished man that the teachings of Nietzsche made Leopold a murderer.[67]

---

66. Quoted in Ratner-Rosenhagen, *American Nietzsche,* 145.

67. Quoted in Konoval, "The Scopes Trial," 553.

Creating Creationism

Even when Bryan was called to the stand to give his supposedly "expert" testimony on Scripture and science, his objections to evolution were not based upon the idea of a young earth, or that evolution contradicted the so-called "literal" meaning of Scripture. Rather, despite Bryan bumbling through his answers and revealing a rather naïve take both upon biblical hermeneutics and the broader theological concerns involved, it remained clear that his concerns were steadfast in their resolve to target the broader naturalistic and anti-theistic interpretations that had become attached to evolution. More specifically, Bryan sought to advance the ideas of equality, to critique eugenics, and curb what he saw as the dangerous precedent set by those like Loeb and Leopold. Such worldviews Bryan—mistakenly, though not without cause—saw as a necessary part of the theory and its implications. In doing so, he was ironically championing many goals that Darwin himself cherished and actually saw as implications of his theory. Bryan also leveled criticism at what he saw as the scientific deficiencies of evolution, which had a great deal more plausibility in his time, as the so-called "Neo-Darwinian Synthesis" had yet to occur, and Darwinism even to many secular scientists appeared to be on the ropes, if not wholly discredited. But if one is to isolate Bryan's "religious" critiques, on his view, evolution excluded miracles, the soul's immortality, the virgin birth, and the resurrection of Christ. Evolution touched upon none of these things, of course, except insofar as it had, as a worldview, been extended far beyond the confines of the theory as scientific. Nonetheless, memories of the trial still often represent Bryan as a typical six-day creationist, hell-bent on promoting a young earth and a literal view of Scripture that has immediate domain over scientific pronouncements. As was common at the time, Bryan in fact believed in an old earth, where the "days" of creation in Genesis represented epochs of indeterminate length rather than twenty-four-hour periods. In doing so he was not a strange outlier but in impeccable "orthodox" company, as we shall see momentarily.

Just as much as before the trial started, perhaps even more important were works that curated the image of the event as the twentieth century continued. A key turning point in how the memory of the Scopes trial was managed was the 1931 publication of Frederick Lewis Allen's *Only Yesterday: An Informal History of the 1920's*. Despite many noting that its closeness to the events in turn could serve to be detrimental, in that it had left little room for objective reflection, it nonetheless quickly became the standard reference text to understand history in America through the tumultuous 1920s, what Roderick Nash has termed "the nervous generation."[68] One of the major fallouts of Allen's representation was that everyone was epitomized in one of three conflicting

68. Nash, *The Nervous Generation*. Nash continues, and notes one of the legacies of Allen's work was the representation of 1920's America as nearly unanimous in its attempt to shed older Victorian values. "No one has done more to shape the conception of the American 1920's than Frederick Lewis Allen," he says. Allen was also, as such, the first to represent Bryan's personal defeat as symbolic for the fall of Fundamentalism at large (5–8).

groups: the science-hating fundamentalists, the science-loving and freethinking Skeptics, and Modernists who sought a third way between two extremes. Allen opens his book in a fairly sardonic tone, in fact, trying what seems to be a gambit to disarm his readers. "Further research will undoubtedly disclose errors and deficiencies in this book," he casually remarks. Indeed, given how close it is to events, many may even find his work amusing, he theorizes, as different interpretations than their own have been "woven into a pattern that at least masquerades as history."[69] Nonetheless, even this posture of self-deprecation could hardly prepare one for the truly astounding reductions that occur. Tapping into the wellspring of Draper and White's warfare narratives, which very neatly fit onto Allen's threefold classification of Fundamentalists, Skeptics, and Modernists, Allen casts the whole thing into a mold of two very clear adversaries by combing modernists and skeptics into a single allegiance against the militant fundamentalism of William Jennings Bryan. What occurred at the trial was "a battle between Fundamentalism . . . and twentieth-century skepticism (assisted by modernism) on the other."[70] In setting these terms, Lewis created the precedent for directly applying Draper and White's work to the trial that lasted well into the twentieth century, even as late as the 1980s, which just goes to show the type of staying power warfare had accumulated by that point.[71] Allen's work in this way also created what appeared to be a continuous line of creationist opposition to evolution that repeated the same themes again and again. Even today, many like philosopher Philip Kitcher reflexively draw a straight line between, say, Wilberforce and later six-day creationists, with whom many of us are more familiar.[72] Such equations may seem straightforward and perhaps even harmless. And yet for the reader of Kitcher's work and those like it, an entire alternative history has now been conjured by this sleight of hand. On the other hand, many fundamentalists and evangelicals have likewise identified six-day creationism as the defense of orthodoxy, and the latest in a "long tradition of evangelical anti-evolutionism" when such alternative histories are simply not the case.[73] Indeed, "we will not fail to notice . . . that evolutionists were represented in the pages of the . . . manifesto *The Fundamentals* published between 1901–1915. This in itself points to a radical disjunction between earlier pluralistic fundamentalism and its later, more caustic counterpart."[74]

---

69. Allen, *Only Yesterday,* xiii–xiv.

70. Allen, *Only Yesterday*, 202.

71. As the historian Barry Hankins writes of Larson's pathbreaking work on the Scopes trial, *Summer for the Gods* (which eventually won Larson a Pulitzer prize), it "could not have been written between 1930 and 1980" not only because of the continued power of caricature. But, in addition, it was not until the "rise of the Christian Right in the 1980's" that anyone took fundamentalist Christianity seriously enough to reevaluate it. See: Hankins, *Jesus and Gin*, 104–5.

72. Kitcher, *Abusing Science*, 1–3.

73. Livingstone, *Darwin's Forgotten Defenders*, ix.

74. Livingstone, *Darwin's Forgotten Defenders*, xii. And the prior fundamentalism, often relying upon Baconianism, itself represented a rather intense mutation and difference from the theologies

Though briefly, in chapter 4 we saw that "Darwinism" and "evolution" could describe a rather large array of positions not immediately reconcilable with one another. And in broader terms as was mentioned, Darwinism and evolutionism cannot be equated. Darwin's was not the first evolutionary theory. Nor does it constitute a singular thing now. As such the Darwin industry over the last forty years has become increasingly sensitive to the plasticity and function that such terms have had through the history of their use. This accelerated again when it became apparent in the 1980s that there was broad disagreement regarding the contents and implications of the theory,[75] with 2016 marking yet another of these reevaluations when the meeting of the Royal Society was convened under the theme "New Trends in Evolutionary Biology," calling into question the centrality of natural selection.[76] Such are not indications of the collapse of Darwinism as a theory, as some have taken it. Rather these are disagreements about what constitutes its primary essence and central core. Such granularity to the term *creation*, or *creationist*, however, has been much slower to emerge. Ronald Numbers with some embarrassment notes that even in his "450-page book on the history of modern creationism" (now longer as it has been revised in a second edition) "failed to address the issue of when [creationism and its cognates] first came into use."[77]

Looking into the different nuances of how the term has been used is incredibly important, for "creationism" has meant quite different things at different times. For example, in the mid-nineteenth century when Darwin published the *Origin*, "creationist" referred not to God's creation of the world, but rather to a debate involving the origins of one's soul—naming those who believed God specially created the soul in each individual, rather than it being passed on in an almost quasi-genetic manner from the parents (traducianism). Moreover, as Numbers writes in his big book on the history of creationism:

> Contemporary readers who associate creationism with the teachings of the so-called scientific creationists will no doubt be surprised by the small number of nineteenth-century creationist writers who subscribed to a recent creation in six literal days and the even greater rarity of those who attributed the fossil record to the Noachian flood. Creationists of the Victorian era generally assimilated the findings of historical geology to such an extent that today they

---

that preceded it (Noll, *America's God*, 93–113; Holifield, *Theology in America*, ch. 8; the injection of the concept of "worldview" which originated in German Idealism and migrated into evangelicalism also changed the content an manner of discourse considerably as well, as recorded adeptly in Worthen, *Apostles of Reason*).

75. Hull, "Darwinism as a Historical Entity."

76. For some work on the broadening concept of evolution (and whether it should even be called "Darwinian"), see Nobel, *The Music of Life*; Morris, *Life's Solutions*; Marshall, *Evolution 2.0*; Shapiro, *Evolution*; C. Cunningham, *Darwin's Pious Idea*; Prum, *The Evolution of Beauty*; Morange, *The Misunderstood Gene*.

77. Numbers, *Darwinism Comes to America*, 49.

seem intellectually closer to the theistic evolutionists of their time than to the scientific creationists of the late twentieth century. . . . Even Princeton Seminary's rock-ribbed Charles Hodge (1797–1878), who concluded that Darwinism was atheism because it banished God from the world and enabled one "to account for design without referring it to the purpose or agency of God," conceded the great antiquity of the earth and gave his imprimatur to Guyot's and Dana's interpretation of the days of Genesis as geological epochs.[78]

Such conclusions of an enormous array of work will no doubt confuse many, for in the twentieth century a young earth above nearly all other doctrines (even, sadly, that of Christ himself) has come to represent Christian tradition and orthodoxy. Perhaps the most notorious example brought up of Scripture being used to produce a young earth—that of Archbishop James Ussher (1581–1656)—is deeply misunderstood. Widely popularized by being printed in the best-selling Scofield Reference Bible, a King James translation that innovated by printing commentary alongside the sacred text rather than in a separate volume, Ussher's dating was mistaken by many to be the sole authoritative interpretation of the meaning of the genealogies in Scripture. Far from a crude literalist, Ussher's work in fact represented a huge feat of classical scholarship. For Ussher did not just use Scripture, but rather a deep working knowledge of Persian, Greek, Roman, and even Chinese history, all of which themselves concluded in a relatively young age of the earth. "By far the greater part of Ussher's evidence, like that of other chronologists, came not from the Bible but from ancient *secular* records."[79] Ussher was, in other words, using the best scientific and historical data available at the time in dialogue with Scripture to come up with the figure of 4004 BC for the starting date of the world. And, while it "has since become the most famous, but at the same time was just one of dozens of rival calculations . . . the huge variation reflected the uncertainties of the complex textual analyses that the science of chronology required. It was a notoriously disputatious science; [and] there was certainly no rigidly orthodox line about the date of Creation or of anything else."[80]

Part of the controversy over rejection of a young earth was not necessarily because it overturned Ussher's chronology or a "literal" sense of Scripture. It was rather because many geologists advocating for much lengthier geological and cosmological time-scales did so in order to advocate for the philosophical and theological position of an eternal universe.[81] The idea that God created the world *ex nihilo* (out of nothing) since the time of its formulation in early Christianity had run up against Aristotle's notion of the eternality of the universe, and it remains something of a perennial issue.[82] In seventeenth- and eighteenth-century Europe, such issues were taken up again

78. Numbers, *The Creationists*, 16, 26.

79. Rudwick, *Earth's Deep History*, 14.

80. Rudwick, *Bursting the Limits of Time*, 116.

81. Rudwick, *Bursting the Limits of Time*, 115–16.

82. On "ex nihilo" creation, see May, *Creatio Ex Nihilo*.

and transposed into the terms of geology.[83] Many were scandalized at the deist James Hutton's denial of the historicity of the Noahic flood, for example, but this was more often than not because behind it lay "a radically ahistorical vision of the earth's revolutions; and behind that in turn lay an eternalism that denied that the cosmos had an ultimately divine foundation."[84] It should be remarked that such eternalism was also quite different than modern geological theories, since human civilization was thought to itself be as eternal as the rocks upon which they stood. Again, with the deletion of theology from our memories this will appear like a science and religion clash, when in fact "this was certainly not a straightforward struggle of enlightened Reason against religious Dogma. There were strong 'ideological' issues at stake on *both* sides of the argument."[85] In the same way, the Christian contributions to geology and the discovery of deep time, evolution, and the like have likewise often gone unnoticed. We might recall from chapter 3 that the concept of history as we know it was a theological and philosophical gift of Judeo-Christianity. In this same vein, the world's leading authority on the history of geology Martin Rudwick notes that the view of nature as itself historical and subject to narratives telling of changes both contingent and cumulative had as "one major source—even arguably *the* major source . . . [the] Judeo-Christian scriptures," and theologies.[86]

By the nineteenth century, even among circles where Ussher's chronology was popular, conservative biblical scholars and Hebraists suggested that it was in error, or at the very least hardly the only or even the most obvious way to interpret the genealogies in context.[87] Moreover, while taking the days of creation as twenty-four-hour days was certainly present throughout the Christian tradition, it was hardly the only or even the predominant focus of Genesis interpretation.[88] Today a "literal" interpretation, coupled with (something akin to) Ussher's chronology, "implies a stance toward geology, biology, and science that ancient interpreters could not have imagined, much less chosen. . . . [For these and other reasons] present-day literalism is a new phenomenon."[89] Indeed, while Charles Lyell is popularly represented as singlehandedly introducing the deep geological age of the earth in distinction from biblical literalism, in point of fact churchmen had accepted the long age of the earth since at least 1790. This is not to decide the issue of interpreting Genesis one way or another— it is, however, to point out that historically speaking literal six-day creationism and all of its accoutrements is not just a modern phenomenon, but one that can largely be sequestered to the contexts of the latter half of the twentieth century and beyond,

83. Cutler, "Steno and the Problem of Deep Time," 143–48.

84. Rudwick, *Bursting the Limits of Time*, 334.

85. Rudwick, *Earth's Deep History*, 28.

86. Rudwick, *Earth's Deep History*, 4.

87. Numbers, "The Most Important Biblical Discovery of Our Time."

88. A. Brown, *The Days of Creation*.

89. A. Brown, *The Days of Creation*, 296.

when the issue became conflated with another storm in the evangelical teacup—that over the authority of Scripture. The two became conflated in the eyes of many, and so to speak of anything other than a narrowly defined "literal" sense came to be seen by quite a few—though certainly not all— as a denial of Scripture itself.[90]

There were, nonetheless, concerted theological arguments against Darwinism, or evolution at large. The danger of generalization is ever present when the attempt to place all of these under the heading of "creationism" is undertaken, however. Today the general notion one hears is that "Darwin defeated theology" by overturning design arguments displaying how God perfectly fit creatures into their environments, or how the complexity of a creature could not be explained as a random assemblage of parts. As we have already seen, while there is a morsel of truth to this, Darwinism and evolutionary theory were often just as theologically minded as the design arguments against which they were providing an alternative vision of providence. But, more to the point, while important throughout Christian history "design applied to the earth [and to its inhabitants was] but one of the violins in a mighty orchestra" that was natural and revealed theology.[91] There is a grand and indeed orchestral tradition of natural theology (always accompanied, it must be admitted, by a few off-key instruments that never quite fit, or tried to monopolize the whole song). Such natural theologies, including arguments for God, not only accompanied science but in many instances guided it, partnered with it, and were in turn informed and shaped by it. What tends to be forgotten in the "Darwin killed design" notions of history associated with the term *creationism* is the fact that by the late eighteenth through the nineteenth century, not just the centrality placed on design arguments, but the very notion of what constituted design arguments at all had drastically altered. In fact, the clash in the nineteenth century and "the conviction of American Protestants that Darwin's theory undermined natural theology becomes comprehensible only when that theory is placed within the context of mid-nineteenth-century Christian apologetics," especially in the United States but also in Britain.[92] This is extremely important because it was precisely these changes that were undone by Darwinian alternatives, in many senses because "Providence came to mean solely God's beneficence in constructing the world so that it conduced to the good."[93] That is to say, a particular slenderized variation of design came to bear nearly the whole weight of God. It was hardly a surprise when such a slender foundation could not hold so much complex weight. If "Darwinism had developed against a theological background shaped by alternative

90. For this story see Worthen, *Apostles of Reason.*

91. Glacken, *Traces,* 377. Cf. 355–550 for a detailed survey on the history of physico-theology in Christianity.

92. J. Roberts, *Darwinism and the Divine,* 7–8.

93. J. Turner, *Without God,* 40; Fergusson, *Providence,* 109–239.

approaches to natural theology," writes historian and scientist Alister McGrath, a "different outcome would have resulted."[94]

The success of Newtonian mechanics had shifted arguments in Christianity from a more philosophical or metaphysical orientation (whether utilizing Aristotelian or Platonic philosophy, for example) to one whose reliance on physical and mechanical pictures of nature grew unwieldy over time. In nature (where law reigned), the providence of God was traditionally thought to be demonstrable. In history, God's acts were thought to be discerned only with the eyes of faith, a theological vision that knit the world together providentially. As John Calvin wrote, "what seems to us contingence, faith will recognize as the secret impulse of God."[95] Newtonianism demanded an extension of lawlikeness to natural history, or what would later be called biology. This was an extension that was traditionally seen as inappropriate theologically speaking— and one that had some bad historical timing. In the period "from the seventeenth to nineteenth centuries, the nature–history boundary became blurred, as natural history was gradually transformed from a discipline that was concerned primarily with atemporal spatial and taxonomic relations, to a genuinely historical discipline that was concerned with organic change over time." The problem with this was that "theological interpreters of evolutionary biology continued to carry with them a set of expectations about the perspicuity of God's providential activity *in nature* that were no longer appropriate [based on the tradition of Christianity] for a discipline that had become essentially *historical*."[96] Or, in other words, "whereas God's providence in history was 'secret,' 'hidden,' 'inaccessible,' 'not discernable,' in the natural world such providence was 'demonstrable,' 'very manifest,' and convincing." Such clear distinctions between these two realms of knowledge regarding God "was to become blurred when in the nineteenth century living things became subject to historical change."[97] In this respect, many theologians were slow to realize that the transposition that had happened also meant a reevaluation of which types of theological judgments could apply where, and how those could be expected to perform in larger theological demonstrations and arguments. In a realm previously reserved exclusively for the eyes of faith, a sort of

94. McGrath, *Darwinism and the Divine*, 5; 217–47.

95. Quoted in Harrison, "Evolution, Providence, and the Problem of Chance," 272.

96. Harrison, "Evolution, Providence, and the Problem of Chance," 261.

97. Harrison, "Evolution, Providence, and the Problem of Chance," 273. Cf. Osler, *Divine Will and the Mechanical Philosophy*, 236: "The unpredictable actions of an omnipotent God have been replaced by unpredictable variations, which respond to selective pressures in ways only empirical study can determine. Contingency, rather than reason is the basis for explanation . . . evolution does not have a predictable and determinate course of a kind that can be known by a priori and deductive methods. The metaphysical assumptions underlying this style of evolutionary science can be traced back to a voluntarist interpretation of the biblical worldview. Although theological language has dropped out of scientific discourse, contemporary styles of science are historically linked to the dialectic of the absolute and ordained powers of God. The interplay between necessity and contingency in the world is now constructed entirely in naturalistic terms, but it grew from roots embedded in an earlier, theological understanding."

reasoned and informed theological "sight" would allow theological inferences and descriptions of the world to be produced. Now, however, the realm of biology was scoured for lawlike proofs via theological reasoning that was not originally designed for such things. As Peter Harrison summarizes the matter:

> If the history of nature were understood to be more akin to human history at the time Darwin published the *Origin*, no [theologian] would have expected to see conspicuous instances of purpose at every moment. What the doctrine of providence would dictate, then, would be that *in spite of* the prominence of apparent chance events, the faithful should be prepared to accept [or even perceive] that God's hidden purposes were being fulfilled. Viewed in this light, the purpose behind the successions of species in the Darwinian framework would be no more obvious than the logic of the rise and fall of kingdoms in human history. Why so few read Darwinian evolution in this fashion is owing to a combination of [ironically contingent] historical factors—foremost among them the fact that the full implications of natural history's transformations into a temporal enterprise were not fully recognized at the time. Hence, the fateful collision between the expectation of design and the apparent randomness of the Darwinian theory.[98]

Moreover, to view design as being overcome in this period is also to reduce our examples to a narrow band that were being used within apologetics as a genre. Beyond this "common context" design functioned in a much broader way as simply the doxological manner in which popular science was often communicated and received.[99] As historian Alister McGrath speaks of the broad transition in design arguments, whereas earlier design arguments were arguments noting "order implies an orderer (an argument *from* design)" meaning that the eyes of faith could see and even unpack God's acts in contingent events, this transitioned into arguments that had to lead *to* conclusions of design.[100] And so, he recognizes that there was a large-scale shift in the method and the content of the arguments being put forward. "From 1690 . . . English natural theology increasingly became concerned with finding 'evidence of design' rather than [theologically describing the] observation of order. 'Physico theology' increasingly became identified with this quest for design, [now] framed specifically in terms of the notion of contrivance."[101] And this was a newly defined level of contrivance whose magnified level of specificity and granularity of detail, along with the burden of proof needed to evidence design, became inversely related to the broader level of sophistication of the philosophical, theological, and metaphysical lattice of arguments and ideas within which design was occurring.[102] As one historian has aptly

98. Harrison, "Evolution, Providence, and the Problem of Chance," 277.

99. Topham, "Beyond the 'Common Context.'"

100. McGrath, *Darwinism and the Divine*, 63.

101. McGrath, *Darwinism and the Divine*, 53.

102. Hanby, *No God, No Science*, 105–296.

put it, today it takes "an effort of historical imagination to realize that theology helped make [scientific knowledge] natural," and appear as not just the gold star of knowledge, but in a sort of quixotic self-marginalization the only type of knowledge claim. "It was, after all, theologians and ministers [in the nineteenth century] who had . . . insisted that knowledge of God's existence and benevolence could be pinned down as securely as the structure of a frog's anatomy—and by roughly the same method. It was they who obscured the difference between natural and supernatural knowledge . . . by the mid-nineteenth century [many] had, really, no effectual model of [theological] knowledge except science."[103]

Chance in the tradition—say, for Aquinas—could have a relatively positive relation to theological interpretation. The eyes of faith could read chance as but a surface level description of what at a deeper, ontological level was providence at play. But, the new stress of arguments *to* design meant that these arguments had to have an almost entirely negative view of chance, explaining it away as illusion or, perhaps, a misunderstanding of what was in reality *demonstrable* lawlikeness, now understood as contrivance, or direct design. Moreover, to bolster this search for providence within the realm where once only the eyes of faith could pierce, divine attributes and activities were given increasingly "physical meaning" and such "transformations of metaphysical axioms into prescriptions for the natural world were extremely common in early modern science."[104] Increasingly seen as an object within the scope of scientific investigation, any hesitance or qualification in our ability to speak of God or discern God's purposes and work was discarded for increasingly straightforward clarity about exactly what God was doing in the world, what it entailed, and how to detect it. In line with the theme of this book, not only did this colonize the prior tradition so it looked like everyone previous was made into the image of nineteenth-century physico-theology efforts, more importantly "the question of creation ceased to be about creation in its proper sense and becomes instead a question of *manufacture*." Creation is no longer understood as a question of "ontological constitution but is rather misinterpreted as a question of temporal origins in a series of causes and effects which culminate in the manufactured artifact."[105] Very early on, writes historian Jon Roberts, "Protestant thinkers opted to eschew ontological refection concerning the relationship between God as Being Itself and the beings that inhabit the universe in favor of an analysis of the relationship between Creator and creation grounded on the discoveries disclosed by science concerning the structure and behavior of phenomena. . . . this greatly contributed to . . . the impoverishment of the [prior Christian] religious vision of the

---

103. Turner, *Without God*, 93.

104. Brooke, *Science and Religion*, 99. Cf. Grant, *The Foundations of Modern Science*, 125–26: Harrison, *The Bible, Protestantism, and the Rise of Natural Science*, 159–60; Brague, *The Law of God*, 236–37.

105. Hanby, *No God, No Science*, 35.

world."[106] And that included Protestant visions just prior to many of these changes as well. The physico-theology of the nineteenth century, whatever family resemblance it may have, is quite distinct to arguments like Thomas Aquinas's that came before it.[107] Not only are these prior arguments more nestled in a sophisticated theology and metaphysics, there is a constant qualification regarding the nature and limits of what they can achieve. The rosy-eyed physico-theology often had no such reservations, at least in terms of the very physical information displaying contrivance for which they were searching. In this way, as one scholar has eloquently summarized the matter, "Once God regained transparency or even a body, he was all the easier to identify and so kill."[108] The entire order of theology in this tradition, in other words, was naturalized, brought down to the same level of explanation as scientific description. Just so, a sharp natural vs. supernatural dichotomy that was present in a new or at least completely exaggerated way than ever before in Christian history became quite evident.[109] As was mentioned in chapter 4, sometime in the middle third of the nineteenth century, some observers began to suspect that "every new conquest achieved by science involved the loss of a domain to religion."[110] This was in many ways the result of the changes just mentioned.

In the early 1740s, for example, a now-forgotten craze swept Europe that many viewed as threatening to undermine the very foundations of religion. In its wake, a panic ensued. The rich elite who heard of it were both horrified and fascinated; the Paris Academy of Science was in uproar; and salons everywhere were abuzz with the latest gossip of a new afront to the Almighty. Amid rumors of increasing atheism, the revival of skepticism and Epicureanism, and the *philosophes* like Voltaire and Condorcet reorganizing knowledge in their militantly anti-Christian *Encyclopedias*, this mania sweeping Europe was nonetheless what, for a time, occupied the idle hands of those ranging from the merely curious to the impish, tantalized by dabbling with blasphemy. The cause of this scandal? Hydras and starfish. An unexpected adversary for God, to be sure. But a discovery had been made by Abraham Trembley, who reported to his amazement that if one were to cut the hydra in half, the poor creature put to the knife would heal by, in fact, becoming two whole creatures. Like a fever, polypchopping swept through Europe, and the only cure was more chopping. Hardly able to believe the reports, one R. A. Réamur performed such dismemberings so many times it was likened to Jesus' multiplication of fishes and loaves. Starting with a few polyps, Réamur ended up with hundreds as he sent his findings to the University of Paris and the Academy of Sciences.[111]

---

106. J. Roberts, *Darwinism and the Divine*, xix; cf. Gilson, *The Spirit of Medieval Philosophy*, 79.

107. Feser, *Neo-Scholastic Essays*, 147–93.

108. Funkenstein, *Theology and the Scientific Imagination*, 116.

109. J. Turner, *Without God*, 38.

110. Numbers, "Aggressors, Victims, and Peacemakers," 20.

111. Brooke, *Science and Religion*, 234.

Such concerns today, even by the most zealous of Christian, may come across as quaint if not outright misguided. Surely the existence of the soul is not something that can or cannot be proven by scientific means but, as qualitatively different, is patient only of philosophical or theological reflection that may or may not incorporate scientific findings as premises, illustrations, or just general data to be interpreted? Yet, concerns such as this attended the design arguments because more and more they placed a unique burden upon physical demonstrations in the world that were not only unreasonable, but quite different than the nearly 1,600 years of theology that preceded them: "the [design] arguments . . . are constructed as if sixteen hundred years of Christianity had never occurred. Both treat the atheistic question as if religion had nothing to say to this issue short of categorical faith, that before this radical challenge to its cognate claims religion stood empty-handed."[112] The design arguments of William Paley so famously brought low by Darwin in a sense represents the historical culmination of this trajectory:

> Paley, in effect, had thrown down a challenge. No natural law, comparable to, or derivable from, other genuine natural laws, he tacitly claimed, would ever be found to explain organic function and adaptedness, including the morphological co-adaptedness of parts to one another and the ecological fittingness of organisms to their niches.[113]

The change here is evident: whereas before the arguments involved the difference between an atheistic vs. a theological interpretation of the laws of nature, and the extent the complexity and adaptedness of life served as evidence of a higher ordering wisdom. Now, the picture is becoming more and more one where natural laws are held up in antithesis to design, that is, God's direct action. Such boldness, coupled with more "Baconian" methods of induction where the examples were thought to quite clearly speak for themselves, is why one so often sees the unusual strategy in treatises on design in the nineteenth century of simply piling up examples of design and letting the heap speak for itself. But this was not to last. Even before alternative evolutionary natural theologies (even of the Darwinian variety, as was mentioned in chapter 4), many realized design arguments both theologically and scientifically had been asked to bear a burden far greater than their newly anemic forms could manage: "natural theology [of the physical sort we briefly described] was not so much destroyed by science as eased out of scientific culture by a growing irrelevance."[114] Eased out, we should remind ourselves, by a culture that was not just secularists but densely populated by theologians and churchmen as well. It was, we might say, a research trajectory that eventually found an end as alternatives were sought.[115] So, we have to be care-

112. Buckley, "The Newtonian Settlement," 92.

113. Depew and Weber, *Darwinism Evolving,* 102.

114. Brooke, *Science and Religion,* 298.

115. This is not to comment on the viability of Intelligent Design or other current movements.

ful what is meant here. The point here is "not that science undermined the design arguments . . . quite the contrary. It was rather that religious apologists were asking too much of it . . . [design] simply became too blunt an instrument [both theologically and scientifically] to yield precise information at the rock face of research."[116] This growing sense of irrelevance came both from within theology and from without. From within, we must remind ourselves again that the alternative evolutionary theories on offer, including Darwin's, had robust Christian representation.[117] Many, moreover, sensed a sort of inanity at the lengths many would go to try to correlate every feature of the universe to some purposeful end, especially apart from a unifying metaphysics like Thomas Aquinas had previously provided. Alfred Russell Wallace "grew impatient" for example, when some praised the soft-scar on a coconut as a wise contrivance of design, which allowed the embryonic shoot to emerge instead of being trapped within. Far from equating a denial of design with a denial of God, here Wallace thought the extremity of the design argument insulted God—"it was like praising an architect for remembering to put a door in his house."[118] Perhaps the most humorous of these arguments in circulation was that the cushion of the buttocks was evidence of God's providential care of man. Now surely on one level we can appreciate the point at an

---

This is intended as a corrective to the usual historical observation of what occurred in the nineteenth century. For a brief summary on ID, see: Numbers, *The Creationists*, 373–398; Ruse, *Darwin and Design*, 313–338; the best representatives of the ID literature are no doubt Meyer, *Darwin's Doubt*; Behe, *A Mousetrap for Darwin*, which is a long awaited sequel to Behe's *Darwin's Black Box*, which has proven itself to be the most seminal and popular piece of ID literature. ID is important because (in theory) it is separate from scientific creationism, both in the sense that it does not appeal to the Bible for its arguments (accepting, for example, the long age of the earth without blushing). William Dembski, "The Design Argument" 335, rightly notes that, for example, the "design argument must be distinguished from a prior metaphysical commitment to design," which certainly as we are arguing at large in this chapter, marks a major shift. However, as the 2006 "Intelligent Design" trials in Dover, Pennsylvania reveal, it is not as clearly distinct as it ostensibly wants to be. Barbara Forrest and her team of researchers discovered that ID had perpetrated a "find and replace" word-search in a popular creationist text, substituting "Intelligent Design" for "Creator" and its cognates after a 1987 court decision ruling Creationism as religious and unable to be taught in the classroom (see Forrest and Gross, *Creationism's Trojan Horse*, esp. 25–34). On the trial, which is sometimes referenced as "Scopes Part II" see Lebo, *The Devil In Dover*; Miller, *Only A Theory*, 88–134. We must sadly bracket this discussion, which, while largely diminishing in the twenty-first century, remains a fascinating debate if for no other reason than it embodies and perpetuates many of the historiographic themes that were present from the beginning with the X-Club and Darwiniana generally. Regardless of how much one wants to make about the textual discovery of Forrest et. al., its result was a widespread reinforcing of the opinion that ID is merely a perpetuation of the continuing creationist tradition.

116. Brooke, *Science and Religion*, 195.

117. As David Kohn, in his essay "Darwin's Ambiguity," 238, reminds us: "Time and again, while damning the narrowness of special creation, and by direct implication providential theology, [Darwin] appeals to [what he wants to represent as] a higher, nobler, more enlarged and enlightened theological perspective. For Darwin in the Origin, the laws of nature implied that there was order in the universe . . . But his open position was not that of an atheist. He can say the laws of nature are impressed on matter by [the] Creator . . . God was an implication of nature's order. And evolution by natural selection was an explanation of natural order that the highest, honest religious mind ought not despise."

118. Brooke, *Science and Religion*, 299.

aesthetic level that every good thing is a gift from God (even if there is a healthy dose of sarcasm here)—but as an earnest argument one is left rather speechless before it. In a way this is part of the problem. Design arguments fulfilled any number of different genre functions aside from demonstration and argument[119]: edification, doxology, even being the mainstream way of communicating science at all (the sugar that made the medicine go down, as it were[120]). In an environment that increasingly rang of war, all of these genre-functions of design were pressed into the single image of demonstration and proof. And it was thus that countless straw men sent eyes rolling or were put to the sword.[121]

From without, of course, we have already seen some of the moves Huxley was making to remove the last veneer of theology from the whole edifice of science. "Design" as such references a particular sequence of physico-theology that collapsed the metaphysical into the physical, equating the ability to discern theological "meaning" in the world with one's ability to describe an organism exhaustively in terms of perfect physical pre-adaptation to fit an environment: "Paley's natural theology . . . is not the antithesis to modern naturalism. It *is* modern naturalism in its theological guise."[122] It was design as perfect pre-adaptation, especially as exemplified by Paley, which were the theories threatened by Darwinian natural selection.[123] It is largely through Paley's influence upon Darwin that the problem of "pre-adaptation"—the fit between 'biological insides and environmental outsides"—would become, in Stephen J. Gould's words, the "primary problem of evolution."[124] But far from representative of the tradition, this was a surprisingly parochial, time-bound mutation within the broader ambit of the tradition.[125] On the other end, it was quite separate in many ways from the creationism of the twentieth century, as the Victorian arguments were often more about physicalizing theological ideas instead of demanding that a literal interpretation of Genesis had to be scientifically applied to the natural world. Once an orchestra of broader theological and metaphysical concerns, a lonely horn that represented the last of an already deserted "physico-theology" section was now blaring for all it was worth. And so, it was little wonder that friend and foe alike often wanted to stop it up.

---

119. Mandelbrote, "What Was Physico-Theology For?" 67–77.

120. On some of the variety of natural theology, see: Topham, "Beyond the Common Context"; Topham, "Science, Natural Theology, and the Practice of Christian Piety," 37–66.

121. For some prehistory on the reduction of natural theology into a purely apologetic function for God's existence, cf. Buckley, *At the Origins of Modern Atheism*, 37–144; Leech, *The Hammer of the Cartesians*, 229–240.

122. Hanby, *No God, No Science*, 170.

123. Brooke, *Science and Religion*, 377.

124. Gould, *The Structure of Evolutionary Theory*, 188.

125. For a good survey of evangelical beliefs pre-Darwin, see Roberts, *Evangelicals and Science*, 33–112.

Conclusion: Re-Creating Creationism—Scopes and Beyond

Creationist responses to Darwinism were vast, and it is in fact difficult to speak of "creationism" as a single tradition without taking into account a myriad of nuances, even down to the geographical situations in which they arose.[126] Where design arguments were in fact overturned by Darwinism, this represented a small subsection of Christian tradition, and one that was overcome more often than not by theistic interpretations of Darwinism or evolutionary theory that were working on alternative ways to conceptualize God's providence. "Darwin's blow to the evidence of design, widely regarded nowadays as the coup de grace to natural theology, actually provoked an explosion of theistic literature." In fact, such was this explosion that "we might call the period from the 1880s to 1910 the 'heyday of theism.'"[127] Moreover, as was mentioned in chapter 4, David Livingstone notes how varied responses were even among Calvinists, depending upon their locations in Edinburgh, Belfast, or elsewhere.[128] And the relevant contexts only multiply from there. Just as design arguments in Victorian Britain, for example, were geared to uphold the particular political and economic situation while simultaneously defending against what was perceived as French radicalism and materialism and cannot be understood outside that context,[129] so too later fundamentalist reworkings of the arguments from design received a curious ally in big business, which found the reduction of the ability for individual choices to shape the world in Darwinism uncongenially for their business models. Evangelicals "found an ideological haven in an alternate science of humanity that relied on a cluster of ideas orbiting business."[130] This in turn fueled fundamentalism as an ostensibly non-denominational, parachurch alliance, and reinforced ideas like the design argument that allowed proponents to circumvent theological argumentation that relied upon more robust, confessionally driven theology.[131]

Ultimately the Scopes trial, while in one sense an event in the Christian engagement with science, must be appropriately qualified as distinct from what came before and after it as an event wholly "contingent on a complex interplay of timing, nation, and culture."[132] And yet, again, these necessary qualifications and distinctions were rewritten to create a continuous history of conflict—and this from multiple sides like

126. See: McGrath, *Nature*, 81–134.

127. Gundlach, *Process and Providence*, 171; J. Roberts, *Darwinism and the Divine*, 120–121; Brooke, *Science and Religion*, 261–306.

128. Livingstone, *Dealing With Darwin*.

129. F. Turner, *Contesting Cultural Authority*, 101–30.

130. Gloege, *Guaranteed Pure*, 3.

131. This has not stopped such non-denominational theology from trying to appropriate past theologies as on their side. Arguably, however, this often severely skews the meaning and intent of past theology and scriptural readings. See Allert, *Early Christian Readings of Genesis One*, 51–160 for examples.

132. Moran, *American Genesis*, 3.

Allen's *Only Yesterday*, where he made "Bryan's personal humiliation at Dayton into a decisive defeat for Fundamentalism as a whole." Indeed, it was a defeat for "old time religion" as a whole.[133] From the fundamentalist vantage point, "because of their unprecedented success in pushing evangelicals to choose between young- and old-Earth histories, after the early 1960's the flood geologists increasingly dictated the terms of debate over origins."[134] This campaign was so vocal, so recognizable that "many, ignorant of history, equated it with traditional biblical creationism."[135] As we have seen, this was so successful that until recently it has been completely forgotten by nearly everyone that "theology provided epistemic aid to evolution by its use of unbroken law and perhaps also shaped the content of evolution by endorsing a naturalized means of biological change."[136] In addition, "the success of these efforts," has also led us to forget that another "discontinuity with the past" was that "until the 1960s the use of the Bible to create scientific views, so-called Bible-only or Bible-dependent science, was a practice largely rejected by a large number of evangelicals," who opted for a broader and more circumspect use of natural theological and philosophical argumentation that nested scripture as a piece of broader argument, rather than the direct, unmediated scripture-to-world argumentation that fundamentalism has made so familiar of late.[137] As Numbers notes, the rise of the title "creation science" signaled a "major tactical shift among six-day creationists" in that they now moved to a "teach the controversy" strategy, whereby both scientific creationism and evolution were held up side by side as equally capable of either scientific confirmation or repudiation.[138] Indeed by the last decades of the twentieth century the richness of Christian theology and its diverse opinions of an even more diverse array of scientific practices seemed to shrivel in the popular consciousness as Young-Earth Creationists "virtually co-opted the creationist label," in the public mind.[139]

On the other hand, in the 1959 centenary of Darwin, as we have mentioned in chapter 4, an explicitly materialist vision of Darwinism was retroactively christened as the true and proper legacy of Darwin. And this despite the fact, as Robert J. Richards puts it, "Darwin's theory both before and after Malthus remained a theory of evolutionary progress."[140] Richards notes that it is only through the historiographical decisions of Darwin's followers that he becomes a "neo-Darwinian" in terms of rejecting any notion of progression.[141] As another Darwin historian puts it, "It might be queried

133. Larson, *Summer*, 226–27.

134. Numbers, *Darwinism Comes to America*, 130.

135. Numbers, *Darwinism Comes to America,* 109.

136. Dilley, "Darwin's Use of Theology," 7.

137. Rios, *After the Monkey Trial*, 115.

138. Numbers, *The Creationists*, 269.

139. Numbers, *Darwinism Comes to America,* 57.

140. Richards, *The Meaning of Evolution*, 114.

141. Richards, *The Meaning of Evolution*, 136 and 143.

whether the absolute prohibition on the concession to any idea of progress is not itself a product of materialist ideology."[142]

Perhaps nothing in the "cycle of self-justification that permeates accounts of science and religion"[143] has affected the legacy of the Scopes trial in terms of religion vs. science than the production of *Inherit the Wind*. Initially performed on Broadway for three years over 806 performances starting in 1955, it was ultimately translated into twenty languages and became a film in 1960 directed by Stanley Kramer. Written in the era of McCarthyism, *Inherit the Wind* was never meant as journalism, but rather as a (quite literal) theatrical replication of the Scopes trial in order to vocalize the dangers of authoritarianism and the repression of free speech so rampant at the time. The whole trial became set up like a sinister witch hunt. Bryan is represented as a yokel who damns all science as "Godless." And in general, the whole atmosphere is one that is threatening to tilt into religious totalitarianism if not stopped. Despite its rather generous license with history, *Inherit the Wind* has remained remarkably resilient in terms of being seen as an accurate report of the trial. Indeed, in the *Corpus for American English* detailing the use of the term "Scopes' Trial" *Inherit the Wind* accounts in total for *26 percent* of all references to the trial.[144] When this is taken into consideration along with all the other various competing memories of the trial that have their war banners held high, it is little wonder that the warfare thesis has dug itself so deep. In the minds of many, it was literally onstage before the eyes of America, perhaps even the world.[145] It is for this and other reasons that "despite a shelf of scholarly studies on Fundamentalism, antievolutionism, and Bryanism, the Scopes trial remains a grotesquely misunderstood event—largely the result, I think, of its ability to serve so many competing interests."[146]

142. Bowler, *Monkey Trials*, 222.

143. Shapiro, *Trying Biology*, 4.

144. Barczewska, "Scopes," 270.

145. Larson, *Summer*, 242–44. Larson remarks "both the play and the movie proved remarkably durable . . . despite the critics." Curiously, "Ever since *Inherit the Wind* appeared, conservative Christians have displayed a greater interest in countering the popular impression created by it [rather than] by the trial" itself (245). Though well intentioned, this no doubt reinforced, rather than actually debunked, the effect of *Inherit the Wind*. The myth, as such, lives on.

146. Numbers, *Darwinism Comes to America*, 76.

—Conclusion—

QUIETED BY ANTICIPATION, THE only sound ticking over the room was a staccato of chalk striking blackboard. A room full of eyes fixated on the writer through pipe smoke and tweedy suits. Nature herself was apparently eager to witness this demonstration, as the winter air pressed through iron grilles and thin glass, filling the moments between chalk ticks with an almost frozen stillness. Angular, rune-like things scrawled and scratched into white-stark life on the board's open black face, line after line. There were few alive who could read the mathematics that was undoubtedly, to its author, a poem; as it went, most were now present in the room. After a short while, the ticking stopped, and the lattice of heavenly things hung there now for all to see. Rustling his mustache, a typically unkempt Einstein—whose hair sprung like a cloud over his own tweed suit—took a step back to look at his creation.

It was about a decade now, after what was later called his "year of miracles" in 1905, in which he had written and published five epoch-making papers on physics. He was currently in Göttingen among friends and fellow physicists, having just completed his theory of general relativity. But there was a problem, and a question just asked by a colleague seemed to bite through the brimming excitement like the November cold. As beautiful as these rune-like things were, whatever their glory—could they be true? Did they scratch into the fabric of reality as well as the midnight of the black board?

The mathematics were sound, yes, but empirical evidence and experiments were lagging behind. Yet Einstein was unperturbed. So beautiful to him was this poem of the heavenly spheres, the lithe symbols flowing to and fro on the board may as well have been a chorus of angels cantillating revelations to which not even Moses was privy. He looked thoughtful for a moment at the question, then somewhat stern. What would happen if such a beautiful theory were false? "I would feel sorry for the Good Lord," said Einstein, finally. Though its hierophanic language escapes most of us, to those who understood it, such an equation was too beautiful to be untrue.

Luckily the Lord apparently had other plans. Soon after arriving at Cambridge, another physicist named Arthur Stanley Eddington (previously a secretary at the Royal Astronomical Society) received Einstein's 1915 paper on relativity. As it happened, this itself was no small feat, for to reach Eddington's hands it had to be smuggled across Berlin through the entrenched battlefronts of World War I. In these papers

Einstein's math predicted that space curved by gravity would bend incoming light. Eddington sought to prove this by observation. As Larry Witham remarks, "this made him the Indiana Jones of physics."[1] In 1919, Eddington became a visage of Harrison Ford's later character, decked head to foot in typical British safari gear as he led an expedition of assistants to Principe, West Africa. There he and his team would observe in the wilderness a solar eclipse, allowing them to view the trajectories of passing starlight. The moment came; the light, bent. Einstein's poem was indeed a tune the universe sang, and good thing too: as the rumors of violent storms increased and the stakes rose, one of the assistants asked a senior scientist regarding the consequences of failure in Principe. "Then Eddington will go mad," remarked his colleague dryly, "and we will have to come home."[2]

Not often mentioned, Eddington's famous expedition was made possible because, as a Quaker, Eddington gained exemption from the war as a conscientious objector on religious grounds. Moreover, his religious background also made him a firm believer in the unity and ecumenicism of science, which crossed international borders and rivalries, and made possible the tricky passage of Einstein's papers and the eventual trip to Principe. The experimental confirmation at Principe was so electric that it in many ways it "helped reunite a world torn apart by war."[3]

Our work here is done to achieve the same end, though in its own small way. For in the world of religion and science we are still living amidst a world torn by the warfare thesis. While deconstructing many of the historical misunderstands that have gone into the thesis and continue to linger in our consciousness does not solve all our problems, or prove Christianity true, or that God is real, it does help us precisely by clearing the decks. While the rhetoric of "everything you know is wrong" can be quite trying when used as marketing, often in the areas of the history of science and religion one does truly wish we could all start over. If nothing else the story of this book has been told to let us sit more comfortably in complexity; to spark a sudden curiosity that might make us travel back to the sources, or to look again at what we thought we knew. For the warfare thesis often came about because so many thought they could boldly stride forward without taking a true inventory of that which lay around them. Perhaps it is time now to reevaluate the landscape, in the hopes that we might begin again.

Sadly, however, the notion of warfare still connected to popular science communication has has not gone away. If anything, it has increased. As Karl Giberson and Mariano Artigas record in their book *Oracles of Science: Celebrity Scientists versus God and Religion*, one is completely overwhelmed by the total number of books sold by their main subjects.[4] Between Richard Dawkins, Stephen Jay Gould, Stephen Hawk-

---

1. Witham, *The Measure of God*, 121.
2. Witham, *The Measure of God*, 122.
3. Ellis et al., "90 Years On," 13.
4. Giberson and Artigas, *Oracles of Science*.

ing, Carl Sagan, Steven Weinberg, and Edward O. Wilson, the number of books sold literally swells into the billions (that is indeed billion, with a "b"). We can pause here for a moment as we collectively sit back in our chairs to let that sink in. Hawking by himself, with his *A Brief History of Time* had by Giberson and Artigas's publication date sold one copy for every 750 people on earth. That was 2007. Hawking has since sadly passed on, no doubt spurring this number to ever new heights. These are the "oracles of science" for the public. Even where warfare is not explicit (in the genial Gould, for example) the contorted and typical notions of the Scientific and Darwinian Revolutions as a separation and overcoming of religion are still present. For Gould, science and religion literally have nothing to say to one another. In broader terms, a 1986 study by Paul Vitz, a professor at New York University, indicated that statistically the vast majority of elementary and high school textbooks went to extraordinary lengths to avoid reference to religion. Pilgrims were defined as "people who take long trips" and fundamentalists as those who "follow the values or traditions of an earlier period." Of 300 important events in America listed in one textbook, three had even oblique reference to religion. A world history textbook left out the Protestant Reformation. The fate of religion appears to be warfare or deletion; often both simultaneously.[5]

And we do not need to stop with these influential figures. One could point to the history of journalism, for example. Christian contributions to the advancing technology and techniques of public journalism in the United States, to beat our dead horse, "has largely been marginalized, to the extent that it has been addressed at all." But more than this, the history of journalism in the states, whether conservative or liberal, can also be told as one where "key persons in journalism," influenced at least in part by the new standards of science put forward by the X-Club, "especially publishers and editors, and also journalism professionalizers from the ranks of the universities and the active press," all "actively sought to minimize and ultimately undermine traditional religion."[6] Not only this, but taking a key from the X-Club, Comte, and others, journalism in these early stages actively presented itself as "the ideal successor to religion, because it alone could provide the appropriate guidance for both individuals and society."[7]

Unfortunately, especially after the tragic terrorist attacks of 9/11 and with the rise of the so-called "New Atheism," the warfare thesis found a new audience whose readiness was white hot and forged in the ashes and melted steel of the twin towers. As we just saw, the elements that had aided its creation and reception never left, anyway. In many ways this environment that once again primed reception of the warfare narrative was very similar to the blood-soaked atmosphere of the post-Civil War context priming the initial reception of Draper and White. Frenzy again filled the air, and freely lent its energies to strife of all kinds. Such was the energy of this new situation

---

5. This summary of Vitz's study is taken from Olasky and Smith, *The Prodigal Press*, 3.

6. Flory, "Promoting a Secular Standard," 397.

7. Flory, "Promoting a Secular Standard," 413.

that many could use it to leap over all the historiographical gains of the twentieth century, and land squarely back onto the pages of Draper and White's work. The late physicist Victor Stenger, for example, dismissed recent historical scholarship overturning White and Draper's stories with a wave of the hand, noting "historians have not been particularly careful or accurate in their criticisms of Draper and White."[8] The case, as we have seen, is actually quite the opposite. But, as we have also seen, most people have not let this stop them. At any rate, he continues, the history of their relations matters little when placed in the shadow of the contemporary shenanigans of American fundamentalist Christianity.

The evolutionary biologist Jerry Coyne, in his attempt to provide a winsome corrective to some of his fellow New Atheists, concedes that in fact Draper and White "did make some errors and omit some countervailing observations." But these, he says, hardly amount to anything approaching an invalidation of the central thesis of either book: that religion is a poor man's science, which has rightfully been sent off with its tail between its legs.[9] Indeed, it is humorous to note that one of the examples of the intrinsic absurdity of faith that Coyne uses is Tertullian's paradox. As does philosophy professor and atheist proselytizer Peter Boghossian.[10] Strangely enough, Coyne cites the original Tertullian quote correctly, but then glosses it with an explanation that depends entirely upon the fabricated version by Voltaire.[11] Coyne's book had the particular misfortune of coming out the same year as Peter Harrison's *The Territories of Science and Religion*, which is already considered something of a modern classic debunking the historical notion of warfare and in general a methodological *tour de force* in the realm of the history of science. Coyne's imprecise use of history (to put it politely) is placed into a stunning contrast by this coincidence. But undoing the contortions we have recorded is a major uphill battle. Moreover, even within Christianity, theology as a discipline is in a state of woeful disrepair and disrepute among the populace. The deeply humanistic, multidisciplinary mastery traditionally required of the theologian is not only difficult to find these days, but often actively chastised by the American believer as of a piece with the deceptions of "experts" and the contrivances of the world's learned. Yet, with brilliant and accessible works like Tom Holland's *Dominion*, Alan Jacob's *The Year of Our Lord 1943*, or Seb Falk's *The Light Ages*, all hope is not lost. Instead of hard-nosed apologetics (which will always have its place), histories that simply have Christian theology and practice restored for historical retelling in all its majesty and clumsiness can, I think, be a simple balm that aids in healing the long neglected wounds of a war that never happened.

Our story, to reiterate, is not that conflict never existed. It did, and in abundance. Even on occasion between what we might call "scientific evidence" and what we might

8. Stenger, *God and the Folly of Faith*, 33.

9. Coyne, *Faith vs. Fact*, 4–5.

10. Boghossian, *A Manual for Creating Atheists*, 34–35.

11. Coyne, *Faith vs. Fact*, 68.

call "religious" or "theological" reasoning. However, these were always local conflicts, and to describe them in terms of "science vs. religion" not only does a major disservice to the complexity of what was actually occurring, it also deeply ignores the deep interpenetration (or perhaps even nonexistence) of our supposedly clear-cut categories in the historical episodes in question. The strange tale of how the conflict of science and Christianity was written into history has sunk so deeply, though, that one wonders if it can ever be expunged. What is worse, because of humanity's ability to shape themselves to the categories they use, the warfare thesis has begun to incarnate itself into reality. Many in the twentieth century—whatever their allegiance—wave these battle flags, and in some sense a fiction becomes truth. Of course, in reality, things still are never so neat and simple. Part of the solution is to revisit some of the episodes we thought we knew. We have of course already done this. The question is not truly whether these episodes embody the warfare of science and religion—they do not. The real question is, once these pet examples have been exploded, where to then? The true next step will not be an apologetic cascade of works, but rigorous intellectual and practical history that proceeds to tell the stories of the rise of the world as we know it with the complex roles of Christianity firmly put back into their proper places as part of the grand adventure of human knowledge and practice. Easier said than done, but adventures are never supposed to be simple, or easy.

—Bibliography—

Abrams, M. H. *Natural Supernaturalism: Tradition and Revolution in Romantic Literature.* New York: W. W. Norton, 1971.

Aechtner, Thomas. "Galileo Still Goes to Jail: Conflict Model Persistence Within Introductory Anthropological Materials." *Zygon* 50 no. 1 (2015) 209–26.

Agamben, Giorgio. *The Omnibus Homo Sacer.* Stanford: Stanford University Press, 2017.

Agassi, Joseph. "Kuhn's Way." *Philosophy of the Social Sciences* 32 no. 3 (2002) 394–430.

Allen, Frederick Lewis. *Only Yesterday: An Informal History of the 1920's.* New York: Wiley, 1997.

Allert, Craig D. *Early Christian Readings of Genesis One: Patristic Exegesis and Literal Interpretation.* Downers Grove, IL: InterVarsity, 2018.

Altschuler, Glenn C. *Andrew D. White: Educator, Historian, Diplomat.* Ithaca, NY: Cornell University Press, 1979.

Appleby, Joyce, Lynn Hunt, and Margaret Jacob. *Telling the Truth About History.* New York: W. W. Norton, 1995.

Aquinas, Thomas. *The Summa Theologia in a Single Volume.* Claremont, CA: Coyote Canyon, 2018.

Ariew, Robert. *Descartes Among the Scholastics.* Leiden: Brill, 2011.

Ariew, Robert, and Peter Barker. "Duhem and Continuity in the History of Science." *Revue Internationale de Philosophie* 46 (1992) 323–43.

Aristotle. *De Caelo.* Translated by C. D. C. Reeve. Indianapolis: Hackett, 2020.

———. *The Metaphysics.* Translated by John H. McMahon. New York: Dover, 2018.

———. *Politics.* Translated by Carnes Lord. Chicago: University of Chicago Press, 2013.

Arsdall, Anne Van. "Rehabilitating Medieval Medicine." In *Misconceptions About the Middle Ages,* edited by Stephen J. Harris and Bryon L. Grigsby, 135–41. New York: Routledge, 2008.

Artigas, Mariano, et al. *Negotiating Darwin: The Vatican Confronts Evolution, 1877–1902.* Baltimore: Johns Hopkins University Press, 2006.

Asad, Talal. "Reading A Modern Classic: W. C. Smith's *The Meaning and End of Religion.*" *History of Religions* 40 (2001) 205–22.

Asúa, Miguel de. "The 'Conflict Thesis' and the Positivist History of Science: A View from the Periphery." *Zygon* 53 no. 4 (November 2018) 1131–48.

Auden, W. H. "Heresies—A Review of *Pagans and Christians in an Age of Anxiety* by E. R. Dodds." *The New York Review of Books,* February 17, 1966. https://www.nybooks.com/articles/1966/02/17/heresies/.

Augustine. *On Genesis: The Works of St. Augustine.* New York: New City, 2002.

Bacon, Francis. *The New Organon.* Edited by Lisa Jardine and Michael Silverthorne. Cambridge: Cambridge University Press, 2008.

Backus, Irena. *Leibniz: Protestant Theologian*. Oxford: Oxford University Press, 2016.

Bagnall, Roger S. "Alexandria: Library of Dreams." *Proceedings of the American Philosophical Society* 146 no. 4 (December 2002) 348–62.

Bakke, O. M. *When Children Became People: The Birth of Childhood in Early Christianity*. Minneapolis: Fortress, 2015.

Boethius, Ancius. *The Consolations of Philosophy*. Translated by Victor Watts. New York: Penguin, 1999.

Balthasar, Hans Urs Von. *The Theology of Karl Barth*. San Francisco: Ignatius, 1992.

Bannister, Robert C. *Sociology and Scientism: The American Quest for Objectivity, 1880–1940*. Chapel Hill, NC: University of North Carolina Press, 1987.

Barbour, Ian. *Myths, Models, and Paradigms*. New York: HarperCollins, 1974.

Barczewska, Shala. "The 1925 Scopes Trial as Discursive Event: Does Reference to the 1925 Trial Affect Our View of Teachers in the Current Debate Over Evolution?" *Token: A Journal of English Linguistics* 2 (2013) 265–87.

Barker, Peter. "The Role of Religion in Lutheran Responses to Copernicus." In *Rethinking the Scientific Revolution*, edited by Margaret J. Osler, 59–89. Cambridge: Cambridge University Press, 2000.

Baring, Edward. *Converts to the Real: Catholicism and the Making of Continental Philosophy*. Cambridge, MA: Harvard University Press, 2019.

Barnett, S. J. *The Enlightenment and Religion: The Myths of Modernity*. Manchester, UK: Manchester University, 2003.

Barr, James. "The Ecological Controversy and the Old Testament." *Bulletin of the John Rylands Library* 55 (1972) 9–32.

Barth, Karl. *Protestant Theology in the Nineteenth Century: Its Background and History*. Translated by Brian Cozens and John Bowden. Grand Rapids: Eerdmans, 2001.

Bartlett, Robert. *The Natural and the Supernatural in the Middle Ages*. Cambridge: Cambridge University Press, 2008.

Barton, Ruth. "Evolution: The Whitworth Gun in Huxley's War for the Liberation of Science from Theology." In *The Wider Domain of Evolutionary Thought*, edited by David Oldroyd and Ian Langham, 261–87. Dordrecht: Reidel, 1983.

———. "Huxley, Lubbock, and a Half Dozen Others: Professionals and Gentlemen in the Formation of the X-Club, 1851–1864." *Isis* 89 no. 3 (September 1998) 410–44.

———. "'An Influential Set of Chaps': The X-Club and Royal Society Politics, 1864–1885." *British Journal for the History of Science* 23 (1990) 53–81.

———. "John Tyndall, Pantheist: A Re-reading of the Belfast Address." *Osiris* 3 (1987) 111–34.

———. *The X-Club: Power and Authority in Victorian Science*. Chicago: University of Chicago Press, 2018.

Becker, Carl L. *The Heavenly City of Eighteenth Century Philosophers*. New Haven, CT: Yale University Press, 2003.

Bede. *On the Nature of Things and On Times*. Translated by Calvin B. Kendall. Liverpool: Liverpool University Press, 2011.

Behe, Michael J. *Darwin's Black Box: The Biochemical Challenge to Evolution*. New York: Free, 1996.

———. *A Mousetrap for Darwin: Michael J. Behe Answers His Critics*. Seattle: Discovery Institute, 2020.

Beiser, Friedrich. *The Fate of Reason: German Philosophy from Kant to Fichte.* Cambridge, MA: Harvard University Press, 1987.

Bellarmine, Robert. "Cardinal R. Bellarmine to P. Foscarini, 12 April 1615." In *The Galileo Affair: A Documentary History,* edited by Maurice Finocchiaro, 67–9. Oakland, CA: University of California Press, 1989.

Biddick, Kathleen. *The Shock of Medievalism.* Durham, NC: Duke University Press, 1998.

Biagioli, Mario. *Galileo Courtier: The Practice of Science in the Culture of Absolutism.* Chicago: University of Chicago Press, 1993.

Bidna, David B. *We The People: A History of the United States.* London: Heath, 1977.

Bildhauer, Bettina, and Robert Mills, eds. *The Monstrous Middle Ages.* Toronto: University of Toronto Press, 2003.

Black, Winston. *The Middle Ages: Facts and Fictions.* Santa Barbara, CA: ABS-Clio, 2019.

Blackburn, Simon. *The Oxford Dictionary of Philosophy.* Oxford: Oxford University Press, 1996.

Blackwell, Richard. "Galileo Galilei." In *Science & Religion: A Historical Introduction,* edited by Gary Ferngren, 105–17. Baltimore: Johns Hopkins University Press, 2002.

Blakely, Jason. *Alasdair MacIntyre, Charles Taylor, and the Demise of Naturalism.* Notre Dame, IN: University of Notre Dame Press, 2016.

Blumenberg, Hans. *The Legitimacy of the Modern Age.* Cambridge, MA: MIT Press, 1985.

Boer, Roland, and Christina Petterson. *Idols of Nations: Biblical Myth at the Origins of Capitalism.* Minneapolis: Fortress, 2014.

Boghossian, Peter. *A Manual for Creating Atheists.* Chapel Hill, NC: Pitchstone, 2013.

Boorstin, Daniel. *The Discoverers: A History of Man's Search to Know His World and Himself.* New York: Vintage, 1985.

Bordoni, Stefano. *When Historiography Met Epistemology: Sophisticated Histories and Philosophies of Science in French-Speaking Countries in the Second Half of the Nineteenth Century.* Leiden: Brill, 2017.

Bowersock, Glen W. *Fiction into History: Nero to Julian.* Berkeley, CA: University of California Press, 1994.

———. *Julian the Apostate.* Cambridge, MA: Harvard University Press, 1997.

———. "The Vanishing Paradigm of the Fall of Rome." *Bulletin of the American Academy of Arts and Sciences* 49 (May 1996) 29–43.

Bowler, Peter J. *The Eclipse of Darwinism: Anti-Darwinian Evolutionary Theories in the Decades Around 1900.* Baltimore: Johns Hopkins University Press, 1983.

———. *Evolution: The History of an Idea.* 3rd ed. Berkeley, CA: University of California Press, 2003.

———. *Monkey Trials & Gorilla Sermons: Evolution and Christianity from Darwin to Intelligent Design.* Cambridge, MA: Harvard University Press, 2007.

———. *The Non-Darwinian Revolution: Reinterpreting a Historical Myth.* Baltimore: Johns Hopkins University Press, 1988.

———. *Reconciling Science and Religion: The Debate in Early Twentieth-Century Britain.* Chicago: University of Chicago Press, 2001.

———. "Revisiting the Eclipse of Darwinism." *Journal of the History of Biology* 38 (2005) 19–32.

Bozeman, Theodore Dwight. *Protestants in an Age of Science: The Baconian Ideal and Antebellum American Religious Thought.* Chapel Hill, NC: University of North Carolina Press, 1977.

Brague, Remi. *On the God of the Christians (And on One or Two Others).* Notre Dame, IN: St. Augustine's, 2013.

———. *The Kingdom of Man: The Genesis and Failure of a Modern Project.* Notre Dame, IN: University of Notre Dame Press, 2018.

———. *The Law of God: The Philosophical History of an Idea.* Chicago: University of Chicago Press, 2007.

———. *The Legend of the Middle Ages: Philosophical Explorations of Medieval Christianity, Judaism, and Islam.* Chicago: University of Chicago Press, 2011.

———. *The Wisdom of the World: The Human Experience of the Universe in Western Thought.* Chicago: University of Chicago Press, 2004.

Bregman, Jay. *Synesius of Cyrene: Philosopher-Bishop.* Berkeley, CA: University of California Press, 1982.

Bremmer, Jan N. *The Rise of Christianity Through the Eyes of Gibbon, Harnack, and Rodney Stark.* San Bernardino, CA: Barkhuis Groningen, 2010.

Brewer, Daniel. *The Enlightenment Past: Reconstructing Eighteenth-Century French Thought.* Cambridge: Cambridge University Press, 2008.

Broers, Michael. *The Politics of Religion in Napoleonic Italy: The War Against God 1801–1814.* London: Routledge, 2002.

Brooke, John Hedley. "The Relationship Between Darwin's Science and His Religion." In *Darwinism and Divinity,* edited by John Durant, 40–75. New York: Oxford University Press, 1985.

———. "Religious Belief and the Content of the Sciences." *Osiris* 16 (2001) 3–28.

———. *Science and Religion: Some Historical Perspectives.* Cambridge: Cambridge University Press, 1991.

———. "The Wilberforce-Huxley Debate: Why Did It Happen?" *Science & Christian Belief* 13 no. 2 (2001) 127–41.

Brooke, John Hedley, and Geoffrey Cantor. *Reconstructing Nature: The Engagement of Science and Religion.* Oxford: Oxford University Press, 1998.

Brown, Andrew. *The Days of Creation: A History of Christian Interpretation of Genesis 1:1–2:3.* Dorset: Deo, 2014.

Brown, Nancy Marie. *The Abacus and the Cross: The Story of the Pope Who Brought the Light of Science to the Dark Ages.* New York: Basic, 2010.

Brown, Peter. *The Rise of Western Christendom: Triumph and Diversity—AD 200–1000.* Oxford: Wiley-Blackwell, 2013.

———. *Society and the Holy in Late Antiquity.* Berkeley, CA: University of California Press, 1982.

———. *The World of Late Antiquity: AD 150–750.* New York: W. W. Norton, 1989.

Brown, William P. *The Seven Pillars of Creation: The Bible, Creation, and the Ecology of Creation.* Oxford: Oxford University Press, 2010.

Buckley, Michael. *Denying and Disclosing God: The Ambiguous Progress of Modern Atheism.* New Haven, CT: Yale University Press, 2004.

———. *At the Origins of Modern Atheism.* New Haven, CT: Yale University Press, 1987.

———. "The Newtonian Settlement and the Origins of Atheism." In *Physics, Philosophy, and Theology: A Common Quest for Understanding,* edited by Robert J. Russell et al., 81–102. Notre Dame, IN: University of Notre Dame Press, 1988.

———. "The Study of Religion and the Rise of Atheism: Conflict or Confirmation?" In *Fields of Faith: Theology and Religious Studies for the Twenty-First Century,* edited by David F. Ford et al., 3–24. Cambridge: Cambridge University Press, 2005.

Bullen, Barrie. "Walter Pater's 'Renaissance' and Leonardo Da Vinci's Reputation in the Nineteenth Century." *The Modern Language Review* 74 no. 2 (1979) 268–80.

Bulman, William J., and Robert G. Ingram, eds. *God in the Enlightenment.* Oxford: Oxford University Press, 2016.

Burckhardt, Jacob. *The Civilization of the Renaissance in Italy: An Essay.* New York: Penguin, 1990.

Burtt, E. A. *The Metaphysical Foundations of Modern Science: The Scientific Thinking of Copernicus, Galileo, Newton, And Their Contemporaries.* Atlantic Highlands: Humanities, 1952.

Butterfield, Herbert. *Christianity and History.* New York: Charles Scribner's Sons, 1949.

———. *The Origins of Modern Science.* New York: Free, 1957.

———. *The Whig Interpretation of History.* New York: W. W. Norton, 1965.

Bynum, Caroline Walker. *The Resurrection of the Body in Western Christianity 200–1336.* New York: Columbia University Press, 1996.

Cahan, David, ed. *From Natural Philosophy to the Sciences: Writing the History of Nineteenth-Century Science.* Chicago: University of Chicago Press, 2003.

Calder, W. M., III. "The Spherical Earth in Plato's *Phaedo.*" *Phronesis* 3 (1958) 121–25.

Caldwell, Bruce. *Beyond Positivism: Economic Methodology in the Twentieth Century.* New York: Routledge, 1994.

Callaway, Kutter, and Barry Taylor. *The Aesthetics of Atheism: Theology and Imagination in Contemporary Culture.* Minneapolis: Fortress, 2019.

Cameron, Averil. "Blame the Christians: A Review of Catherine Nixey, *A Darkening Age.*" *Christian Theology* (September 2017). https://davidtinikashvili.wordpress.com/2017/11/29/cameron-blame-the-christians-review/.

———. *Byzantine Matters.* Princeton, NJ: Princeton University Press, 2014.

Cantor, Georg. "What Shall We Do with the 'Conflict Thesis'?" In *Science and Religion: New Historical Perspectives,* edited by Thomas Dixon, Geoffrey Cantor, and Stephen Pumfrey, 282–98. Cambridge: Cambridge University Press, 2010.

Cantor, Norman F. *Inventing the Middle Ages: The Lives, Works, and Ideas of the Great Medievalists of the Twentieth Century.* New York: Quill, 1991.

Cantor, Geoffrey, and Sally Shuttleworth, eds. *Science Serialized: Representation of the Sciences in Nineteenth-Century Periodicals.* Cambridge, MA: MIT Press, 2004.

Casanova, José. "Beyond European and American Exceptionalisms." In *Predicting Religion: Christian, Secular, and Alternative Futures,* edited by Grace Davie et al., 17–29. Aldershot: Ashgate, 2003.

Cashdollar, Charles D. *The Transformation of Theology 1830–1890: Positivism and Protestant Thought in Britain and America.* Princeton, NJ: Princeton University Press, 2014.

Casson, Lionel. *Libraries in the Ancient World.* New Haven, CT: Yale University Press, 2002.

Cassirer, Ernst. *The Individual and the Cosmos in Renaissance Philosophy.* Philadelphia: University of Pennsylvania Press, 1972.

———. *The Philosophy of the Enlightenment.* Princeton, NJ: Princeton University Press, 1968.

Cavanaugh, William. *The Myth of Religious Violence: Secular Ideology and the Roots of Modern Conflict.* Oxford: Oxford University Press, 2009.

Cavanaugh, William, and James K. A. Smith, eds. *Evolution and The Fall*. Grand Rapids: Eerdmans, 2017.

Chadwick, Owen. *Catholicism and History: The Opening of the Vatican Archives*. Cambridge: Cambridge University Press, 2009.

———. *The Secularization of the European Mind in the 19th Century*. Cambridge: Cambridge University Press, 1975.

Chapman, Alister, et al. *Seeing Things Their Way: Intellectual History and the Return of Religion*. Notre Dame, IN: University of Notre Dame Press, 2009.

Chapman, Allan. *Stargazers: Copernicus, Galileo, the Telescope, and the Church—The Astronomical Renaissance 1500–1700*. Oxford: Lion, 2014.

Chapp, Larry S. *God of Covenant and Creation: Scientific Naturalism and Its Challenge to the Christian Faith*. Edinburgh: T&T Clark, 2011.

Chesterton, G. K. *The Everlasting Man*. San Francisco: Ignatius, 1996.

Cicero. *The Republic and the Laws*. Translated by Nial Rudd. Oxford: Oxford University Press, 2009.

Clark, Elizabeth A. *History, Theory, Text: Historians and the Linguistic Turn*. Cambridge, MA: Harvard University Press, 2004.

Clark, J. C. D. "Secularization and Modernization: The Failure of a 'Grand Narrative.'" *The Historical Journal* 55 no. 2 (2012) 161–194.

Claggett, Marshall. "The Impact of Archimedes on Medieval Science." In *Archimedes in the Middle Ages: Volume I —The Arabo-Latin Tradition*, 1–14. Madison, WI: University of Wisconsin-Madison Press, 1964.

Clayton, Philip. *The Problem of God in Modern Thought*. Grand Rapids: Eerdmans, 1998.

———. "The Religious Spinoza." In *The Persistence of the Sacred in Modern Thought*, edited by Chris Firestone and Nathan Jacobs, 66–86. Notre Dame: University of Notre Dame Press, 2012.

Clouser, Roy A. *The Myth of Religious Neutrality: An Essay on the Hidden Role of Religious Belief in Theories*. Notre Dame, IN: University of Notre Dame Press, 2005.

Coffey, John, and Alister Chapman. "Intellectual History and the Return of Religion." In *Seeing Things Their Way: Intellectual History and the Return of Religion*, edited by John Coffey, Alister Chapman, and Brad Gregory, 1–22. Notre Dame, IN: University of Notre Dame Press, 2009.

Cohen, H. Floris. *The Scientific Revolution: A Historiographical Inquiry*. Chicago: University of Chicago Press, 1994.

Cohen, I. Bernard, ed. *Puritanism and the Rise of Modern Science: The Merton Thesis*. New Brunswick, NJ: Rutgers University Press, 1990.

———. *Revolutions in Science*. Cambridge, MA: The Belknap Press of Harvard University Press, 1985.

Cohen, Jeffrey Jerome. "Monster Culture (Seven Theses)." In *Monster Theory: Reading Culture*, edited by Jeffrey Jerome Cohen, 3–25. Minneapolis: University of Minnesota Press, 1996.

Cohn, Norman. *Europe's Inner Demons: The Demonization of Christians in Medieval Christendom*. Rev. ed. Chicago: University of Chicago Press, 1993.

Cole, Andrew, and D. Vance Smith, eds. *The Legitimacy of the Middle Ages: On the Unwritten History of Theory*. Durham, NC: Duke University Press, 2010.

Colish, Marcia L. *Medieval Foundations of the Western Intellectual Tradition 400–1400*. New Haven, CT: Yale University Press, 1997.

Colon, Ferdinando. *The Life of the Admiral Christopher Columbus by His Son Ferdinand.* New Brunswick, NJ: Rutgers University Press, 1959.

Corduan, Winfred. *In the Beginning God: A Fresh Look At The Case For Original Monotheism.* Nashville: B&H, 2013.

Cormack, Lesley B. "Flat Earth or Round Sphere: Misconceptions of the Shape of the Earth and the Fifteenth-Century Transformation of the World." *Ecumene* 1 (1994) 363–85.

———. "Myth 3: That Medieval Christians Taught the Earth Was Flat." In *Galileo Goes to Jail and Other Myths About Science and Religion*, edited by Ronald Numbers, 28–34. Cambridge, MA: Harvard University Press, 2007.

Coyne, Jerry. *Faith vs. Fact: Why Science and Religion Are Incompatible.* New York: Penguin, 2016.

Coyne, Ryan. *Heidegger's Confessions: The Remains of Saint Augustine in 'Being and Time' and Beyond.* Chicago: University of Chicago Press, 2016.

Critchley, Simon. *Continental Philosophy: A Very Short Introduction.* Oxford: Oxford University Press, 2001.

Croce, Benedetto. "We Cannot Help But Call Ourselves Christians." In *My Philosophy and Other Essays on the Moral and Political Problems of Our Times,* translated by E. F. Caritt, 37–47. London: George Allen & Unwin, 1949.

Crombie, A. C. *Medieval and Early Modern Science: Science in the Middle Ages V–XVII Centuries.* 2 volumes. New York: Doubleday Anchor, 1959.

———. "Review: The Appreciation of Ancient and Medieval Science During the Renaissance 1450–1600 by George Sarton." *The British Journal for the Philosophy of Science* X no. 38 (August 1959) 164–65.

———. *Robert Grosseteste and the Origins of Experimental Science 1100–1700.* Oxford: Oxford University Press, 1955.

Cunningham, Andrew. *The Anatomical Renaissance: The Resurrection of the Anatomical Projects of the Ancients.* New York: Routledge, 1997.

———. *Before Science: The Invention of the Friar's Natural Philosophy.* New York: Routledge, 1996.

———. "Getting the Game Right: Some Plain Words on the Identity and Invention of Science." *Studies in the History and Philosophy of Science* 19 no. 3 (September 1988) 365–89.

———. "How the *Principia* Got Its Name: Or, Taking Natural Philosophy Seriously." *History of Science* 29 no. 4 (1991) 377–92.

———. "The Identity of Natural Philosophy: A Response to Edward Grant." *Early Science and Medicine* 5 no. 3 (2000) 259–78.

Cunningham, Andrew, and Perry Williams. "De-Centring the 'Big Picture': The Origins of Modern Science and the Modern Origins of Science." *The British Journal for the History of Science* 26 no. 4 (1993) 407–32.

Cunningham, Conor. *Darwin's Pious Idea: Why Ultra-Darwinists and Creationists Both Get It Wrong.* Grand Rapids: Eerdmans, 2010.

Cusa, Nicolas. *On Learned Ignorance: A Translation of de docta ignorantia.* Translated by Jasper Hopkins. New York: Arthur Banning, 1985.

Cutler, Alan H. "Nicolas Steno and the Problem of Deep Time." In *The Revolution in Geology from the Renaissance to Enlightenment,* edited by Gary B. Rosenberg, 143–48. Denver: Geological Society of America, 2009.

Damascius. *The Philosophical History*. Translated by Polymnia Athanassiadi. New York: Apamea, 1999.

Daniels, George H. *American Science in the Age of Jackson*. Tuscaloosa, AL: University of Alabama Press, 1994.

Danielson, Dennis R. "The Great Copernican Cliché." *American Journal of Physics* 69 no. 10 (April 2001) 1029–35.

———. "Myth 6: That Copernicanism Demoted Humans from the Center of the Cosmos." In *Galileo Goes to Jail and Other Myths About Science and Religion*, edited by Ronald Numbers, 50–59. Chicago: University of Chicago Press, 2008.

Darrow, Clarence, and William Jennings Bryan. *The World's Most Famous Court Trial*. Clark, NJ: The Lawbook Exchange, 2010.

Darwin, Charles. *The Autobiography of Charles Darwin 1809–1882: With Original Omissions Restored*. New York: W. W. Norton, 1993.

Daston, Lorraine. "The Naturalistic Fallacy is Modern." *Isis* 105 no. 3 (September 2014) 579–87.

Davies, Paul. *The Goldilocks Enigma: Why Is the Universe Just Right for Life?* Boston: Mariner, 2006.

Davis, Edward B. "Christianity and Early Modern Science: The Foster Thesis Reconsidered." In *Evangelicals and Science in Historical Perspective*, edited by David Livingstone, D. G. Hart, and Mark A. Noll, 75–95. Oxford: Oxford University Press, 1999.

———. "Science and Religious Fundamentalism in the 1920s: Religious Pamphlets by Leading Scientists of the Scopes Era Provide Insight into Public Debates About Science and Religion." *American Scientist* 93 no. 3 (2005) 253–60.

Davis, Kathleen. *Periodization and Sovereignty: How Ideas of Feudalism & Secularization Govern the Politics of Time*. Philadelphia: University of Pennsylvania Press, 2008.

———. "The Sense of an Epoch: Periodization, Sovereignty, and the Limits of Secularization." In *The Legitimacy of the Middle Ages*, edited by Andrew Cole and D. Vance Smith, 39–69. Durham, NC: Duke University Press, 2010.

Dawson, Christopher. *Progress & Religion: A Historical Inquiry*. Washington, DC: Catholic University of America Press, 2001.

Dawson, Gowan. *Darwin, Literature, and Victorian Respectability*. Cambridge: Cambridge University Press, 2007.

Dawkins, Richard. *The Blind Watchmaker: How Evolution Reveals a World Without Design*. New York: W. W. Norton, 2015.

———. *The Devil's Chaplain: Reflections on Hope, Lies, Science and Love*. New York: Mariner, 2003.

Dear, Peter. *Discipline and Experience: The Mathematical Way in the Scientific Revolution*. Chicago: University of Chicago Press, 1995.

———. "Historiography of Not-So-Recent Science." *History of Science* 50 no. 2 (April 2012) 197–211.

de Certeau, Michel. *The Writing of History*. New York: Columbia University Press, 1988.

Decker, Rainer. *Witchcraft & the Papacy: An Account Drawing on the Formerly Secret Records of the Roman Inquisition*. Translated by H. C. Erik Midelfort. Charlottesville, VA: University of Virginia Press, 2009.

Delia, Diana. "From Romance to Rhetoric: The Alexandrian Library in the Classical and Alexandrian Traditions." *The American Historical Review* 97 no. 5 (December 1992) 1449–67.

De Lubac, Henri. *The Drama of Atheist Humanism*. San Francisco: Ignatius, 1983.

———. *The Mystery of the Supernatural*. New York: Herder & Herder, 1998.

Dembski, William A. "The Design Argument." In *Science & Religion: A Historical Introduction*, edited by Gary B. Ferngren, 335–44. Baltimore: Johns Hopkins University Press, 2002.

Dennett, Daniel C. *Darwin's Dangerous Idea: Evolution and the Meaning of Life*. New York: Simon & Schuster, 1995.

Depew, David J., and Bruce H. Weber. *Darwinism Evolving: Systems Dynamics and the Genealogy of Natural Selection*. Cambridge, MA: MIT Press, 1996.

Desmond, Adrian. *Huxley: From Devil's Disciple to Evolution's High Priest*. New York: Basic, 1999.

Desmond, Adrian, and James Moore. *Darwin: The Life of a Tormented Evolutionist*. New York: W. W. Norton, 1991.

———. *Darwin's Sacred Cause: Race, Slavery, and the Quest for Human Origins*. Chicago: University of Chicago Press, 2011.

Dillenberger, John. *Protestant Thought and Natural Science: A Historical Interpretation*. Notre Dame, IN: University of Notre Dame Press, 1989.

Dilley, Stephen. "Charles Darwin's Use of Theology in the Origin of Species." *British Journal for the History of Science* 45 no. 1 (2012) 29–56.

Dirda, Michael. Review of *The Swerve*, by Stephen Greenblatt. *The Washington Post*, September 21, 2011. https://www.washingtonpost.com/entertainment/books/stephen-greenblatts-the-swerve-reviewed-by-michael-dirda/2011/09/20/gIQA8WmVmK_story.html?utm_term=.72a1f6ab50f5.

Dixon, Thomas. *From Passions to Emotions: The Creation of a Secular Psychological Category*. Cambridge: Cambridge University Press, 2003.

Dixon, Thomas, et al., eds. *Science and Religion: New Historical Perspectives*. Cambridge: Cambridge University Press, 2010.

Dobbs, Betty Jo Teeter. *The Foundation of Newton's Alchemy or, 'The Hunting of the Greene Lyon.'* Cambridge: Cambridge University Press, 1991.

———. *The Janus Faces of Genius: The Role of Alchemy in Newton's Thought*. Cambridge: Cambridge University Press, 1991.

———. "Newton as Final Cause and First Mover." In *Rethinking the Scientific Revolution*, edited by Margaret J. Osler, 25–41. Cambridge: Cambridge University Press, 2000.

Dodds, Michael J. *Unlocking Divine Action: Contemporary Science and Thomas Aquinas*. Washington, DC: Catholic University of America Press, 2012.

Dorrien, Gary. *The Barthian Revolt in Modern Theology: Theology Without Weapons*. Louisville: Westminster John-Knox, 2000.

Draper, John William. *History of the Conflict Between Religion and Science*. London: Henry S. King, 1875.

———. *History of the Intellectual Development of Europe*. New York: Appleton, 1869.

Dreger, Alice. *Galileo's Middle Finger: Heretics, Activists, and One Scholar's Search for Justice*. New York: Penguin, 2016.

Dry, Sara. *The Newton Papers: The Strange and True Odyssey of Isaac Newton's Manuscripts*. Oxford: Oxford University Press, 2014.

Dubuisson, Daniel. *The Western Construction of Religion: Myths, Knowledge, and Ideology*. Translated by William Sayers. Baltimore: Johns Hopkins University Press, 2003.

Duffy, Eamon. *Saints and Sinners: A History of the Popes*. New Haven, CT: Yale University Press, 2006.

Duhem, Pierre. *Medieval Cosmology: Theories of Infinity, Place, Time, Void, and the Plurality of Worlds.* Translated and edited by Roger Ariew. Chicago: University of Chicago Press, 2011.

———. *To Save the Phenomena: An Essay on the Idea of Physical Theory from Plato to Galileo.* Translated by Edmund Dolan. Chicago: University of Chicago Press, 2015.

Dupré, Louis. *The Enlightenment and the Intellectual Foundations of Modern Culture.* New Haven, CT: Yale University Press, 2005.

———. *Passage to Modernity: An Essay in the Hermeneutics of Nature and Culture.* New Haven, CT: Yale University Press, 1993.

Durand, Dana B. "Nicole Oresme and the Medieval Origins of Modern Science." *Speculum* 16 no. 2 (April 1941) 167–85.

Durant, John. "Darwinism and Divinity: A Century of Debate." In *Darwinism and Divinity*, edited by John Durant, 9–39. New York: Basil Blackwell, 1985.

Dzielska, Maria. *Hypatia of Alexandria.* Translated F. Lyra. Cambridge, MA: Harvard University Press, 1995.

Eagleton, Terry. *Culture and the Death of God.* New Haven, CT: Yale University Press, 2014.

Eco, Umberto. "The Force of Falsity." In *Serendipities: Language and Lunacy*, 1–23. Orlando: Harvest, 1999.

———. *Postscript to* The Name of the Rose. New York: Harcourt, 1989.

Edelstein, Dan. *The Enlightenment: A Genealogy.* Chicago: University of Chicago Press, 2010.

Edmonds, David. *The Murder of Professor Schlick: The Rise and Fall of the Vienna Circle.* Princeton, NJ: Princeton University Press, 2020.

Efron, Noah. "Sciences and Religions: What It Means to Take Historical Perspectives Seriously." In *Science and Religion: New Historical Perspectives*, edited by Thomas Dixon, Geoffrey Cantor, and Stephen Pumfrey, 247–62. Cambridge: Cambridge University Press, 2010.

Ellis, Richard, et al. "90 Years On—The 1919 Eclipse Expedition at Principe." *Astronomy & Geophysics* 50 no. 4 (August 2009) 12–15.

England, Richard. "Natural Selection, Teleology, and Logos: From Darwin to the Oxford Neo-Darwinists, 1859–1909." *Osiris* 16 (2001) 270–287.

Erdozain, Dominic. "A Heavenly Poise: Radical Religion and the Making of the Enlightenment." *Intellectual History Review* 27 no. 1 (2017) 71–96.

———. *The Soul of Doubt: The Religious Roots of Unbelief from Luther to Marx.* Oxford: Oxford University Press, 2016.

Falque, Emmanuel. *The Loving Struggle: Phenomenological and Theological Debates.* New York: Rowman & Littlefield, 2018.

Fara, Patricia. *Newton: The Making of Genius.* New York: Columbia University Press, 2004.

Farr, A. D. "Early Opposition to Obstetric Anesthesia." *Anesthesia* 35 (1980) 896–907.

———. "Religious Opposition to Obstetric Anesthesia: A Myth?" *Annals of Science* 40 (1983) 159–77.

Farrell, Frank B. *How Theology Shaped Twentieth-Century Philosophy.* Cambridge University Press, 2019.

Fasolt, Constantin. *The Limits of History.* Chicago: University of Chicago Press, 2003.

Faulkner, William. *Requiem for a Nun.* New York: Vintage, 2011.

Feigl, Herbert. "Beyond Peaceful Coexistence." In *Historical and Philosophical Perspectives of Science*, edited by R. H. Stuewer, 3–11. Minneapolis: University of Minnesota Press, 1970.

Feldhay, Rivka. *Galileo and the Church: Political Inquisition or Critical Dialogue?* Cambridge: Cambridge University Press, 1995.

Fergusson, David. *The Providence of God: A Polyphonic Approach.* Cambridge: Cambridge University Press, 2019.

Fergusson, Wallace. *The Renaissance in Historical Thought.* Boston: Houghton Mifflin, 1948.

Ferngren, Gary B. *Medicine & Health Care in Early Christianity.* Baltimore: Johns Hopkins University Press, 2009.

———, ed. *Science and Religion: A Historical Introduction.* Baltimore: Johns Hopkins University Press, 2002.

Ferris, Timothy. *Coming of Age in the Milky Way.* New York: Harper Perennial, 2003.

Feser, Edward. *Aristotle's Revenge: The Metaphysical Foundations of Physical and Biological Science.* Scheelscheid, Germany: Editiones Scholasticae, 2019.

———. *Five Proofs of the Existence of God: Aristotle, Plotinus, Augustine, Aquinas, Leibniz.* San Francisco: Ignatius, 2017.

———. *Neo-Scholastic Essays.* South Bend, IN: St. Augustine's, 2015.

Feyerabend, Paul. *Against Method.* London: Verso, 2010.

Finnegan, Diarmid A. "The Spatial Turn: Geographical Approaches in the History of Science." *Journal of the History of Biology* (2008) 369–88.

Finocchiaro, Maurice A. *Retrying Galileo: 1633–1992.* Berkeley, CA: University of California Press, 2005.

———. *The Trial of Galileo: Essential Documents.* Indianapolis: Hackett, 2014.

Firestone, Chris L., and Nathan A. Jacobs. *The Persistence of the Sacred in Modern Thought.* Notre Dame, IN: University of Notre Dame Press, 2012.

Fleming, Donald. *John William Draper and The Religion of Science.* London: Octagon, 1972.

Fleming, John V. *The Dark Side of the Enlightenment: Wizards, Alchemists, and Spiritual Seekers in the Age of Reason.* New York: W. W. Norton, 2013.

Flew, Antony. *There Is a God: How the World's Most Notorious Atheist Changed His Mind.* New York: HarperOne, 2007.

Flory, Richard W. "Promoting a Secular Standard: Secularization and Modern Journalism, 1870–1930." In *The Secular Revolution: Power, Interests, and Conflict in the Secularization of American Public Life,* edited by Christian Smith, 395–433. Berkeley, CA: University of California Press, 2003.

Force, James E. "Newton's God of Dominion: The Unity of Newton's Theological, Scientific, and Political Thought." In *Essays on the Context, Nature, and Influence of Isaac Newton's Theology,* edited by James Force and Richard Popkin, 75–102. Dordrecht: Springer, 1990.

Forrest, Barbara, and Paul R. Gross. *Creationism's Trojan Horse: The Wedge of Intelligent Design.* New York: Oxford University Press, 2004.

Foster, M. B. "The Christian Doctrine of Creation and the Rise of Natural Science." *Mind* 43 (1934) 446–68.

———. "Christian Theology and Modern Science of Nature (I)." *Mind* 44 (1935) 439–66.

———. "Christian Theology and Modern Science of Nature (II)." *Mind* 45 (1936) 1–27.

———. "Greek and Christian Ideas of Nature." *The Free University Quarterly* 6 (1959) 122–27.

Francis, Mark. *Herbert Spencer and the Invention of Modern Life.* Ithaca, NY: Cornell University Press, 2007.

Frazer, James George. *The Golden Bough: A Study in Magic and Religion.* Cambridge: Cambridge University Press, 2012.

Freely, John. *Before Galileo: The Birth of Modern Science in Medieval Europe*. New York: Overlook Duckworth, 2012.

Freeman, Charles. *The Closing of the Western Mind: The Rise of Faith and the Fall of Reason*. New York: Knopf, 2003.

Frei, Hans. "Appendix A: Theology in the University." In *Types of Christian Theology*, 95–133. New Haven, CT: Yale University Press, 1992.

———. *The Eclipse of the Biblical Narrative: A Study in Eighteenth and Nineteenth Century Hermeneutics*. New Haven, CT: Yale University Press, 1974.

French, Roger. *Medicine Before Science: The Rational and Learned Doctor from the Middle Ages to the Enlightenment*. Cambridge: Cambridge University Press, 2003.

Frend, W. H. C. "Edward Gibbon (1737–1794) and Early Christianity." *Journal of Ecclesiastical History* 45 (1994) 661–772.

Freud, Sigmund. *The Complete Psychological Works of Sigmund Freud: The Introductory Letters of Psychoanalysis*. New York: Vintage, 2001.

———. *The Future of an Illusion*. New York: W. W. Norton, 1989.

Friedman, Michael. *Reconsidering Logical Positivism*. Cambridge: Cambridge University Press, 1999.

Fuller, Randall. *The Book that Changed America: How Darwin's Theory of Evolution Ignited a Nation*. New York: Penguin, 2018.

Funkenstein, Amos. *Theology and the Scientific Imagination: From the Middle Ages to the Seventeenth Century*. Princeton, NJ: Princeton University Press, 1986.

Fyfe, Aileen. *Science and Salvation: Evangelical Popular Science Publishing in Victorian Britain*. Chicago: University of Chicago Press, 2004.

Gadamer, Hans-Georg. *Truth and Method*. New York: Bloomsbury Academic, 2004.

Galilei, Galileo. *Dialogue Concerning Two Chief World Systems*. Translated and edited by Stillman Drake. New York: Modern Library, 2001.

———. "Letter to the Grand Duchess Christina (1615)." In *The Galileo Affair: A Documentary History*, vol. 1, edited by Maurice A. Finocchiaro, 101–2. Berkeley, CA: University of California Press, 1989.

Garcia, Robert K., and Nathan L. King. "Introduction." In *Is Goodness Without God Good Enough? A Debate on Faith, Secularism, and Ethics*, edited by Robert K. Garcia and Nathan L. King, 1–24. Lanham, MD: Rowman and Littlefield, 2009.

Garroutte, Eva Marie. *Language and Cultural Authority: Nineteenth-Century Science and the Colonization of Religious Discourse*. PhD diss., Princeton University, 1993.

———. "The Positivist Attack on Baconian Science and Religious Knowledge in the 1870s." In *The Secular Revolution: Power, Interests, and Conflict in the Secularization of American Public Life*, edited by Christian Smith, 197–215. Berkeley, CA: University of California Press, 2003.

Garwood, Christine. *Flat Earth: The History of an Infamous Idea*. New York: Thomas Dunne, 2007.

Gatti, Hilari. *Giordano Bruno and Renaissance Science*. Ithaca, NY: Cornell University Press, 2002.

Gaukroger, Stephen. *Descartes: An Intellectual Biography*. Oxford: Oxford University Press, 1997.

———. *The Emergence of a Scientific Culture: Science and the Shaping of Modernity 1210–1685*. Oxford: Oxford University Press, 2019.

Geertz, Clifford. *Works and Lives: The Anthropologist as Author*. Stanford, CA: Stanford University Press, 1989.

Gellius, Allus. *The Attic Nights of Allus Gellius*. New York: Forgotten, 2019.

Gibbon, Edward. *The Decline and Fall of the Roman Empire*. Vol. 5. London: The Folio Society, 1999.

Giberson, Karl, and Mariano Arigas. *Oracles of Science: Celebrity Scientists Versus God and Religion*. Oxford: Oxford University Press, 2009.

Giberson, Karl W., and Donald A. Yerxa. *Species of Origin: America's Search for a Creation Story*. Lanham, MD: Rowman and Littlefield, 2002.

Gies, Frances, and Joseph Gies. *Cathedral, Forge, and Waterwheel: Technology and Invention in the Middle Ages*. New York: Harper Perennial, 1995.

Gillis, John. "The Future of European History." *Perspectives* 34 no. 4 (April 1996) 6.

Gillespie, Michael Allen. *Nihilism Before Nietzsche*. Chicago: University of Chicago Press, 1996.

———. *The Theological Origins of Modernity*. Chicago: University of Chicago Press, 2009.

Gillespie, Neal C. *Charles Darwin and the Problem of Creation*. Chicago: University of Chicago Press, 1979.

Gillispie, Charles Coulston. *Genesis and Geology: A Study in the Relations of Scientific Thought, Natural Theology, and Social Opinion in Great Britain, 1790–1850*. Cambridge, MA: Harvard University Press, 1996.

Gilley, Sheridan. "The Huxley-Wilberforce Debate: A Reconsideration." In *Religion and Humanism*, edited by Keith Robbins, 325–40. Studies in Church History 17. Oxford: Blackwell, 1981.

Gilley, Sheridan, and Ann Loades. "Thomas Henry Huxley: The War Between Science and Religion." *The Journal of Religion* 61 no. 3 (July 1981) 285–308.

Gilson, Etienne. *Being and Some Philosophers*. 2d ed. Toronto: The Pontifical Institute for Medieval Studies, 1952.

———. *From Aristotle to Darwin and Back Again: A Journey in Final Causality, Species, and Evolution*. San Francisco: Ignatius, 1984.

———. *History of Christian Philosophy in the Middle Ages*. Washington, DC: Catholic University of America Press, 2019.

———. *The Philosopher and Theology*. Providence: Cluny Media LLC, 2020.

———. *The Spirit of Medieval Philosophy (The Gifford Lectures 1931–1932)*. Translated by A. H. C. Downes. Notre Dame, IN: University of Notre Dame Press, 2009.

Gimple, Jean. *The Medieval Machine: The Industrial Revolution of the Middle Ages*. New York: Penguin, 1977.

Gingerich, Owen. *The Book Nobody Read: Chasing the Revolutions of Nicolaus Copernicus*. New York: Walker, 2004.

———. "The Censorship of *De Revolutionibus*." In *The Eye of Heaven: Ptolemy, Copernicus, Kepler*, 269–85. New York: American Institute of Physics, 1993.

———. "Crisis vs. Aesthetic in the Copernican Revolution." In *The Eye of Heaven: Ptolemy, Copernicus, Kepler*, 193–204. New York: American Institute of Physics, 1993.

Glacken, Clarence J. *Traces on the Rhodian Shore: Nature and Culture in Western Thought from Ancient Times to the End of the Eighteenth Century*. Berkeley, CA: University of California Press, 1967.

Gleiser, Marcelo. *The Dancing Universe: From Creation Myths to the Big Bang*. Hanover, NH: Dartmouth University Press, 2005.

Gloege, Timothy E. W. *Guaranteed Pure: The Moody Bible Institute, Business, and the Making of Modern Evangelicalism*. Chapel Hill, NC: University of North Carolina Press, 2015.

Goetz, Stewart, and Charles Taliaferro. *Naturalism*. Grand Rapids: Eerdmans, 2008.

Gould, Stephen Jay. "Late Birth of the Flat Earth. In *Dinosaur in a Haystack: Reflections in Natural History*, 38–53. Cambridge, MA: Harvard University Press, 2011.

———. *Life's Grandeur: The Spread of Excellence from Plato to Darwin*. New York: Jonathan Cape, 1996.

———. *The Structure of Evolutionary Theory*. Cambridge, MA: The Belknap Press of Harvard University Press, 2002.

Grafton, Anthony. *The Footnote: A Curious History*. Cambridge, MA: Harvard University Press, 1997.

Graney, Christopher M. *Setting Aside All Authority: Giovanni Battista Riccioli and the Science Against Copernicus in the Age of Galileo*. Notre Dame, IN: University of Notre Dame Press, 2015.

Grant, Edward. *The Foundations of Modern Science in the Middle Ages: Their Religious, Institutional, and Intellectual Contexts*. Cambridge: Cambridge University Press, 1996.

———. *God & Reason in the Middle Ages*. Cambridge: Cambridge University Press, 2001.

Greenblatt, Stephen. *The Swerve: How the World Became Modern*. New York: W. W. Norton, 2012.

Gregory, Alan P. R. *Science Fiction Theology: Beauty and the Transformation of the Sublime*. Waco, TX: Baylor University Press, 2015.

Gregory, Brad S. *The Unintended Reformation: How A Religious Revolution Secularized Society*. Cambridge, MA: The Belknap Press of Harvard University Press, 2012.

Gregory, Frederick. "Continental Europe." In *The Warfare Between Science and Religion: The Idea That Wouldn't Die*, edited by Jeff Hardin et al., 89. Baltimore: Johns Hopkins University Press, 2018.

———. *Nature Lost? Natural Science and the German Theological Traditions of the Nineteenth Century*. Cambridge, MA: Harvard University Press, 1992.

Griffiths, Richard. "Very, Very Frightening." *The Daily Telegraph*, November 5, 1994, 6.

Gross, Alan G. *The Scientific Sublime: Popular Science Unravels the Mysteries of the Universe*. New York: Oxford University Press, 2018.

Gundlach, Bradley J. *Process and Providence: The Evolution Question at Princeton, 1845–1929*. Grand Rapids: Eerdmans, 2013.

Haas, Christopher. *Alexandria in Late Antiquity: Topography and Social Conflict*. Baltimore: Johns Hopkins University Press, 1996.

Hadden, Jeffrey K. "Toward Desacralizing Secularization Theory." *Social Forces* 65 (March 1987) 587–611.

Hadot, Pierre. *What Is Ancient Philosophy?* Cambridge, MA: The Belknap Press of Harvard University Press, 2004.

Hahn, Roger. "Laplace and the Mechanistic Universe." In *God & Nature: Historical Essays on the Encounter Between Christianity and Science*, edited by Ronald Numbers and David Lindberg, 256–276. Berkley: University of California Press, 1986.

———. "Laplace and the Vanishing Role of God in the Physical Universe." In *The Analytic Spirit: Essays in the History of Science in Honor of Henry Guerlac*, edited by Harry Woolf, 85–95. Ithaca: Cornell University Press, 1981.

Hahn, Scott and Jeffrey Morrow. *Modern Biblical Criticism as a Tool of Statecraft (1700–1900)*. Steubenville: Emmaus Academic, 2020.

Hampton, Monte Harrell. *Storm of Words: Science, Religion, and Evolution in the Civil War Era*. Tuscaloosa, AL: University of Alabama Press, 2014.

Hanby, Michael. *No God, No Science? Theology, Cosmology, Biology*. Oxford: Wiley-Blackwell, 2013.

Hankins, Barry. *Jesus and Gin: Evangelicalism, The Roaring Twenties, and Today's Culture Wars*. New York: St. Martin's, 2010.

Hannam, James. *The Genesis of Science: How the Christian Middle Ages Launched the Scientific Revolution*. Washington, DC: Regnery, 2011.

Hardin, Jeff, et al., eds. *The Warfare Between Science & Religion: The Idea that Wouldn't Die*. Baltimore: Johns Hopkins University Press, 2018.

Harries, Karsten. *Infinity and Perspective*. Cambridge, MA: MIT Press, 2001.

Harrison, Peter. "Adam Smith and the History of the Invisible Hand." *Journal of the History of Ideas* 72 no. 1 (January 2011) 29–49.

———. "Angels on Pinheads and Needles' Points." *Notes and Queries* 63 no. 1 (March 2016) 45–47.

———. *The Bible, Protestantism, and the Rise of Natural Science*. Cambridge: Cambridge University Press, 1998.

———, ed. *The Cambridge Companion to Science and Religion*. Cambridge: Cambridge University Press, 2010.

———. "Evolution, Providence, and the Problem of Chance." In *Abraham's Dice: Chance and Providence in the Monotheistic Traditions*, edited by Karl Giberson, 260–90. Oxford University Press, 2016.

———. *The Fall of Man and the Foundations of Science*. Cambridge: Cambridge University Press, 2008.

———. "'I Believe Because It Is Absurd': The Enlightenment Invention of Tertullian's *Credo*." *Church History* 86 (2017) 339–64.

———. "Is Science-Religion Conflict Always A Bad Thing?" In *Evolution and The Fall*, edited by William Cavanaugh and James K. A. Smith, 204–26. Grand Rapids: Eerdmans, 2017.

———, ed. *Narratives of Secularization*. New York: Routledge, 2017.

———. *'Religion' and the Religions in the English Enlightenment*. Cambridge: Cambridge University Press, 2002.

———. "Religion, Innovation, and Secular Modernity." In *Religion and Innovation: Antagonists or Protagonists?*, edited by Donald A. Yerxa, 74–87. New York: Bloomsbury, 2015.

———. "Religion, Scientific Naturalism, and Historical Progress." In *Religion and Innovation: Antagonists or Protagonists?*, edited by Donald A. Yerxa, 87–100. New York: Bloomsbury, 2015.

———. "'Science' and 'Religion': Constructing the Boundaries." *Journal of Religion* 86 no. 1 (January 2006) 81–106.

———. "Science and Secularization." *Intellectual History Review* 27 (January 2017) 47–70.

———. "Subduing the Earth: Genesis 1, Early Modern Science, and the Exploitation of Nature." *The Journal of Religion* 79 no. 1 (January 1999) 86–109.

———. *The Territories of Science and Religion*. Chicago: University of Chicago Press, 2015.

———. "Voluntarism and Early Modern Science." *History of Science* 40 no. 1 (2002) 63–89.

———. "Was There a Scientific Revolution?" *European Review* 15, no. 4 (2007) 445–57.

———, ed. *Wrestling with Nature: From Omens to Science*. Chicago: University of Chicago Press, 2011.

Hart, David Bentley. *Atheist Delusions: The Christian Revolution and Its Fashionable Enemies.* New Haven, CT: Yale University Press, 2009.

———. *The Beauty of the Infinite: The Aesthetics of Christian Truth.* Grand Rapids: Eerdmans, 2003.

———. *The Experience of God: Being, Consciousness, Bliss.* New Haven, CT: Yale University Press, 2013.

———. "Hypatia Reassembled." In *The Dream Child's Progress and Other Essays,* by David Bentley Hart, 91–97. Brooklyn, NY: Angelico, 2017.

Harrison, Peter, and Jon Roberts, eds. *Science Without God? Rethinking the History of Scientific Naturalism.* Oxford: Oxford University Press, 2019.

Hasler, August. *How the Pope Became Infallible: Pius IX and the Politics of Persuasion.* New York: Doubleday, 1981.

Hauerwas, Stanley. *The State of the University: Academic Knowledges and the Knowledge of God.* Malden, MA: Blackwell, 2007.

Heilbron, John L. *Galileo.* Oxford: Oxford University Press, 2012.

———. *Sun in the Church: Cathedrals as Solar Observatories.* Cambridge, MA: Harvard University Press, 1999.

Herodotus. *The Histories.* Translated by Aubrey de Selencourt. New York: Penguin, 2003.

Hesketh, Ian. "From Copernicus to Darwin to You: History and the Meaning(s) of Evolution." In *Rethinking History, Science, and Religion: An Exploration of Conflict and the Complexity Principle,* edited by Bernard Lightman, 190–205. Pittsburgh: University of Pittsburgh Press, 2019.

———. *Of Apes and Ancestors: Evolution, Christianity, and the Oxford Debate.* Toronto: University of Toronto Press, 2009.

Higgit, Rebekah. *Recreating Newton: Newtonian Biography and the Making of Nineteenth-Century History of Science.* Pittsburgh: University of Pittsburgh Press, 2007.

Hill, Lisa. "The Hidden Theology of Adam Smith." *European Journal of the History of Economic Thought* 8 (2011) 1–29.

Hilton, Boyd. *The Age of Atonement: The Influence of Evangelicalism on Social and Economic Thought 1785–1865.* Oxford: Oxford University Press, 1993.

Hinde, John R. *Jacob Burckhardt and the Crisis of Modernity.* New York: McGill-Queen University Press, 2000.

Hitchens, Christopher. "Equal Time: No Tax-Exempt Status for Churches that Refuse to Distribute Pro-Evolution Propaganda!" *Slate* 23 (August 2005). https://slate.com/news-and-politics/2005/08/equal-time-for-evolutionists.html.

Hoff, Johannes. *The Analogical Turn: Rethinking Modernity with Nicholas of Cusa.* Grand Rapids: Eerdmans, 2013.

Holifield, E. Brooks. *Theology in America: Christian Thought from the Age of the Puritans to the Civil War.* New Haven, CT: Yale University Press, 2003.

Holland, Tom. *Dominion: How the Christian Revolution Remade the World.* New York: Basic, 2019.

Holsinger, Bruce. *The Premodern Condition: Medievalism and the Making of Theory.* Chicago: University of Chicago Press, 2005.

Holt-Jenson, Arild. *Geography: Its History and Concepts: A Student's Guide.* 2d ed. London: Sage, 1988.

Hooykaas, Reijer. *Religion and the Rise of Modern Science.* Edinburgh: Scottish Academic, 1972.

Horkheimer, Max, and Theodore Adorno. *Dialectic of Enlightenment: Philosophical Fragments.* Stanford, CA: Stanford University Press, 2007.

Howard, Thomas Albert. *God and the Atlantic: America, Europe, and the Religious Divide.* Oxford: Oxford University Press, 2011.

———. *The Pope and the Professor: Pius IX, Ignaz von Dollinger, and the Quandary of the Modern Age.* Oxford: Oxford University Press, 2017.

———. *Protestant Theology and the Making of the Modern German University.* Oxford: Oxford University Press, 2006.

———. *Religion and the Rise of Historicism: W. M. L. de Wette, Jacob Burckhardt, and the Theological Origins of Nineteenth-Century Historical Consciousness.* Cambridge: Cambridge University Press, 1999.

Howell, Kenneth J. *God's Two Books: Copernican Cosmology and Biblical Interpretation in Early Modern Science.* Notre Dame, IN: University of Notre Dame Press, 2002.

Howsam, Leslie. "An Experiment with Science for the Nineteenth-Century Book Trade: *The International Scientific Series.*" *The British Journal for the History of Science* 33 no. 2 (June 2000) 187–207.

Huizinga, Johan. *The Autumn of the Middle Ages.* Translated by Rodney J. Payton and Ulrich Mammitzsch. Chicago: University of Chicago Press, 1996.

Hull, David. "Darwinism as a Historical Entity: A Historiographical Proposal." In *The Darwinian Heritage,* edited by David Kohn, 773–812. Princeton, NJ: Princeton University Press, 1985.

———. *The Metaphysics of Evolution.* Albany, NY: State University of New York Press, 1989.

Hunter, Ian. "Secularization: Process, Program, and Historiography." *Intellectual History Review* 27 no. 1 (2017) 7–29.

Hunter, James Davison. *To Change the World: The Irony, Possibility, and Tragedy of Christianity in the Late Modern World.* Oxford: Oxford University Press, 2010.

Hunter, Michael. *Boyle: Between God and Science.* New Haven, CT: Yale University Press, 2010.

Hutchings, David. "Demonic Cheese-Donkeys and Immortal Peacocks: Augustine Does Science." *Christianity Today,* January 26, 2018. https://www.christianitytoday.com/ct/2018/january-web-only/demonic-cheese-donkeys-and-immortal-peacocks-augustine-does.html.

Huxley, Thomas Henry. "Agnosticism." In *Collected Essays,* vol. 5, 224. London: Macmillan, 1895.

———. "Science and Religion." *The Builder* 18 (1859) 35.

———. *Science and the Hebrew Tradition.* London: Macmillan, 1904.

Hyman, Gavin. *A Short History of Atheism.* New York: I. B. Tauris, 2010.

Iggers, Georg G. "The Image of Ranke in American and German Thought." *History and Theory* 2 no. 1 (1962) 17–40.

Iliffe, Rob. *Priest of Nature: The Religious Worlds of Isaac Newton.* Oxford: Oxford University Press, 2017.

Inglis, John. *Spheres of Philosophical Inquiry and the Historiography of Medieval Philosophy.* Leiden: Brill, 1998.

Inman, Daniel. *The Making of Modern English Theology: God and the Academy at Oxford.* Minneapolis: Fortress, 2014.

Irvine, William. *Apes, Angels, and Victorians: A Joint Biography of Darwin and Huxley.* London: Weidenfield and Nicolson, 1956.

Irving, Washington. *The Life and Voyages of Christopher Columbus.* New York: Forgotten, 2017.

Israel, Jonathan L. *Radical Enlightenment: Philosopy and the Making of Modernity 1650–1750.* New York: Oxford University Press, 2001.

Jaeger, Werner. *Early Christianity and the Greek Paidea.* Cambridge, MA: Harvard University Press, 1961.

Jakab, Peter. "An Extraordinary Journey: Leonardo Da Vinci and Flight." *Smithsonian: National Air and Space Museum.* August 22, 2013. https://airandspace.si.edu/stories/editorial/leonardo-da-vinci-and-flight.

Jaki, Stanley. "Introduction." In *Prémices Philosophique: Présentées Avec Une Introduction En Anglais Par S. L. Jaki* by Pierre Duhem, i–xvi. Leiden: Brill, 1987.

———. *Origins of Science and the Science of Origins.* Edinburgh: Scottish Academic, 1979.

———. *Reluctant Heroine: The Life and Work of Helene Duhem.* Edinburgh: Scottish Academic, 1992.

———. *The Road of Science and the Ways to God.* Chicago: University of Chicago Press, 1978.

———. *The Savior of Science.* Grand Rapids: Eerdmans, 2000.

———. *Science and Creation: From Infinite Cycles to an Oscillating Universe.* Lanham, MD: University Press of America, 1990.

———. *Scientist and Catholic: Pierre Duhem.* Fort Royal, VA: Christendom, 2004.

———. *Uneasy Genius: The Life and Work of Pierre Duhem.* Dordrecht: Martinus Nijhoff, 1987.

James, Frank. "On Wilberforce and Huxley." *Astronomy & Geophysics* 46 (February 2005) 1.9.

Jardine, Nick. "Epistemology of the Sciences." In *The Cambridge History of Renaissance Philosophy,* edited by C. B. Schmitt and Q. Skinner, 685–712. Cambridge: Cambridge University Press, 1991.

———. "Whigs and Stories: Herbert Butterfield and the Historiography of Science." *History of Science* 41 (2003) 125–40.

Jefferson, Thomas. *Notes on the State of Virginia.* New York: Penguin Classics, 1998.

Jenkins, Philip. *The New Anti-Catholicism: The Last Acceptable Prejudice.* New York: Oxford University Press, 2003.

Johnson, Monte, and Catherine Wilson. "Lucretius and the History of Science." In *The Cambridge Companion to Lucretius,* edited by Stewart Gillespie and Philip Hardie, 131–48. Cambridge: Cambridge University Press, 2007.

Jones, Gareth Steadman. "History: The Poverty of Empiricism." In *Ideology in Social Science: Readings in Critical Theory,* edited by Robin Blackburn, 96–118. Glasgow: William Collins Sons, 1972.

Jordan, Mark D. *Convulsing Bodies: Religion and Resistance in Foucault.* Stanford, CA: Stanford University Press, 2015.

Josephson-Storm, Jason Ã. *The Myth of Disenchantment: Magic, Modernity, and the Birth of the Human Sciences.* Chicago: University of Chicago Press, 2017.

Jung, Carl G. *Psychological Types.* Translated by Gerhard Adler and R. F. C. Hull. Princeton, NJ: Princeton University Press, 1971.

Jüngel, Eberhard. *God as the Mystery of the World: On the Foundation of the Theology of the Crucified One in the Dispute Between Theism and Atheism.* Grand Rapids: Eerdmans, 1983.

Kaiser, Christopher B. "Calvin, Copernicus, and Castellio." *Calvin Theological Journal* 21 no. 1 (April 1986) 5–31.

Kamen, Henry. *The Spanish Inquisition: A Historical Revision.* 4th ed. New Haven, CT: Yale University Press, 2014.

Kantorowicz, Ernst. *The King's Two Bodies: A Study in Medieval Political Theology.* Princeton, NJ: Princeton University Press, 1957.

Karlowicz, Darius. *Socrates and Other Saints: Early Christian Understanding of Reason and Philosophy.* Translated by Artur Rosman. Eugene, OR: Cascade, 2017.

Keas, Michael Newton. *Unbelievable: 7 Myths About the History and Future of Science and Religion.* Delaware: ISI, 2019.

Kelley, Donald R. *History and the Disciplines: The Reclassification of Knowledge in Early Modern Europe.* Rochester, NY: University of Rochester, 1997.

Kemp, Kenneth W. *The War That Never Was: Evolution and Christianity.* Eugene, OR: Cascade, 2020.

Kenny, Anthony. *A New History of Western Philosophy.* Oxford: Oxford University Press, 2010.

Keynes, Randal. *Darwin, His Daughter, and Human Evolution.* New York: Riverhead, 2002.

Kohn, David. "Darwin's Ambiguity: The Secularization of Biological Meaning." *British Journal of the History of Science* 22 (1989) 215–39.

Kim, Kwangsu. "Adam Smith: Natural Theology and Its Implications for His Method of Social Inquiry." *Review of Social Economy* 55 (1997) 312–36.

Kinnaman, David, and Gabe Lyons. *Unchristian: What a New Generation Really Thinks About Christianity . . . And Why It Matters.* Grand Rapids: Baker, 2007.

King, Nathan L., and Robert K. Garcia, eds. *Is Goodness Without God Good Enough?: A Debate on Faith, Secularism, and Ethics.* Lanham, MD: Rowman & Littlefield, 2009.

Kirsch, Jonathan. *God Against the Gods: The History of the War Between Monotheism and Polytheism.* New York: Penguin, 2005.

Kitcher, Philip. *Abusing Science: The Case Against Creationism.* Cambridge, MA: MIT Press, 1986.

Klaaren, Eugene M. *Religious Origins of Modern Science.* Grand Rapids: Eerdmans, 1977.

Knight, John Allan. *Liberalism versus Postliberalism: The Great Divide in Twentieth-Century Theology.* Oxford: Oxford University Press, 2012.

Konoval, Brandon. "What Does Dayton Have to Do With Sils-Maria? Nietzsche and the Scopes Trial." *Perspectives on Science* 22 no. 4 (2014) 545–73.

Kors, Alan. *Atheism in France 1650–1729, Volume One: The Orthodox Sources of Unbelief.* Princeton, NJ: Princeton University Press, 1990.

Krell, David Farrell. *The Tragic Absolute: German Idealism and the Languishing of God.* Bloomington, IN: Indiana University Press, 2005.

Kuhn, Thomas. *The Essential Tension: Selected Studies in Scientific Tradition and Change.* Chicago: University of Chicago Press, 1977.

———. "The History of Science." In *The Essential Tension:Selected Studies in Scientific Tradition and Change,* 105–26. Chicago: University of Chicago Press, 1977.

———. *The Road Since Structure: Philosophical Essays 1970–1993.* Edited by James Conant and John Haugeland. Chicago: University of Chicago Press, 2002.

———. *The Structure of Scientific Revolutions: 50th Anniversary Edition.* Chicago: University of Chicago Press, 2012.

Kuklick, Bruce. *Churchmen and Philosophers: From Jonathan Edwards to John Dewey*. New Haven, CT: Yale University Press, 1985.

Küng, Hans. *Does God Exist? An Answer for Today*. New York: Doubleday, 2013.

Kupfer, Marcia. "Reflections on the Ebstorf Map: Cartography, Theology, and *Delectio Speculationis*." In *Mapping Medieval Geographies: Geographical Encounters in the Latin West and Beyond, 300–1600*, edited by Keith D. Lilley, 110–18. Cambridge: Cambridge University Press, 2013.

Lai, Tyrone. "Nicholas of Cusa and the Finite Universe." *Journal of the History of Philosophy* 11 no. 2 (1973) 161–67.

Lakoff, George, and Mark Johnson. "Conceptual Metaphor in Everyday Language." *The Journal of Philosophy* 77 no. 8 (1980) 453–86.

Larsen, Timothy. *Crisis of Doubt: Honest Faith in Nineteenth-Century England*. Oxford: Oxford University Press, 2006.

———. *The Slain God: Anthropologists & the Christian Faith*. Oxford: Oxford University Press, 2014.

Larson, Edward J. "Law and Society in the Courtroom: Introducing the Trials of the Century." *University of Missouri-Kansas City Law Review* no. 68 (2000) 543–48.

———. *Summer for the Gods: The Scopes Trial and America's Continuing Debate Over Science and Religion*. New York: Basic, 2006.

Lavan, Luke A. "The End of the Temples: Toward a New Narrative?" In *The Archaeology of Late Antique 'Paganism,'* edited by Luke A. Lavan and Michael Mulryan, xv–lxv. Leiden: Brill, 2011.

LeBlanc, Steven A. *Constant Battles: Why We Fight*. New York: St. Martin's Griffin, 2004.

Lebo, Lauri. *The Devil in Dover: An Insider's Story of Dogma v. Darwin in Small-Town America*. New York: New, 2016.

Leech, D. *The Hammer of the Cartesians: Henry More's Philosophy of Spirit and the Origins of Modern Atheism*. Leiden: Brill, 2013.

Legaspi, Michael C. *The Death of Scripture and the Rise of Biblical Studies*. Oxford: Oxford University Press, 2010.

Le Goff, Jacques. *Medieval Civilization 400–1500*. Oxford: Blackwell, 1999.

———. *Must We Divide History Into Periods?* New York: Columbia University Press, 2015.

———. "The Several Middle Ages of Jules Michelet." In *Time, Work, and Culture in the Middle Ages*, 3–28. Chicago: University of Chicago Press, 1980.

———. *Time, Work, and Culture in the Middle Ages*. Translated by Jacques Le Goff. Chicago: University of Chicago Press, 1980.

Lehner, Ulrich. *The Catholic Enlightenment: The Forgotten History of a Global Movement*. Oxford: Oxford University Press, 2018.

Leighton, Albert C. *Transport and Communication in Early Medieval Europe AD 500–1100*. Exeter: David & Charles, 1972.

Leslie, John. *Universes*. New York: Routledge, 1989.

Levering, Matthew. *Engaging the Doctrine of Creation: Cosmos, Creatures, and the Wise and Good Creator*. Grand Rapids: Baker Academic, 2017.

———. *Predestination: Biblical and Theological Paths*. Oxford: Oxford University Press, 2011.

———. *Proofs of God: Classical Arguments from Tertullian to Barth*. Grand Rapids: Baker Academic, 2016.

Lewis, C .S. "*De Descriptione Temporum*." In *Selected Literary Essays*, 1–15. Cambridge: Cambridge University Press, 1969.

———. *The Discarded Image: An Introduction to Medieval and Renaissance Literature*. Cambridge: Cambridge University Press, 2012.

———. *Selected Literary Essays*. Cambridge: Cambridge University Press, 1969.

———. *Studies in Words*. Cambridge: Cambridge University Press, 2013.

———. *The Voyage of the Dawn Treader*. New York: Scholastic, 1987.

Lienesch, Michael. *In the Beginning: Fundamentalism, the Scopes Trial, and the Making of the Antievolution Movement*. Chapel Hill, NC: University of North Carolina Press, 2007.

Lightman, Bernard. "Does the History of Science and Religion Change Depending on the Narrator? Some Atheist and Agnostic Perspectives." *Science and Christian Belief* 24 (October 2012) 149–68.

———. "The Evolution of the Evolutionary Epic." In *Victorian Popularizers of Science: Designing Nature for New Audiences*, 219–294. Chicago: University of Chicago Press, 2007.

———. "The International Scientific Series and the Communication of Darwinism." *Journal of Cambridge Studies* 5 no. 4 (2010) 27–38.

———. *The Origins of Agnosticism*. Baltimore: Johns Hopkins University Press, 1987.

———, ed. *Rethinking History, Science, and Religion: An Exploration of Conflict and the Complexity Principle*. Pittsburgh: University of Pittsburgh Press, 2019.

———. "Scientists as Materialists in the Periodical Press: Tyndall's Belfast Address." In *Science Serialized: Representation of the Sciences in Nineteenth-Century Periodicals*, edited by Geoffrey Cantor and Sally Shuttlesworth, 199–238. Cambridge, MA: MIT Press, 2004.

———. "Spencer's American Disciples: Fiske, Youmans, and the Appropriation of the System. In *Global Spencerism: The Communication and Appropriation of a British Evolutionist*, edited by Bernard Lightman, 123–48. Leiden: Brill, 2015.

———. "The Theology of Victorian Scientific Naturalists." In *Science Without God? Rethinking the History of Scientific Naturalism*, edited by Peter Harrison and Jon H. Roberts, 235–54. Oxford: Oxford University Press, 2019.

———. *Victorian Popularizers of Science: Designing Nature for New Audiences*. Chicago: University of Chicago, 2010.

———. "The Victorians." In *The Warfare Between Science and Religion: The Idea That Wouldn't Die*, edited by Jeff Hardin et al., 65–84 Baltimore: Johns Hopkins University Press, 2018.

Lincoln, Bruce. *Theorizing Myth: Narrative, Ideology, and Scholarship*. Chicago: University of Chicago Press, 2000.

Lindberg, David C. "Galileo, The Church, and the Cosmos." In *When Science and Christianity Meet*, edited by David C. Lindberg and Ronald L. Numbers, 33–60. Chicago: University of Chicago, 2003.

Lindberg David C., and Ronald L. Numbers, eds. *God & Nature: Historical Essays on the Encounter Between Christianity and Science*. Berkley: University of California Press, 1986.

———. *When Science & Christianity Meet*. Chicago: University of Chicago Press, 2003.

Lindberg, David C., and Robert S. Westman, eds. *Reappraisals of the Scientific Revolution*. Cambridge: Cambridge University Press, 1990.

Livingstone, David N. *Adam's Ancestors: Race, Religion, and the Politics of Human Origins.* Baltimore: Johns Hopkins University Press, 2011.

———. *Darwin's Forgotten Defenders: The Encounter Between Evangelical Theology and Evolutionary Thought.* Grand Rapids: Eerdmans, 1987.

———. *Dealing with Darwin: Place, Politics, and Rhetoric in Religious Engagements with Evolution.* Baltimore: Johns Hopkins University Press, 2014.

———. "Myth 17: That Huxley Defeated Wilberforce in Their Debate Over Evolution and Religion." In *Galileo Goes to Jail and Other Myths About Science and Religion,* edited by Ronald L. Numbers, 152–61. Chicago: University of Chicago Press, 2008.

———. *Putting Science in its Place: Geographies of Scientific Knowledge.* Chicago: University of Chicago Press, 2013.

———. "Re-Placing Darwinism and Christianity." In *When Science & Christianity Meet,* edited by David C. Lindberg and Ronald L. Numbers, 183–202. Chicago: University of Chicago Press, 2003.

Livingstone, David, D. G. Hart, and Mark A. Noll, eds. *Evangelicals and Science in Historical Perspective.* Oxford: Oxford University Press, 1999.

Lloyd, Genevieve. *Providence Lost.* Cambridge, MA: Harvard University Press, 2008.

Logan, Peter Melville. *Victorian Fetishism: Intellectuals and Primitives.* New York: SUNY Press, 2009.

Lorenz, Chris. "Can Histories Be True? Narrativism, Positivism, and the Metaphorical Turn." *History and Theory* 3 no. 37 (1998) 309–29.

Louth, Andrew. *Greek East and Latin West: The Church AD 681–1071.* Crestwood, NY: St. Vladimir's Seminary Press, 2007.

Löwith, Karl. *Meaning in History: The Theological Implications of the Philosophy of History.* Chicago: University of Chicago Press, 1957.

Lucas, J. R. "Wilberforce and Huxley: A Legendary Encounter." *History Journal* 22 (1979) 313–30.

Lundin, Roger. *Believing Again: Doubt and Faith in a Secular Age.* Grand Rapids: Eerdmans, 2009.

Lupton, Julia Reinhard. *The Afterlives of the Saints: Hagiography, Typology, and Renaissance Literature.* Stanford, CA: Stanford University Press, 1996.

Mach, Ernst. *The Science of Mechanics: A Critical and Historical Account of Its Development.* Columbia, SC: Bibliolife, 2019.

Macleod, Roy. "The X-Club: A Scientific Network in Late-Victorian England." *Notes and Records of the Royal Society* 24 (1970) 305–22.

MacIntyre, Alasdair. *After Virtue: A Study in Moral Theory.* 3rd ed. Notre Dame, IN: University of Notre Dame Press, 2007.

———. *Three Rival Versions of Moral Inquiry: Encyclopedia, Genealogy, and Tradition.* Notre Dame, IN: University of Notre Dame Press, 1990.

———. *Whose Justice? Which Rationality?* Notre Dame, IN: University of Notre Dame Press, 1988.

Maiocchi, Roberto. "Pierre Duhem's *Aim and Structure of Physical Theory: A Book Against Conventionalism.*" *Synthese* 83 no. 3 (1990) 395.

Manaugh, Geoff. "Why Catholics Built Secret Astronomical Features into Churches to Help Save Souls." *Atlas Obscura,* November 15, 2016. https://www.atlasobscura.com/articles/catholics-built-secret-astronomical-features-into-churches-to-help-save-souls.

Manchester, William. *A World Lit Only By Fire: The Medieval Mind and the Renaissance—The Portrait of An Age*. New York: Little, Brown, 1993.

Mandelbrote, Scott. "What Was Physico-Theology For?" In *Physico-Theology: Religion and Science in Europe, 1650–1750*, edited by Ann Blair and Kaspar Von Greyer, 67–77. Baltimore: Johns Hopkins University Press, 2020.

Manfasani, John. "Review of *The Swerve: How the Renaissance Began* (Review no. 1283)." *Reviews In History*, July 2012. http://www.history.ac.uk/reviews/review/1283.

Manuel, Frank Edward. *Isaac Newton, Historian*. Cambridge, MA: Harvard University Press, 1963.

Marenbon, John. *Pagans and Philosophers: The Problem of Paganism from Augustine to Leibniz*. Princeton, NJ: Princeton University Press, 2015.

Marion, Jean-Luc. *On Descartes' Metaphysical Prism: The Constitution and Limits of Onto-Theo-Logy*. Chicago: University of Chicago Press, 1999.

Marsden, George M. *The Soul of the American University: From Protestant Establishment to Established Nonbelief*. Oxford University Press, 1996.

———. *Understanding Fundamentalism and Evangelicalism*. Grand Rapids: Eerdmans, 1991.

Marsden, Richard. "Game of Thrones: Imagined World Combines Romantic and Grotesque Visions of the Middle Ages." *The Conversation*, October 17, 2018. https://theconversation.com/game-of-thrones-imagined-world-combines-romantic-and-grotesque-visions-of-middle-ages-10541#comment_1752980.

Marshall, Perry. *Evolution 2.0: Breaking the Deadlock Between Darwin and Design*. Dallas: BenBella, 2015.

Martin, R. N. D. *Pierre Duhem: Philosophy and History in the Work of a Believing Physicist*. Chicago: Open Court, 1991.

Martinich, A. P. *The Two Gods of Leviathan: Thomas Hobbes on Religion and Politics*. Cambridge: Cambridge University Press, 2003.

Masuzawa, Tomoko. *The Invention of World Religions: Or, How European Universalism Was Preserved in the Language of Pluralism*. Chicago: University of Chicago Press, 2005.

Matthews, David. *Medievalism: A Critical History*. New York: D. S. Brewer, 2015.

Matthews, Steven. *Theology and Science in the Thought of Francis Bacon*. Aldershot: Ashgate, 2008.

Matthewes, Charles, and Christopher McKnight Nichols. *Prophesies of Godlessness: Predictions of America's Imminent Secularization from the Puritans to the Present Day*. Oxford: Oxford University Press, 2008.

May, Gerhard. *Creation Ex Nihilo: The Doctrine of 'Creation out of Nothing' in Early Christian Thought*. New York: T&T Clark, 1994.

McAllister, James W. "Truth and Beauty in Scientific Reason." *Synthese* 78 (1989) 25–51.

McCarraher, Eugene. *The Enchantments of Mammon: How Capitalism Became the Religion of Modernity*. Cambridge, MA: The Belknap Press of Harvard University, 2019.

McClory, Robert. *Power and the Papacy: The People and Politics Behind the Doctrine of Papal Infallibility*. Chicago: Triumph, 1997.

McCumber, John. *Time in the Ditch: American Philosophy and the McCarthy Era*. Evanston, IL: Northwestern University Press, 2001.

McGrath, Alister E. *C. S. Lewis—A Life: Eccentric Genius, Reluctant Prophet*. Wheaton, IL: Tyndale House, 2016.

———. *Darwinism and the Divine: Evolutionary Thought and Natural Theologians*. Oxford: Wiley-Blackwell, 2011.

————. *A Fine-Tuned Universe: The Quest for God in Science and Theology*. Louisville, KY: Westminster John Knox, 2009.

————. *A Scientific Theology Volume 1: Nature*. New York: T&T Clark, 2001.

————. *A Scientific Theology Volume 2: Reality*. New York: T&T Clark, 2002.

————. *A Scientific Theology Volume 3: Theory*. New York: T&T Clark, 2003.

————. *The Territories of Human Reason: Science and Theology in an Age of Multiple Rationalities*. Oxford: Oxford University Press, 2019.

————. *The Twilight of Atheism: The Rise and Fall of Disbelief in the Modern World*. New York: Doubleday, 2006.

McGrath, S. J. *The Early Heidegger and Medieval Philosophy: Phenomenology for the Godforsaken*. Washington, DC: Catholic University of America Press, 2014.

McGuckin, John Anthony. *Saint Cyril of Alexandria and the Christological Controversy*. New York: St. Vladimir's Seminary Press, 2010.

McKenna, John. "John Philoponus: Sixth Century Alexandrian Grammarian, Christian Theologian, and Scientific Philosopher." *Quodlibet Journal* 5 no. 1 (January 2003) 1–36.

McLellan, David. "Religion and Socialism in Europe." In *Religion in Europe: Contemporary Perspectives,* edited by Sean Gill et al., 150–69. Kampen, the Netherlands: Pharos, 1994.

McMahon, Darrin M. *Divine Fury: A History of Genius*. New York: Basic, 2013.

————. *Enemies of the Enlightenment: The French Counter-Enlightenment and the Making of Modernity*. Oxford: Oxford University Press, 2002.

McMullin, Ernan. "Galileo on Science and Scripture." In *The Cambridge Companion to Galileo,* edited by Peter Machamer, 271–347. Cambridge: Cambridge University Press, 1998.

————. "Medieval and Modern Science: Continuity or Discontinuity?" *International Philosophical Quarterly* 5 no. 1 (1965) 103–29.

Meador, Keith G. "'My Own Salvation': The Christian Century and Psychology's Secularizing of American Protestantism." In *The Secular Revolution: Power, Interests, and Conflict in the Secularization of American Public Life,* edited by Christian Smith, 269–310. Berkeley, CA: The University of California Press, 2012.

Melzer, Arthur M. *Philosophy Between the Lines: The Lost History of Esoteric Writing*. Chicago: University of Chicago Press, 2014.

Mercer, Christia. "Descartes' Debt to Teresa of Avila Or: Why We Should Work with Women in the History of Philosophy." *Philosophical Studies* 174 no. 10 (2017) 2539–55.

————. "Descartes Is Not Our Father." *The New York Times,* September 25, 2017. https://www.nytimes.com/2017/09/25/opinion/descartes-is-not-our-father.html.

————. *Leibniz' Metaphysics: Its Origins and Development*. Cambridge: Cambridge University Press, 2001.

Merchant, Carolyn. *The Death of Nature: Women, Ecology, and the Scientific Revolution*. New York: HarperCollins, 1980.

Meyer, Stephen C. *Darwin's Doubt: The Explosive Origin of Animal Life and the Case for Intelligent Design*. San Francisco: Harper One, 2014.

Milbank, John. *Theology and Social Theory: Beyond Secular Reason*. 2d ed. Malden, MA: Blackwell, 2006.

Miles, Laura Saetvit. "Stephen Greenblatt's *The Swerve* Racked Up Prizes—And Completely Misled You About The Middle Ages." *Vox,* July 20, 2016. https://www.vox.com/2016/7/20/12216712/harvard-professor-the-swerve-greenblatt-middle-ages-false.

Miller, Kenneth R. *Only a Theory: Evolution and the Battle for America's Soul*. New York: Viking, 2008.

Miller, Timothy S. *The Birth of the Hospital in the Byzantine Empire*. Baltimore: Johns Hopkins University Press, 1997.

Minich, Thomas Joseph. *Bulwarks of Unbelief: A Phenomenology of Atheism and Divine Absence in Late Modernity*. PhD diss., University of Texas at Dallas, 2019.

Molinaro, Ursula. "A Christian Martyr in Reverse, Hypatia: 370–415 A.D." *Hypatia: A Journal for Feminist Philosophy* 4 (1989) 6–8.

Moore, James R. "1859 And All That: Remaking the Story of Evolution and Religion." In *Charles Darwin, 1809–1882: A Centennial Commemorative*, edited by Roger G. Chapman and Cleveland T. Duval, 167–94. Wellington: Nova Pacifica, 1982.

———. "Of Love and Death: Why Darwin 'Gave Up Christianity.'" In *History, Humanity, and Evolution: Essays for John C. Greene*, edited by James R. Moore, 195–229. Cambridge: Cambridge University Press, 1989

———. *The Post-Darwinian Controversies: A Study of the Protestant Struggle to Come to Terms with Darwinism in Great Britain and America, 1870–1900*. Cambridge: Cambridge University Press, 1981.

———. "Review: Creation and the Problem of Charles Darwin." *British Journal for the History of Science* 14 no. 2 (1981) 189–200.

———. "Speaking of 'Science' and 'Religion'—Then and Now." *History of Science* 30 (1992) 311–23.

———. "Telling Tales: Evangelicals and the Darwin Legend." In *Evangelicals and Science in Historical Perspective*, edited by David Livingstone, D. G. Hart, and Mark A. Noll, 220–234. Oxford: Oxford University Press, 1999.

———. "Theodicy and Society: The Crisis of the Intelligentsia." In *Victorian Faith in Crisis: Essays on Continuity and Change in Nineteenth Century Religious Belief*, edited by Richard J. Helmstadter and Bernard Lightman, 153–86. London: Palgrave-Macmillan, 1990.

Moore, Michael E. *Nicholas of Cusa and the Kairos of Modernity: Cassirer, Gadamer, Blumenberg*. New York: Punctum, 2013.

Moran, Jeffrey P. *American Genesis: The Evolution Controversies from Scopes to Creation Science*. Oxford: Oxford University Press, 2012.

Morange, Michel. *The Misunderstood Gene*. Cambridge, MA: Harvard University Press, 2001.

Morgan, Luke. *The Monster in the Garden: The Grotesque and the Gigantic in Renaissance Landscape Design*. Philadelphia: University of Pennsylvania Press, 2016.

Moritz, Joshua M. *Science and Religion: Beyond Warfare Toward Understanding*. Winona, MN: Anselm Academic, 2016.

———. *The Role of Theology in the History and Philosophy of Science*. Leiden: Brill, 2017.

Morris, Simon Conway. *Life's Solution: Inevitable Humans in a Lonely Universe*. Cambridge: Cambridge University Press, 2003.

———. *The Runes of Evolution: How the Universe Became Self-Aware*. Conshohocken, PA: Templeton, 2015.

Mulhall, Stephen. *Philosophical Myths of the Fall*. Princeton: Princeton University Press, 2005.

Mullin, Robert Bruce. "Science, Miracles, and the Prayer-Gauge Debate." In *When Science & Christianity Meet*, edited by David C. Lindberg and Ronald Numbers, 203–24. Chicago: University of Chicago Press, 2008.

Murphy, Francesca Aran. *Art and Intellect in the Philosophy of Etienne Gilson.* Columbia, MO: University of Missouri Press, 2004.

———. *God is Not a Story: Realism Revisited.* Oxford: Oxford University Press, 2007.

Nabokov, Vladimir. *Pale Fire.* New York: Vintage, 1962.

Nash, Roderick. *The Nervous Generation: American Thought 1917–1930.* Chicago: Ivan R. Dee, 1990.

Nebelsick, Harold P. *Circles of God: Theology and Science from the Greeks to Copernicus.* Edinburgh: Scottish Academic, 1985.

Needham, Joseph. "Human Laws and the Laws of Nature in China and the West." *Journal of the History of Ideas* no. 12 (1951) 3–32, 194–230.

Neem, Johann N. "The Early Republic: Thomas Jefferson's Philosophy of History and the Future of American Christianity." In *Prophecies of Godlessness: Predictions of America's Immanent Secularization from the Puritans to the Present Day*, edited by Charles Matthewes and Christopher McKnight Nichols, 35–52. Oxford: Oxford University Press, 2008.

Neiman, Susan. *Evil in Modern Thought: An Alternative History of Philosophy.* Princeton, NJ: Princeton University Press, 2002.

Nelson, Robert H. *Economics as Religion: From Samuelson to Chicago and Beyond.* State College, PA: Pennsylvania State University, 2001.

———. *Reaching for Heaven on Earth: The Theological Meaning of Economics.* Lanham, MD: Rowman & Littlefield, 1991.

Nesteruk, Alexei V. *Light from the East: Theology, Science, and the Eastern Orthodox Tradition.* Minneapolis: Fortress, 2000.

———. *The Sense of the Universe: Philosophical Explications of Theological Commitment in Modern Cosmology.* Minneapolis: Fortress, 2015.

———. *Universe as Communion: Towards a Neo-Patristic Synthesis of Theology and Science.* New York: T&T Clark, 2012.

Netz, Reviel, and William Noel. *The Archimedes Codex: How A Medieval Prayer Book is Revealing the True Genius of Antiquity's Greatest Scientist.* Philadelphia: De Capo, 2007.

Nicolaides, Efthymios. "Eastern Orthodox Christians." In *The Warfare Between Science & Religion: The Idea That Wouldn't Die*, edited by Jeff Hardin et al., 123–42. Baltimore: Johns Hopkins University Press, 2019.

———. *Science and Eastern Orthodoxy: From the Greek Fathers to the Age of Globalization.* Baltimore: Johns Hopkins University Press, 2011.

Nietzsche, Friedrich. *Genealogy of Morals and Ecce Homo.* Translated and edited by Walter Kaufmann. New York: Vintage, 1989.

———. "On Truth and Lies in an Unmoral Sense." In *The Portable Nietzsche*, translated and edited by Walter Kaufmann, 42–47. New York: Penguin, 1977.

Nisbet, Robert A. *The Sociological Tradition.* New Brunswick, NJ: Transaction, 1993.

Nixey, Catherine. *A Darkening Age: The Christian Destruction of the Classical World.* London: Houghton Mifflin, 2017.

Nobel, Dennis. *The Music of Life: Biology Beyond Genes.* Oxford: Oxford University Press, 2008.

Noll, Mark. *America's God: From Jonathan Edwards to Abraham Lincoln.* Oxford: Oxford University Press, 2005.

———. *The Civil War as a Theological Crisis.* Chapel Hill, NC: University of North Carolina Press, 2015.

Noll, Mark, and David Livingstone. "Introduction." In *Evangelicals and Science in Historical Perspective,* edited by David Livingstone, D. G. Hart, and Mark A. Noll, 2–17. Oxford: Oxford University Press, 1999.

Nongbri, Brent. *Before Religion: A History of a Modern Concept.* New Haven, CT: Yale University Press, 2013.

Novick, Peter. *That Noble Dream: The 'Objectivity' Question and the American Historical Profession.* Cambridge: Cambridge University Press, 1988.

Nowak, Martin A., and Sarah Coakley. *Evolution, Games, and God: The Principle of Cooperation.* Cambridge, MA: Harvard University Press, 2013.

Numbers, Ronald. "Aggressors, Victims, and Peacemakers: Historical Actors in the Drama of Science and Religion." In *The Religion and Science Debate: Why Does it Continue?,* edited by Harold W. Attridge, 15–54. New Haven, CT: Yale University Press, 2009.

———. *Creation by Natural Law: La Place's Nebular Hypothesis in American Thought.* Seattle: University of Washington Press, 1977.

———. *The Creationists: From Scientific Creationism to Intelligent Design.* Cambridge, MA: Harvard University Press, 2006.

———. *Darwinism Comes to America.* Cambridge, MA: Harvard University Press, 1998.

———, ed. *Galileo Goes to Jail and Other Myths About Science and Religion.* Cambridge, MA: Harvard University Press, 2009.

———. "Introduction." In *Galileo Goes to Jail and Other Myths About Science and Religion,* edited by Ronald Numbers, 1–7. Cambridge, MA: Harvard University Press, 2009.

———. "The Most Important Biblical Discovery of Our Time: William Henry Green and the Demise of Ussher's Chronology." In *Science and Christianity in Pulpit and Pew,* 113–29. Oxford: Oxford University Press, 2007.

———. *Science and Christianity in Pulpit and Pew.* Oxford: Oxford University Press, 2007.

———. "Science and Religion." *Osiris* 1 no. 1 (1985) 59–80.

Numbers, Ronald, and Kostas Kampourakis, eds. *Newton's Apple and Other Myths About Science.* Cambridge, MA: Harvard University Press, 2015.

Nye, Mary Joe. *Michael Polanyi and His Generation: The Origins of the Social Construction of Science.* Chicago: University of Chicago Press, 2011.

Oakley, Francis. "Christian Theology and the Newtonian Science: The Rise of The Concept of Laws of Nature." *Church History* 30 (1961) 433–57.

———. *Natural Law, Laws of Nature, Natural Rights: Continuity and Discontinuity in the History of Ideas.* New York: Continuum, 2005.

———. *Omnipotence, Covenant, and Order: An Excursion into the History of Ideas from Abelard to Leibniz.* Ithaca, NY: Cornell University Press, 1984.

Oakley, Francis, and Daniel O'Connor. *Creation: The Impact of an Idea.* New York: Charles Scribner's Sons, 1969.

Oberman, Heiko. "Reformation and Revolution: Copernicus' Discovery in an Era of Change." In *Dawn of the Reformation: Essays in Late Medieval and Early Reformation Thought,* 179–203. Edinburgh: T&T Clark, 1986.

Obolevitch, Teresa. *Faith and Science in Russian Religious Thought.* Oxford: Oxford University Press, 2019.

O'Hagan, Sean. "Capturing the Light." *The Guardian* (April 2013). https://www.theguardian.com/books/2013/apr/21/capturing-the-light-watson-rappaport-review.

Olasky, Marvin, and Warren Cole Smith. *The Prodigal Press: Confronting the Anti-Christian Bias of the American News Media.* Phillipsburg, NJ: P&R, 2013.

Oliveira, Amélia. "Duhem's Legacy for the Change in Historiography of Science: An Analysis Based on Kuhn's Writing." *Transversal: International Journal of Historiography of Science* 127 no. 2 (2017) 127–39.

Oliver, Simon. *Philosophy, God, and Motion*. London: Routledge, 2013.

O'Neill, Tim. "The Archimedes Palimpsest." *History for Atheists,* September 18, 2017. https://historyforatheists.com/2017/09/the-archimedes-palimpsest/.

———."The Great Myths 5: The Destruction of the Great Library of Alexandria." *History for Atheists,* July 2, 2017. https://historyforatheists.com/2017/07/the-destruction-of-the-great-library-of-alexandria/.

———. "Review: Catherine Nixey, *A Darkening Age*." *History for Atheists* (November 2017). https://historyforatheists.com/2017/11/review-catherine-nixey-the-darkening-age/.

Osborn, Eric. *The Emergence of Christian Theology*. Cambridge: Cambridge University Press, 1993.

———. *Tertullian: First Theologian of the West*. Cambridge: Cambridge University Press, 2003.

Osler, Margaret J. "The Canonical Imperative: Rethinking the Scientific Revolution." In *Rethinking the Scientific Revolution*, edited by Margaret Osler, 3–25. Cambridge: Cambridge University Press, 2000.

———. *Divine Will and the Mechanical Philosophy: Gassendi and Descartes on Contingency and Necessity in the Created World*. Cambridge: Cambridge University Press, 2005.

———. "Mixing Metaphors: Science and Religion or Natural Philosophy and Theology in Early Modern Europe." *History of Science* 36 (1998) 91–113.

———, ed. *Rethinking the Scientific Revolution*. Cambridge: Cambridge University Press, 2000.

———. "Religion and the Changing Historiography of the Scientific Revolution." In *Science and Religion: New Historical Perspectives,* edited by Thomas Dixon, Geoffrey Cantor, and Stephen Pumfrey, 70–86. Cambridge: Cambridge University Press, 2010.

Oslington, Paul. ed., *Adam Smith as Theologian*. New York: Routledge, 2011.

———. *Political Economy as Natural Theology: Smith, Malthus, and their Followers*. New York: Routledge, 2017.

Ospovat, Dov. "'Darwin's Theology', review of Neil Gillespie's *Charles Darwin and the Problem of Creation*." *Science* 207 (1980) 520.

———. *The Development of Darwin's Theory: Natural History, Natural Theology, and Natural Selection, 1838–1859*. Cambridge: Cambridge University Press, 1981.

———. "God and Natural Selection." *Journal of the History of Biology* 13 (1980) 169–94.

Pabst, Adrian. *Metaphysics: The Creation of Hierarchy*. Grand Rapids: Eerdmans, 2012.

Palaver, Wolfgang. *René Girard's Mimetic Theory*. East Lansing, MI: Michigan State University, 2013.

Pannenberg, Wolfhart. *Anthropology in Theological Perspective*. Philadelphia: Westminster, 1985.

———. *Christianity in a Secularized World*. London: SCM, 2012.

———. *The Historicity of Nature*: *Essays on Science and Theology*. Edited by Niels Henrik Gregersen. Conshohocken, PA: Templeton, 2008.

———. *Systematic Theology*. 3 vols. Translated by Geoffrey W. Bromiley. Grand Rapids: Eerdmans, 1991–1994.

———. *Theology and the Philosophy of Science*. Translated by Francis McDonagh. Philadelphia: Westminster, 1976.

————. *Toward a Theology of Nature: Essays on Science and Faith.* Edited by Ted Peters. Louisville: Westminster John-Knox, 1993.

Park, Katharine. "The Life of the Corpse: Division and Dissection in Late Medieval Europe." *Journal of the History of Medicine* 50 (January 1995) 111–32.

————. "Myth 5: That the Medieval Church Prohibited Dissection." In *Galileo Goes to Jail and Other Myths About Science and Religion,* edited by Ronald Numbers, 43–50. Chicago: University of Chicago Press, 2008.

Passmore, John. "Logical Positivism." In *The Encyclopedia of Philosophy*, edited by Paul Edwards, 5:52–57. New York: Macmillan, 1967.

Patapievici, Horia-Roman. "The Discovery of the Middle Ages by Pierre Duhem: The Fate and Meaning of a Truth." In *Meaning and Truth*, edited by Sorin Costreie and Mircea Dumitru, 157–224. Bucharest: Pro Universitaria, 2015.

————. "The 'Pierre Duhem Thesis': A Reappraisal of Duhem's Discovery of Physics in the Middle Ages." *Logos & Episteme* VI no. 2 (2015) 201–18.

Pecora, Vincent P. *Secularization without End: Beckett, Man, Coetzee.* Notre Dame, IN: University of Notre Dame Press, 2015.

Pelikan, Jaroslav. *The Christian Tradition: The History of the Development of Doctrine.* 5 vols. Chicago: University of Chicago Press, 1975–1991.

————. *Christianity and Classical Culture: The Metamorphosis of Natural Theology in the Christian Encounter with Hellenism.* New Haven, CT: Yale University Press, 1993.

————. *What Has Athens to Do with Jerusalem? Timaeus and Genesis in Counterpoint.* Ann Arbor, MI: University of Michigan Press, 1997.

Pernoud, Regine. *Those Terrible Middle Ages! Debunking the Myths.* Translated by Anne Englund Nash. San Francisco: Ignatius, 2000.

Peters, Edward. "The Desire to Know the Secrets of the World." *Journal of the History of Ideas* 62 (2001) 593–610.

————. *Inquisition.* New York: Free, 1988.

Peterson, Derrick. "Galileo Again: Reevaluating Galileo's Conflict with the Church and Its Significance for Today." *Cultural Encounters: A Journal for the Theology of Culture* 13 no 1. (2017) 25–47.

————. "Scribbling in God's Two Books: Some Historical and Normative Reflections on Scripture, Theology, Natural Philosophy, and Science." In *Philosophy and the Christian: The Quest for Wisdom in the Light of Christ,* edited by Joseph Minich and Bradford Littlejohn, 401–60. Richmond, VA: Davenant Institute, 2018.

————. "The War With No Sides: Another Look at the Uses of Scripture and Science in the Reformation Debates About Copernicus and Galileo." *Ad Fontes* 10 no. 2 (June 2018) 6–12.

Pfau, Thomas. *Minding the Modern: Human Agency, Intellectual Tradition, and Responsible Knowledge.* Notre Dame, IN: University of Notre Dame Press, 2013.

Phelps, Lynn A., and Edwin Cohen. "The Wilberforce-Huxley Debate." *Western Speech* 37 (1973) 56–64.

Placher, William C. *The Domestication of Transcendence: How Modern Thinking About God Went Wrong.* Louisville: Westminster John-Knox, 1996.

Popkin, Richard H. *The History of Skepticism: From Savonarola to Bayle.* New York: Oxford University Press, 2003.

———. "The Religious Background of Seventeenth-Century Philosopy." In *The Cambridge History of Seventeenth-Century Philosophy*, edited by Daniel Garber and Michael Ayers, 393–422. Cambridge: Cambridge University Press, 2000.

Principe, Lawrence M. "Transmuting History." *Isis* 98 no. 4 (2007) 779–87.

———. "The Warfare Thesis." In *The Warfare Between Science and Religion: The Idea That Wouldn't Die*, edited by Jeff Hardin et al., 3–27 Baltimore: Johns Hopkins University Press, 2018.

Provan, Iain. *Convenient Myths: The Axial Age, Dark Green Religion, and the World that Never Was*. Waco, TX: Baylor University Press, 2013.

Prum, Richard O. *The Evolution of Beauty: How Darwin's Forgotten Theory of Mate Choice Affects the Animal World*. New York: Anchor, 2018.

Rattansi, M., and J. E. McGuire. "Newton and the 'Pipes of Pan.'" *Notes and Records of the Royal Society of London* 21 (1966) 108–43.

Ratner-Rosenhagen, Jennifer. *American Nietzsche: A History of an Icon and His Ideas*. Chicago: University of Chicago Press, 2011.

Rawlings, Helen. *The Spanish Inquisition*. New York: Wiley-Blackwell, 2005.

Reed, Edward S. *From Soul to Mind: The Emergence of Psychology from Erasmus Darwin to William James*. New Haven, CT: Yale University Press, 1997.

Rees, Martin. *Just Six Numbers: The Deep Forces That Shape the Universe*. New York: Basic, 2000.

Reeve, Michael. "Lucretius in the Middle Ages and Early Renaissance: Transmission and Scholarship." In *The Cambridge Companion to Lucretius,* edited by Stuart Gillespie and Philip Hardie, 205–14. Cambridge: Cambridge University Press, 2007.

Reeves, Eileen. "Augustine and Galileo on Reading the Heavens." *Journal of the History of Ideas* 52 (1991) 563–71.

Richards, Robert J. *The Meaning of Evolution: The Morphological Construction and Ideological Reconstruction of Darwin's Theory*. Chicago: University of Chicago Press, 1992.

———. "Theological Foundations of Darwin's Theory of Evolution." In *Experiencing Nature,* edited P. H. Therman and K. H. Parshall, 61–79. Dordrecht: Klewer Academic, 1997.

———. *The Tragic Sense of Life: Ernst Haeckel and the Struggle Over Evolutionary Thought*. Chicago: University of Chicago Press, 2008.

Rigge, Williams F. "An Historical Examination of the Connection of Calixtus III with Halley's Comet." *Popular Astronomy* 18 (1910) 214–19.

Riley, Patrick. *The General Will Before Rousseau: The Transformation of the Divine into the Civic*. Princeton, NJ: Princeton University Press, 1996.

Riley-Smith, Jonathan, *The Crusades: A History*. 3rd ed. London: Bloomsbury Academic, 2014.

Rios, Christopher M. *After the Monkey Trial: Evangelical Scientists and a New Creationism*. New York: Fordham University Press, 2014.

Riskin, Jessica. *The Restless Clock: A History of the Centuries-Long Argument Over What Makes Living Things Tick*. Chicago: University of Chicago Press, 2016.

Risse, Guenter B. *Mending Bodies, Saving Souls: A History of Hospitals*. Oxford: Oxford University Press, 1999.

Roberts, Jon. *Darwinism and the Divine in America: Protestant Intellectuals and Organic Evolution, 1859–1900*. Notre Dame, IN: University of Notre Dame Press, 1988.

———. "The Idea That Wouldn't Die: The Warfare Between Science and Christianity." *Historically Speaking* 3 no. 3 (February 2003) 21–24.

————. "Psychoanalysis and American Christianity, 1900–1945." In *When Science & Christianity Meet,* edited by David C. Lindberg and Ronald L. Numbers, 225–44. Chicago: University of Chicago Press, 2003.

Roberts, Jon H., and James Turner. *The Sacred & The Secular University.* Princeton, NJ: Princeton University Press, 2000.

Roberts, Michael. *Evangelicals and Science.* Westport, CT: Greenwood, 2008.

Robertson, Jon M. *Christ as Mediator: A Study in the Theologies of Eusebius of Caesarea, Marcellus of Ancyra, and Athanasius of Alexandria.* Oxford: Oxford University Press, 2007.

Rosen, Edward. "Calvin's Attitude toward Copernicus." In *Copernicus and His Successors,* edited by Erna Hilfstein, 161–73. London: Hambledon, 1995.

Ross, Dorothy. "On the Misunderstanding of Ranke and the Origins of the Historical Profession in America." In *Leopold von Ranke and the Shaping of the Historical Discipline,* edited by Georg G. Iggers and James M. Powell, 154–69. Syracuse, NY: Syracuse University Press, 1990.

Ross, Sydney. "Scientist: The Story of a Word." *Annals of Science* 18 no. 2 (1962) 65–85.

Rubenstein, Mary-Jane. *Worlds Without End: The Many Lives of the Multiverse.* New York: Columbia University Press, 2014.

Rudwick, Martin S. *Bursting the Limits of Time: The Reconstruction of Geohistory in the Age of Revolution.* Chicago: University of Chicago Press, 2005.

————. *Earth's Deep History: How It Was Discovered and Why It Matters.* Chicago: University of Chicago Press, 2014.

————. *The Meaning of Fossils: Episodes in the History of Paleontology.* 2d ed. Chicago: University of Chicago Press, 1976.

Ruse, Michael. *Darwin and Design: Does Evolution Have a Purpose?* Cambridge, MA: Harvard University Press, 2003.

————. *Darwinism as Religion: What Literature Tells Us About Evolution.* Oxford: Oxford University Press, 2017.

————. "Removing God from Biology." In *Science Without God? Rethinking the History of Scientific Naturalism,* edited by Peter Harrison and Jon Roberts, 130–147. Oxford: Oxford University Press, 2019.

Russell, Bertrand. *A History of Western Philosophy.* New York: Simon & Schuster, 1967.

Russell, Colin A. "The Conflict Metaphor and Its Social Origins." *Science & Christian Belief* 1 no. 1 (April 1989) 3–26.

————. "The Conflict of Science and Religion." In *Science and Religion: A Historical Introduction,* edited by Gary B. Ferngren, 3–12. Baltimore: Johns Hopkins University Press, 2002.

Russell, Jeffrey Burton. *Inventing the Flat Earth: Columbus and Modern Historians.* Westport, CT: Prager, 1997.

————. "The Myth of the Flat Earth." Lecture given to the American Scientific Affiliate Conference, August 4, 1997. http://www.veritas-uscb.org/library/russell/FlatEarth.html.

Ryrie, Alec. *Unbelievers: An Emotional History of Doubt.* Cambridge, MA: The Belknap Press of Harvard University Press, 2019.

Sadler, Gregory B. *Reason Fulfilled by Revelation: The 1930s Christian Philosophy Debates In France.* Washington, DC: Catholic University of America Press, 2011.

————. "The 1930's Christian Philosophy Debates." *Acta Philosophica* 2 no. 21 (2012) 393–406.

Sagan, Carl. *Cosmos.* New York: Random, 1983.

———. *Pale Blue Dot: A Vision of the Human Future in Space.* New York: Ballantine, 1994.

Sampson, Philip J. *6 Modern Myths About Christianity and Western Civilization.* Downers Grove, IL: InterVarsity, 2001.

Sanders, Elizabeth M. *Genres of Doubt: Science Fiction, Fantasy, and the Victorian Crisis of Faith.* Jefferson, NC: McFarland, 2017.

Santillana, Giorgio de. *The Crime of Galileo.* Chicago: University of Chicago Press, 1955.

Sarton, George. "August Comte, Historian of Science: With a Short Digression on Clotilde de Vaux and Harriet Taylor." *Osiris* 10 (1952) 328–357.

———. "Knowledge and Charity." *Isis* 5 (1923) 5–19.

———. "The History of Science." *The Monist* 26 no. 3 (July 1916) 321–65.

———. *The History of Science and the New Humanism.* New York: George Braziller, 1956.

———. *The Life of Science: Essays on the History of Civilization.* Mishawaka, IN: Palala, 2015.

———. "The Message of Leonardo: His Relation to the Birth of Modern Science." *Scribner's Magazine* 65 no. 5 (1919) 531–40.

———. *The Study of the History of Science.* Cambridge, MA: Harvard University Press, 1958.

Schaefer, Richard. "Andrew Dickson White and the History of a Religious Future." *Zygon* 50 no. 1 (March 2015) 7–27.

Schiebinger, Londa. *The Mind Has No Sex? Women in the Origins of Modern Science.* Cambridge, MA: Harvard University Press, 1991.

Schmidt, Carl. *Political Theology: Four Chapters on the Concept of Sovereignty.* Cambridge, MA: MIT Press, 1985.

Schneider, Nathan. *God In Proof: The Story of a Search From The Ancients To The Internet.* Berkeley, CA: University of California Press, 2013.

Schoepflin, Rennie B. "Myth 14: That the Church Denounced Anesthesia in Childbirth for Biblical Reasons." In *Galileo Goes to Jail and Other Myths About Science and Religion,* edited by Ronald Numbers, 123–31. Chicago: University of Chicago Press, 2008.

Scholasticus, Socrates. *Ecclesiastical History.* Miami, FL: Palala, 2015.

Schreiber, Joan. *America Past and Present.* Teacher's ed. Northbrook, IL: Scott Foresman, 1983.

Schreiner, Susan. *Are You Alone Wise? The Search for Certainty in the Early Modern Era.* Oxford: Oxford University, 2012.

Schukin, Timur. "Matter as Universal: John Philoponus and Maximus the Confessor on the Eternity of the World." *Scrinium: Revue de patrologie d'hagiographie critique et d'historire ecclésiastique* 13 no. 1 (November 2017) 361–82.

Scopes, John. "Reflections—Forty Years After." In *D-Day at Dayton: Reflections on the Scopes Trial,* edited by Jerry R. Tompkins, 25–26. San Francisco: Bolerium, 1965.

Scotti, Dom Paschal. *Galileo Revisited: The Galileo Affair in Context.* San Francisco: Ignatius, 2017.

Secada, Jorge. *Cartesian Metaphysics: The Scholastic Origins of Modern Philosophy.* Cambridge: Cambridge University Press, 2004.

Secord, James A. *Victorian Sensation: The Extraordinary Publication, Reception, and Secret Authorship of Vestiges of the Natural History of Creation.* Chicago: University of Chicago Press, 2000.

Semonin, Paul. *American Monster: How the Nation's First Prehistoric Creature Became a Symbol of National Identity.* New York: New York University Press, 2000.

Shank, J. B. "Between Isaac Newton and Enlightenment Newtonianism: The 'God Question' in the Eighteenth Century." In *Science Without God? Rethinking the History of Scientific*

Naturalism, edited by Peter Harrison and Jon H. Roberts, 77–96. Oxford: Oxford University Press, 2019.

———. *The Newton Wars and the Beginning of the French Enlightenment.* Chicago: University of Chicago Press, 2008.

Shank, Michael A. "Naturalist Tendencies in Medieval Science." In *Science Without God? Rethinking the History of Scientific Naturalism,* edited by Peter Harrison and Jon H. Roberts, 37–57. Oxford: Oxford University Press, 2019.

Shank, Michael F. "Myth 1: That There Was No Scientific Activity Between Greek Antiquity and the Scientific Revolution." In *Newton's Apple and Other Myths About Science,* edited by Ronald Numbers and Kostas Kampourakis, 7–15. Cambridge, MA: Harvard University Press, 2015.

Shapere, Dudley. *Galileo: A Philosophical Study.* Chicago: University of Chicago Press, 1974.

Shapin, Steven. *Never Pure: Historical Studies of Science as if it Was Produced by People with Bodies, Situated in Time, Space, Culture, and Society, and Struggling for Credibility and Authority.* Baltimore: The Johns Hopkins University Press, 2010.

———. *The Scientific Revolution.* Chicago: University of Chicago Press, 1996.

Shapin, Steven, and Simon Schaeffer. *Leviathan and the Air-Pump: Hobbes, Boyle, and the Experimental Life.* Princeton, NJ: Princeton University Press, 2011.

Shapiro, Adam R. "The Scopes Trial Beyond Science and Religion." In *Science and Religion: New Historical Perspectives,* edited by Thomas Dixon, Geoffrey Cantor, and Sephen Pumfrey, 197–219. Cambridge: Cambridge University Press, 2010.

———. *Trying Biology: The Scopes Trial, Textbooks, and the Antievolution Movement in American Schools.* Chicago: University of Chicago Press, 2013.

Shapiro, James A. *Evolution: A View from the 21st Century.* Upper Saddle River, NJ: FT, 2011.

Shea, William R. "Assessing the Relations Between Science and Religion." In *The History of Science and Religion: Historians in Conversation,* edited by Donald A. Yerxa, 11–18. Columbia, SC: University of South Carolina Press, 2009.

Sheehan, Jonathan. *The Enlightenment Bible.* New Jersey: Princeton University Press, 2005.

———. "The Enlightenment, Religion, and the Enigma of Secularization: A Review." *The American Historical Review* 108 no. 4 (October 2003) 1061–1080.

Sheehan, Jonathan, and Dror Wahrman. *Invisible Hands: Self-Organization and the Eighteenth Century.* Chicago: University of Chicago Press, 2015.

Shiner, Larry. *The Invention of Art: A Cultural History.* Chicago: University of Chicago Press, 2001.

Shorto, Russell. *Descartes' Bones: A Skeletal History of the Conflict Between Faith and Reason.* New York: Random, 2008.

Shuger, Debora K. *The Renaissance Bible: Scholarship, Sacrifice, and Subjectivity.* Waco, TX: Baylor University Press, 2010.

Siedentop, Larry. *Inventing the Individual: The Origins of Western Liberalism.* Cambridge, MA: The Belknap Press of Harvard University Press, 2017.

Sloan, Phillip R. "'The Sense of Sublimity': Darwin on Nature and Divinity." *Osiris* 16 no. 2 (2001) 251–270.

Smith, Christian, ed. *Atheist Overreach: What Atheism Can't Deliver.* Oxford: Oxford University Press, 2018.

———. *Moral Believing Animals: Human Personhood and Culture.* Oxford: Oxford University Press, 2009.

———. *Religion: What It Is, How It Works, and Why It Matters* Princeton, NJ: Princeton University Press, 2019.

———. *The Sacred Project of American Sociology.* Oxford: Oxford University Press, 2014.

———. *The Secular Revolution: Power, Interests, and Conflict in the Secularization of American Higher Education.* Berkeley, CA: University of California Press, 2003.

———. "Secularizing American Higher Education: The Case of Early American Sociology." In *The Secular Revolution: Power, Interests, and Conflict in the Secularization of American Public Life,* edited by Christian Smith, 97-153. Berkeley, CA: The University of California Press, 2012.

Smith, James K. A. *How Not To Be Secular: Reading Charles Taylor.* Grand Rapids: Eerdmans, 2014.

Smith, Wilfred Cantwell. *The Meaning and End of Religion.* Minneapolis: Fortress, 1962.

Snobelen, Stephen D. "The Myth of the Clockwork Universe: Newton, Newtonianism, and the Enlightenment." In *The Persistence of the Sacred in Modern Thought,* edited by Chris L. Firestone and Nathan A. Jacobs, 149–84. Notre Dame, IN: University of Notre Dame, 2012.

Sorabji, Richard, ed. *Philoponus and the Rejection of Aristotelian Science.* Indianapolis: Institute for Classical Studies, 2010.

Sorkin, David. *The Religious Enlightenment: Protestants, Jews, and Catholics from London to Vienna.* Princeton, NJ: Princeton University Press, 2011.

Spencer, Nick. *Atheists: The Origin of Species.* New York: Bloomsbury, 2014.

Spitzer, Robert J. *New Proofs for the Existence of God: Contributions of Contemporary Physics and Philosophy.* Grand Rapids: Wm. B. Eerdmans, 2010.

Stanley, Matthew. *Huxley's Church and Maxwell's Demon: From Theistic Science to Naturalistic Science.* Chicago: University of Chicago Press, 2016.

Stark, Rodney. *Bearing False Witness: Debunking Centuries of Anti-Catholic History.* Conshohocken, PA: Templeton, 2016.

———. *For the Glory of God: How Monotheism led to Reformations, Science, Witch-Hunts, and the End of Slavery.* Princeton, NJ: Princeton University Press, 2003.

———. *God's Battalions: The Case for the Crusades.* New York: HarperOne, 2010.

———. *The Triumph of Christianity: How the Jesus Movement Became the World's Largest Religion.* New York: HarperOne, 2012.

———. *The Victory of Reason: How Christianity Led to Freedom, Capitalism, and Western Success.* New York: Random, 2006.

Steiner, George. *Real Presences.* Chicago: University of Chicago Press, 1991.

Stenger, Victor. *God and the Folly of Faith: The Incompatibility of Science and Religion.* New York: Prometheus, 2012.

Stenmark, Mikael. *How to Relate Science and Religion: A Multi-Dimensional Model.* Grand Rapids: Eerdmans, 2001.

Stevenson, Louise. *Scholarly Means to Evangelical Ends: The New Haven Scholars and the Transformation of Higher Learning in America, 1830–1890.* Baltimore: Johns Hopkins University Press, 1986.

Stock, Brian. "Science, Technology, and Economic Progress in the Early Middle Ages." In *Science in the Middle Ages,* edited by David C. Lindberg, 1–51. Chicago: University of Chicago Press, 1978.

Stump, James B. "History of Science Through Koyre's Lenses." *Studies in History and Philosophy of Science* 32 no. 2 (2001) 243–263.

Surin, Kenneth. *Theology and the Problem of Evil*. Eugene, OR: Wipf and Stock, 1986.

Szasz, Ferenc M. "The Scopes Trial in Perspective." *Tennessee Historical Quarterly* 30 (1971) 288–98.

Tanner, Kathryn. *God and Creation in Christian Theology*. Minneapolis: Augsburg Fortress, 2004.

Tattersall, Jill. "Sphere or Disc? Allusions to the Shape of the Earth in Some Twelfth Century and Thirteenth Century Vernacular French Works." *Modern Language Review* 76 no. 1 (January 1981) 31–46.

Taylor, Charles. *The Language Animal: The Full Shape of the Human Linguistic Capacity*. Cambridge, MA: Harvard University Press, 2016.

———. *Modern Social Imaginaries*. Durham, NC: Duke University Press, 2004.

———. *A Secular Age*. Cambridge, MA: The Belknap Press of Harvard University, 2007.

———. *Sources of the Self: The Making of Modern Identity*. Cambridge, MA: Harvard University Press, 1992.

Thistleton, Anthony. *New Horizons in Hermeneutics: The Theory and Practice of Transforming Biblical Reading*. Grand Rapids: Zondervan, 1992.

———. *The Two Horizons: New Testament Hermeneutics and Philosophical Description*. Grand Rapids: Eerdmans, 1980.

Tierney, Brian. *The Idea of Natural Rights: Studies on Natural Rights, Natural Law, and Church Law 115–1625*. Grand Rapids: Eerdmans, 1997.

Tolkien, J. R. R. "*Beowulf*: The Monster and the Critics." In *The Monster and Its Critics and Other Essays*, 5–48. Boston: Houghton Mifflin, 1984.

Topham, Jonathan R. "Beyond the 'Common Context:' The Production and Reading of the Bridgewater Treatises." *Isis* 89 no. 2 (June 1998) 233–262.

———. "Science, Natural Theology, and the Practice of Christian Piety in Early-Nineteenth-Century Religious Magazines." In *Science Serialized: Representation of the Sciences in Nineteenth-Century Periodicals*, edited by Geoffrey Cantor and Sally Shuttleworth, 37–66. Cambridge, MA: MIT Press, 2004.

Torrance, T. F. *Divine Meaning: Studies in Patristic Hermeneutics*. Edinburgh: T&T Clark, 1995.

———. "John Philoponus of Alexandria: Sixth-Century Christian Physicist." In *Theological and Natural Science*, 97–120. Eugene, OR: Wipf and Stock, 2005.

———. *Theological Science*. Oxford: University of Oxford Press, 1969.

———. *Transformation and Convergence in the Frame of Knowledge: Explorations in the Interrelations of Scientific and Theological Enterprise*. Eugene, OR: Wipf and Stock, 1998.

Toulmin, Stephen. *Cosmopolis: The Hidden Agenda of Modernity*. Chicago: University of Chicago Press, 1990.

Toynbee, Arnold J. *A Study of History: An Abridgment of Volumes I–VI by D. C. Somerwell*. Oxford: Oxford University Press, 1987.

Trevor-Roper, Hugh. "The European Witch Craze." In *The Crisis of the Seventeenth Century: Religion, Reformation, and Social Change*, 83–178. Indianapolis: Liberty Fund, 2001.

Trinkaus, Charles. *In Our Image and Likeness: Humanity and Divinity in Italian Humanist Thought*. 2 vols. Notre Dame, IN: University of Notre Dame Press, 1995.

———. "Italian Humanism and Scholastic Theology." In *Renaissance Humanisms*, 3 vols., edited by Albert Rabil, 3:327–44. Philadelphia: University of Pennsylvania Press, 1988.

Truitt, E. R. *Medieval Robots: Mechanism, Magic, Nature, and Art*. Philadelphia: University of Pennsylvania, 2015.

Turner, Frank M. *Contesting Cultural Authority: Essays in Victorian Intellectual Life.* Cambridge: Cambridge University Press, 1993.

———. "The Victorian Conflict Between Science and Religion." In *Contesting Cultural Authority: Essays in Victorian Intellectual Life,* 171–200. Cambridge: Cambridge University Press, 1993.

Turner, James. *Without God, Without Creed: The Origins of Unbelief in America.* Baltimore: Johns Hopkins University Press, 1986.

Tyler, Peter. *The Return to the Mystical: Ludwig Wittgenstein, Teresa of Avila, and the Christian Mystical Tradition.* New York: Continuum, 2011.

Ungureanu, James. "Edward L. Youmans and the 'Peacemakers' in *Popular Science Monthly.*" *Fides et Historia* 51 no. 2 (Fall 2019) 13–32.

———. "Relocating the Conflict Between Science and Religion at the Foundations of the History of Science." *Zygon* 53 no. 4 (December 2018) 1106–30.

———. *Science, Religion, and the Protestant Tradition: Retracing the Origins of Conflict.* Pittsburgh: University of Pittsburgh Press, 2019.

———. "A Yankee at Oxford: John William Draper at the British Association for the Advancement of Science at Oxford, 30 June 1860." *Notes and Records: The Royal Society Journal of the History of Science* 70 (2015) 135–50.

Vickery, John B. *The Literary Impact of "The Golden Bough."* Princeton, NJ: Princeton University Press, *1973.*

Viner, Jacob. *The Role of Providence in the Social Order: An Essay in Intellectual History.* Princeton, NJ: Princeton University Press, 1977.

Vanzo, Alberto, and Peter R. Anstey, eds. *Experiment, Speculation, and Religion in Early Modern Philosophy.* New York: Routledge, 2019.

Waggoner, Paul M. "The Historiography of the Scopes Trial: A Critical Re-Evaluation." *Trinity Journal* 5 (1984) 155–74.

Wallace, William A. "The Enigma of Domingo De Soto." In *Prelude to Galileo: Essays on Medieval and Sixteenth-Century Sources of Galileo's Thought,* 91–109. Dordrecht: D. Reidel, 1981.

———. "Pierre Duhem: Galileo and the Science of Motion." In *Prelude to Galileo: Essays on Medieval and Sixteenth-Century Sources of Galileo's Thought,* 303–20. Dordrecht: D. Reidel, 1981.

Walsham, Alexandra. "The Reformation and 'The Disenchantment of the World' Reassessed." *The Historical Journal* 5 no. 2 (2008) 497–528.

Walton, John. *Ancient Near Eastern Thought and the Old Testament.* 2d ed. Grand Rapids: Baker, 2018.

Ward, Graham. *Cities of God.* New York: Routledge, 2001.

Warfield, B. B. *Evolution, Science, and Scripture: Selected Writings,* edited by Mark Noll and David A. Livingstone. Grand Rapids: Baker Academic, 2000.

Waterman, A.M.C. "Economics as Theology: Adam Smith's Wealth of Nations." *Southern Economic Journal* 68 (2002) 907–21.

Watson, R. A. "Shadow History in Philosophy." *Journal of the History of Philosophy* 31 no. 1 (1993) 95–109.

Watts, Edward J. *Hypatia: The Life and Legend of an Ancient Philosopher.* Oxford: Oxford University Press, 2017.

Weber, Max. "Science as a Vocation." *Daedalus* 48 no. 1 (1958) 111–34.

Webster, Charles. *From Paracelsus to Newton: Magic and the Making of Modern Science.* New York: Dover, 1980.

———. *The Great Instauration: Science, Medicine, and Reform 1626–1660.* London: Duckworth, 1975.

Webster, Richard. *Why Freud Was Wrong: Sin, Science, and Psychoanalysis.* New York: Basic, 1995.

Weckovicz, T. E., and Liebel H. Weckovickz. *A History of Great Ideas in Abnormal Psychology.* Amsterdam: Elsevier, 1990.

Westfall, Richard S. *Essays on the Trial of Galileo.* Notre Dame, IN: University of Notre Dame Press, 1990.

———. *Science and Religion in Seventeenth Century England.* Ann Arbor, MI: University of Michigan Press, 1973.

Westman, Robert S. "The Melanchthon Circle: Rheticus and the Wittenberg Interpretation of Copernican Theory." *Isis* 66 (1975) 165–93.

———. "The Astronomer's Role in the Sixteenth Century: A Preliminary Study." *History of Science* 23 (1980) 105–47.

Wheeler-Barclay, Marjorie. *The Science of Religion in Britain, 1860–1915.* Charlottesville, VA: University of Virginia Press, 2010.

Whewell, William. *Philosophy of the Inductive Sciences Founded Upon Their History.* 2 vols. New York: Forgotten, 2012.

White, Andrew Dickson. *Autobiography of Andrew Dickson White.* 2 vols. New York: Century, 1905.

———. "First of the Course of Scientific Lectures—Prof. White on the 'Battlefields of Science.'" *New York Daily Tribune,* December 18, 1869, 4.

———. *A History of the Warfare of Science With Theology in Christendom.* 2 vols. New York: D. Appleton, 1896.

White, Lynn, Jr. "The Historical Roots of Our Ecological Crisis." *Science* 155 no. 3767 (March 10, 1967) 1203–7.

———. *Medieval Technology and Social Change.* Oxford: Oxford University Press, 1964.

———. "Technology and Invention in the Middle Ages." In *Medieval Religion and Technology: Collected Essays,* 1–21. Berkeley, CA: University of California Press, 2018.

White, Robert. "Calvin and Copernicus: The Problem Reconsidered." *Calvin Theological Journal* 15 (1980) 233–43.

Whitfield, Bryan J. "The Beauty of Reasoning: A Reexamination of Hypatia of Alexandria." *The Mathematics Educator* 6 no. 1 (Summer 1995) 14–21.

Wickham, Chris. *The Inheritance of Rome: Illuminating the Dark Ages 400–1000.* New York: Penguin, 2009.

Wigner, Eugene. "The Unreasonable Effectiveness of Mathematics in the Natural Sciences." *Communications in Pure and Applied Mathematics* 13 vol. 1 (1960) 1–14.

Wilken, Robert Louis. *Christians as the Romans Saw Them.* New Haven, CT: Yale University Press, 2003.

Wilson, Margaret. *Descartes.* London: Routledge, 1978.

Witham, Larry. *By Design: Science and the Search for God.* San Francisco: Encounters, 2003.

———. *The Measure of God: Our Century-Long Struggle to Reconcile Science & Religion: The Story of the Gifford Lectures.* San Francisco: Harper San Francisco, 2005.

———. *The Proof of God: The Debate that Shaped Modern Belief.* New York: Atlas, 2008.

————. *Where Darwin Meets the Bible: Creationists and Evolutionists in America*. Oxford: Oxford University Press, 2002.

Withers, Charles W. J. "Place and the 'Spatial Turn' in Geography and History." *Journal of the History of Ideas* 70 no. 4 (October 2009) 637–58.

Witmer, Andrew. "After the Civil War: Auguste Comte's Theory of History Crosses the Atlantic." In *Prophecies of Godlessness: Predictions of America's Imminent Secularization from the Puritans to the Present Day*, edited by Charles Mathewes and Christopher McKnight Nichols, 95–113. Oxford: Oxford University Press, 2008.

Wolterstorff, Nicholas. "How Philosophical Theology Became Possible Within the Analytic Tradition of Philosophy." In *Analytic Theology: New Essays in the Philosophy of Theology*, edited by Oliver D. Crisp and Michael C. Rea, 155–71. Oxford: Oxford University Press, 2009.

————. "The Migration of the Theistic Arguments: From Natural Theology to Evidentialist Apologetics." In *Rationality, Religious Belief, and Moral Commitment: New Essays in the Philosophy of Religion*, edited by Robert Audi, William Wainwright, and Nicholas Wolterstorff, 38–81. Ithaca, NY: Cornell University Press, 1986.

Woodward, Daniel. "Medieval *Mappaemundi*." In *The History of Cartography: Cartography in Prehistoric, Ancient, and Medieval Europe and the Mediterranean*, vol. 1, edited by J. B. Harley and David Woodward, 286–370. Chicago: University of Chicago Press, 1987.

————. "Reality, Symbolism, Time and Space in Medieval World Maps." *Annals of the Association of American Geographers* 75 (1985) 510–21.

Worrall, John. "'Revolution in Permanence': Popper on Theory-Change in Science." *Royal Institute of Philosophy Supplement* 39 (1995) 75–102.

Worthen, Molly. *Apostles of Reason: The Crisis of Authority in American Evangelicalism*. Oxford: Oxford University Press, 2016.

Worthen, Shana. "The Influence of Lynn White Jr.'s *Medieval Technology and Social Change*." *History Compass* 7 no. 4 (2009) 1201–17.

Wybrow, Cameron. *The Bible, Baconianism, and Mastery Over Nature: The Old Testament and Its Modern Misreading*. New York: Peter Lang, 1992.

————, ed. *Creation, Nature, and Political Order in the Philosophy of Michael Foster (1903–1959): The Classic Mind Articles and Others, with Modern Critical Essays*. Lewiston, NY: E. Mellen, 1992.

Yanni, Carla. *Nature's Museums: Victorian Architecture and the Culture of Display*. New York: Princeton Architectural, 2005.

Yates, Francis A. *Giordano Bruno and the Hermetic Tradition*. Chicago: University of Chicago Press, 1991.

————. *The Rosicrucian Englightenment*. New York: Routledge, 1972.

Yeats, Samuel Butler. *Erewhon and Erewhon Revisited Twenty Years Later*. New York: Dover, 2015.

Yeo, Richard. *Defining Science: William Whewell, Natural Knowledge, and Public Debate in Early Victorian Britain*. Cambridge: Cambridge University Press, 1993.

Yerxa, Donald A., ed. *The History of Science and Religion: Historians in Conversation*. Columbia, SC: University of South Carolina Press, 2009.

Youmans, Edward. "Editor's Table: 'The Conflict of Ages.'" *Popular Science Monthly* 8 (February 1876) 493–94.

Young, Francis M. *Biblical Exegesis and the Formation of Christian Culture*. New York: Hendrikson, 2002.

Young, Robert. *Darwin's Metaphor: Nature's Place in Victorian Culture.* Cambridge: Cambridge University Press, 1985.

Youngson, R. *Scientific Blunders.* New York: Carrol and Graff, 1998.

Zachhuber, Johannes. *Theology as Science in Nineteenth Century Germany: From F. C. Baur to Ernst Troeltsch.* Oxford: Oxford University Press, 2013.

Zammito, John H. *A Nice Derangement of Epistemes: Post-Positivism in the Study of Science from Quine to Latour.* Chicago: University of Chicago Press, 2004.

9 781532 653339